RENEWALS 458-4574
DATE DUE

WITHDRAWN
UTSA LIBRARIES

INTERFACIAL AND CONFINED WATER

INTERFACIAL AND CONFINED WATER

BY
IVAN BROVCHENKO AND ALLA OLEINIKOVA

Physical Chemistry
Technical University of Dortmund
Dortmund, Germany

ELSEVIER

AMSTERDAM • BOSTON • HEIDELBERG • LONDON • NEW YORK • OXFORD
PARIS • SAN DIEGO • SAN FRANCISCO • SINGAPORE • SYDNEY • TOKYO

Elsevier
Radarweg 29, PO Box 211, 1000 AE Amsterdam, The Netherlands
Linacre House, Jordan Hill, Oxford OX2 8DP, UK

First edition 2008

Copyright © 2008 Elsevier B.V. All rights reserved

No part of this publication may be reproduced, stored in a retrieval system or transmitted in any form or by any means electronic, mechanical, photocopying, recording or otherwise without the prior written permission of the publisher

Permissions may be sought directly from Elsevier's Science & Technology Rights Department in Oxford, UK: phone (+44) (0) 1865 843830; fax (+44) (0) 1865 853333; email: permissions@elsevier.com. Alternatively you can submit your request online by visiting the Elsevier web site at http://elsevier.com/locate/permissions, and selecting *Obtaining permission to use Elsevier material*

Notice
No responsibility is assumed by the publisher for any injury and/or damage to persons or property as a matter of products liability, negligence or otherwise, or from any use or operation of any methods, products, instructions or ideas contained in the material herein. Because of rapid advances in the medical sciences, in particular, independent verification of diagnoses and drug dosages should be made

Library of Congress Cataloging-in-Publication Data
A catalog record for this book is available from the Library of Congress

British Library Cataloguing in Publication Data
A catalogue record for this book is available from the British Library

ISBN: 978-0-444-52718-9

For information on all Elsevier publications
visit our website at books.elsevier.com

Printed and bound in Hungary
08 09 10 11 12   10 9 8 7 6 5 4 3 2 1

**Working together to grow
libraries in developing countries**

www.elsevier.com | www.bookaid.org | www.sabre.org

ELSEVIER   BOOK AID International   Sabre Foundation

Library
University of Texas
at San Antonio

Contents

	Preface	vii
1	**Phase diagram of bulk water**	1
2	**Surface transitions of water**	17
	2.1 Surface transitions of fluids	17
	2.2 Layering, prewetting, and wetting transitions of water near hydrophilic surfaces	25
	2.3 Drying transition of water near hydrophobic surfaces . .	50
	2.4 Surface phase diagram of water	61
3	**Surface critical behavior of water**	67
	3.1 Surface critical behavior of fluids	68
	3.2 Surface critical behavior of water	76
4	**Phase diagram of confined water**	91
	4.1 Effect of confinement on the phase transitions	91
	4.2 Phase transitions of confined water	98
	4.3 Capillary condensation and capillary evaporation	114
5	**Water layers at hydrophilic surfaces**	121
	5.1 Percolation transition of hydration water	121
	5.2 Structure of water layers at hydrophilic surfaces	139
6	**Role of interfacial water in biological function**	151

7	**Water in low-hydrated biosystems**	**165**
	7.1 Percolation transition of water in low-hydrated biosystems	165
	7.2 Effect of hydration on the properties of biosystems . . .	194
8	**States of interfacial water in fully hydrated biosystems**	**215**
9	**Summary and outlook**	**233**
	References	**237**
	Index	**303**

Preface

Abundance of water on the earth and in space makes it involved in the processes that are interesting for researches in various fields of science and technology. For better understanding of these processes, it is necessary to characterize water properties in a wide range of thermodynamic conditions. Similar to other substances, water can exist in various phase states with essentially different properties: vapor, liquid, crystalline phases, amorphous phases, glassy states. Therefore, characterization of water properties should be based on the phase diagram, which shows location of the phase transitions in thermodynamic space, i.e. in temperature, pressure, density and other coordinates. Major part of water on the earth is the *bulk* phases: liquid phase in oceans, crystalline forms in polar ice caps and in glaciers, vapor phase in the air. Both on the earth and in space, essential amount of water is affected by the proximity of various boundaries. A bulk three-dimensional phase may be terminated by another phase of the same substance when two phases coexist. For example, a liquid and vapor being at coexistence form liquid–vapor interface. Besides, a boundary may be formed by another substance being in crystalline or amorphous phase, by extended surface of some macromolecular structure, etc.

The *interfacial* water, that is water near a boundary, plays an important role in various biological, geological, technological, and other processes. For example, life is not possible without water, which exists mainly as interfacial water in living organisms. The presence of boundary breaks the translational invariance present in a bulk system. As a consequence, all system properties become local, i.e. dependent on the position of the part of fluid considered relative to the boundary. The phase diagram of any substance including water becomes much more complicated near a boundary, in particular due to the appearance of the *surface transitions*. Besides, the critical behavior of a fluid is strongly modified near a boundary, which strongly affects the fluid properties in a wide temperature range.

Finite extension in one or more spacial directions makes the system to be trapped in the pore geometry, which causes further complications of its phase behavior. The phase diagram of a system, confined between two planar boundaries or within cylindrical boundary, differs from the bulk one. In particular, bulk phase transitions and surface transitions are modified due to the confinement in pores. Structure of the real porous materials is often far from the simple slit or cylindrical geometry. Moreover, various porous media possess a highly disordered structure, and this disorder further complicates the phase behavior of a confined system. On the earth, *confined* water may be found in various geological materials, which possess a porous structure permeable for water. Considerable amount of water in living organisms is confined in cells and their counterparts. Confined water can be often found in porous materials used in technological processes. Essential amount of confined water may be expected in comets, which presumably represent a mixture of dust and ice.

To understand specific properties of interfacial and confined water at various thermodynamic conditions, we have to characterize the phase diagrams of water near surfaces and in various pores. A wide variety of such phase diagrams is expected, as they depend on the strength of the water–wall interaction, heterogeneity, roughness and curvature of a wall, pore size, and shape, etc. Knowledge of these phase diagrams opens a way for the description of the water *density distribution* near the surfaces and in pores, which is crucial for various structural and dynamic water properties. Subsequent analysis of the properties of interfacial and confined water allows understanding of related phenomena. Naturally, the phase behavior and properties of water show some regularities, which are universal for a wide class of fluids or even lattice systems. Knowledge of these universal features allows to distinguish them from the peculiar features, which are connected, first of all, with the strongly anisotropic hydrogen-bonding interactions between water molecules.

In Section 1 of this book, we give a brief description of the phase diagram of bulk water. This includes analysis of the liquid–vapor coexistence curve of water, a possibility to describe it in a universal way, effect of the liquid–vapor critical point on the properties of supercritical water, etc. Besides, we consider some peculiar properties of a liquid bulk water, which appears at ambient and supercooled temperatures. The relation of these properties to the polyamorphism of water and to the liquid–liquid transitions of supercooled water is discussed. Surface phase diagram of water is described in Section 2. Analysis of the surface transitions of water starts with the brief overview of the theoretical, experimental, and simulation results obtained for lattices and simple fluids (Section 2.1). This is followed by the analysis of the surface transitions of water near hydrophilic (Section 2.2) and hydrophobic (Section 2.3) surfaces. Finally, the surface phase diagram of water is presented in Section 2.4. Section 3 is devoted to the surface critical behavior of water, which allows description of water density profiles near various surfaces. This analysis, presented in detail in Section 3.2, is based on the theory of the surface critical behavior and its implementation in simple fluids (Section 3.1). In Section 4, we consider the modifications of the phase diagram of water due to confinement in pores. A brief overview of the general theoretical expectations and the results for simple fluids is given in Section 4.1. Phase transitions of water in various pores are discussed in Section 4.2. Phenomena of capillary evaporation and capillary condensation and characteristic properties of water in pores are briefly described in Section 4.3. Upon adsorbing at hydrophilic surfaces, water may form mono- or bilayers (Section 5). In Section 5.1, we consider a percolation transition of water at hydrophilic surfaces, which results in the formation of water monolayer. Main structural properties of water layers at hydrophilic surfaces are described in Section 5.2. Role of interfacial water in biology is analyzed in Section 6. In this analysis, we show how various forms of biological activity depend on hydration level, temperature, and pressure. To clarify the role of interfacial water in biological function, we consider separately low-hydrated (Section 7) and fully hydrated (Section 8) biosystems. Experimental and simulation studies of the percolation transition of hydration water in biosystems are summarized in Section 7.1. The effect of this transition on various properties of biosystems is analyzed in Section 7.2. For fully hydrated biosystems (Section 8), we analyze the effect of temperature and pressure on the various properties of hydration water, including connectivity of the hydrogen bonds within the hydration shell. The effect of the state of hydration water on the properties of biosystems is discussed. Finally, we summarize the current understanding of the properties of interfacial and confined water and formulate the open questions and controversial problems in Section 9.

In closing, we would like to express our deep gratitude to Alfons Geiger for his hospitality, fruitful collaboration, support of our initiatives in studies of water, and patience. We have greatly profited from collaboration with Roland Winter, Nikolay Smolin, Aljaksei Krukau, and Alexey Mazur. Our view on the properties of interfacial and confined water presented in this book is based on the fundamentals of the theory of phase transitions and critical phenomena in presence of a boundary, and we are greatly indebted to Kurt Binder, Michael Fisher, Gene Stanley, Robert Evans, Pablo Debenedetti, Gerhard Findenegg, and Josef Indekeu for elucidated and encouraging discussions, criticisms, and advices. This book has been made possible by financial support of our researches by Deutsche Forschungsgemeinschaft through the Forschergruppe 436 "Polymorphism, dynamics and functions of water near molecular boundaries," Schwerpunktprogramm 1155 "Molecular modeling and simulations in technology," and Graduiertenkolleg 298 "Structure-Dynamics Relations in Microstructured Systems," and by Bundesministerium fur Bildung und Forschung through the grant 01SF0303.

1 Phase diagram of bulk water

Properties of bulk fluid in distinct phase states differ so strongly that gas, liquid, and solid states are studied in different fields of statistical mechanics. Phase state of a system may be identified based on the phase diagram. Phase diagram of bulk water describes how the phase state of water changes with temperature and pressure. It includes the liquid–vapor, liquid–solid, and solid–solid phase transitions and also hypothesized transitions between amorphous (glassy or liquid) phases of supercooled water. In the solid state, water may form more than 14 crystalline forms [1]. Among these ices, the hexagonal ice is the most abundant. At atmospheric pressure, liquid water freezes into hexagonal ice at 273.15 K. This liquid–solid transition is accompanied by the decrease in water density by about 8%. Other ices can be obtained by increasing the pressure above 2 kbar. In particular, liquid water freezes into one of the high-density ices at ambient temperature when the pressure exceeds 6 kbar.

Major amount of water on the earth exists in the liquid bulk phase, which is close to the coexistence with a bulk water vapor present in the air. Therefore, the liquid–vapor phase transition is of special importance for understanding the water properties in the most of practically important situations. Bulk liquid water coexists with saturated vapor in a wide temperature range of about 374 K from the freezing temperature up to the liquid–vapor critical point. The liquid–vapor coexistence curve of bulk water, i.e. the temperature dependence of the densities of liquid (ρ_l) and vapor (ρ_v) coexisting phases, is shown in Fig. 1 by solid line. A stable bulk vapor phase exists at the densities left to the lower density branch of the coexistence curve. Accordingly, a stable bulk liquid phase exists at the densities right to the high-density branch of the coexistence curve. Water with a density inside two-phase region bounded by the coexistence curve is not thermodynamically stable, and it decomposes into two coexisting phases with the densities ρ_l and ρ_v. By increasing the temperature, two coexisting liquid and vapor phases become more and more similar in their properties until, at the critical point, all differences vanish. Beyond the critical point, only one homogeneous equilibrium water phase can exist, and all changes are continuous and smooth. The coexistence pressure near the melting point is about $6 \cdot 10^{-3}$ of ambient pressure (1 bar),

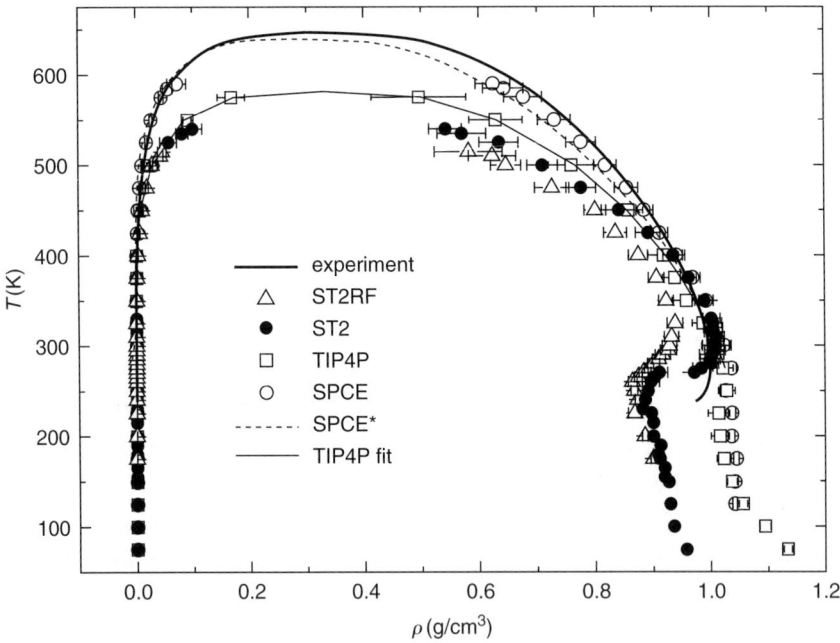

Figure 1: Experimental liquid–vapor coexistence curve of water [3] (thick solid line). Liquid–vapor coexistence curves of several water models: ST2 [6, 10], ST2RF [6], TIP 4P [6], SPCE [6], and SPCE* [11]. Fit of the data for the TIP 4P model to the extended scaling equation with leading asymptotic behavior described by eq. (1) is shown by thin solid line.

and it increases by a factor of ∼36 000 when approaching the critical point (Fig. 2). The liquid–vapor critical point of water is located at the critical temperature $T_c = 647.096$ K, critical pressure $P_c = 22.064$ MPa, and critical density $\rho_c = 0.322$ g/cm^3 [2, 3].

Although the liquid–vapor phase transition of bulk water is well studied experimentally, this is not the case for the phase transitions of interfacial and confined water, which we consider in the next sections. Therefore, studies of the phase transitions of confined water by computer simulation gain a special importance. For meaningful computer simulations, it is necessary to have water model, which is able to describe satisfactorily the liquid–vapor and other phase transitions of bulk water. The coexistence curves of some empirical water models, which represent a water molecule as a set of three to five interacting sites, are shown in Fig. 1. Some model adequately reproduces the location of the liquid–vapor critical point and,

Phase diagram of bulk water

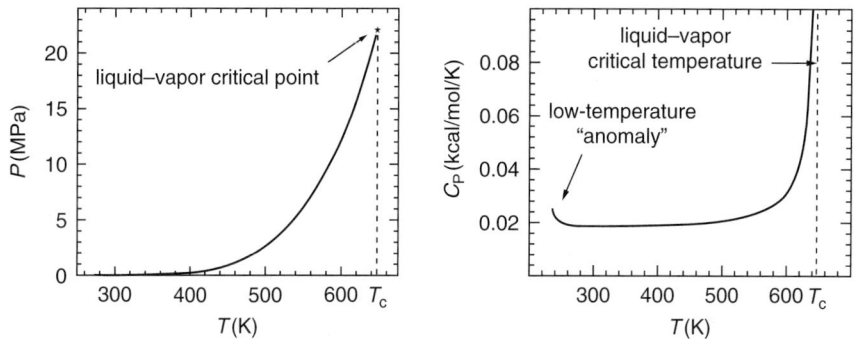

Figure 2: Left: liquid–vapor coexistence curve of water in the pressure–temperature plane from freezing temperature to the critical point [3]. Right: isobaric heat capacity C_P of liquid water along the liquid–vapor coexistence curve [3, 20].

accordingly, the temperature dependence of the liquid water density at high temperatures (SPCE model, see Fig. 1). However, the same model gives extremely low freezing temperature of water of about 214 K [4]. The comparative analysis of various water models can be found in Refs. [5–8]. Generally there is no empirical water model, which adequately describes the whole phase diagram of bulk water. This is not surprising as most of the popular water models were parameterized to fit some of the water properties at some particular thermodynamic conditions. Probably, the phase diagram of water in a wide thermodynamic range cannot be reproduced by a water model with just a few sites [9]. Therefore, there is an urgent need in more adequate water models. The available water models should be used with caution, keeping in mind their limited abilities to reproduce the phase diagram and properties of water quantitatively.

Upon heating, the densities of the coexisting vapor and liquid phases approach each other, and, asymptotically close to the critical temperature, their difference follows the universal power law:

$$\Delta\rho = (\rho_l - \rho_v)/2 = B_0 \tau^\beta, \tag{1}$$

where $\Delta\rho$ is the order parameter of phase transition, $\tau = (T_c - T)/T_c$ is a reduced deviation of temperature from T_c, $\beta \approx 0.326$ [12] is a universal critical exponent, and B_0 is a system-dependent amplitude. The behavior of $\Delta\rho(\tau)$ is shown in a double-logarithmic scale in Fig. 3, where power law (1) is shown by straight dashed lines. In the temperature interval $\sim 130°$ below T_c ($\tau \leq 0.2$), $\Delta\rho(\tau)$ closely follows the asymptotic power law, whereas more complicate description is clearly necessary far away from T_c. In a wider temperature interval, order parameter may be described by the extended scaling equation using Wegner expansion [13]. The temperature dependence of the order parameter $\Delta\rho(\tau)$ of water may be described satisfactorily using several nonasymptotic corrections [11]:

$$\Delta\rho = 2.02\tau^{0.325}\left(1 + 0.396\tau^{0.5} - 1.35\tau + 1.63\tau^{1.5} - 0.693\tau^2 - 0.37\tau^{2.5}\right). \tag{2}$$

The diameter ρ_d of the coexistence curve is the average value of the densities of the coexisting liquid and vapor phases. It is equal to the critical density at $T = T_c$ and changes mainly regulary with τ. In the close vicinity of the critical point, diameter of fluids shows a critical anomaly, which may behave as $\sim \tau^{1-\alpha}$ [14] or $\sim \tau^{2\beta}$ [15], or as superposition of two contributions [16], where $\alpha \approx 0.11$ [12]. Diameter of the coexistence curve of bulk water may be described by the following equation [11]:

$$\rho_d = (\rho_l + \rho_v)/2$$
$$= \rho_c\left(1 + 1.45\tau^{1-\alpha} - 0.10\tau - 0.35\tau^{1.5-\alpha} - 0.13\tau^{2-\alpha} - 1.35\tau^{5.5}\right). \tag{3}$$

Universal behavior of the order parameter as well as of other thermodynamic properties near the critical point originates from the dominant role of the density fluctuations close to the T_c [17, 18]. When approaching the critical point, density fluctuations strongly increase and the correlation length ξ, which describes their growth (extension), diverges as $\xi = \xi_0\tau^{-\nu}$, where $\nu \approx 0.63$ [12] is a universal exponent and $\xi_0 \approx 0.694$ Å for water along the coexistence curve [19]. Under such circumstances, the microscopic details of system structure are not important, and thermodynamic properties depend mainly on the distance to the critical point expressed in terms of temperature, pressure, or density. The thermodynamic domain of universal behavior depends on the property considered, and it does not extend over more than several degrees for shear viscosity, for example. Strictly speaking, the true asymptotic range for the order parameter $\Delta\rho$ is also rather narrow. However, the corrections for nonasymptotic behavior notably compensate each other, providing the behavior close to eq. (1) in the temperature range of dozens and hundreds degrees. For instance, the $\Delta\rho$ of model Lennard-Jones (LJ) fluids, which are used to describe such simple fluids as noble gases, shows behavior close to the $\sim \tau^\beta$ in the whole range of the existence of a liquid phase, i.e. from T_c to freezing temperature (see Fig. 3). This "critical-like" behavior of the order parameter in a wide temperature range is surprising because density fluctuations seem to be negligible far away from the critical point.

When approaching the critical point, all thermodynamic properties of systems behave in anomalous way [18, 21]. In particular, isothermal compressibility and heat capacity diverge, whereas diffusivity and other dynamic properties show a critical slowing down. An example of the critical divergence is shown in Fig. 2 for isobaric heat capacity C_P of liquid water along the coexistence curve. The critical anomalies of various properties are similar when approaching T_c along the coexistence curve or from supercritical temperatures along the critical isochore. At a given supercritical temperature, a property shows remnant of the critical anomaly, which is the largest at some pressure–density point. As a result, there are a number of specific lines in T–ρ plane in supercritical region, which emanate from the critical point and mark some specific

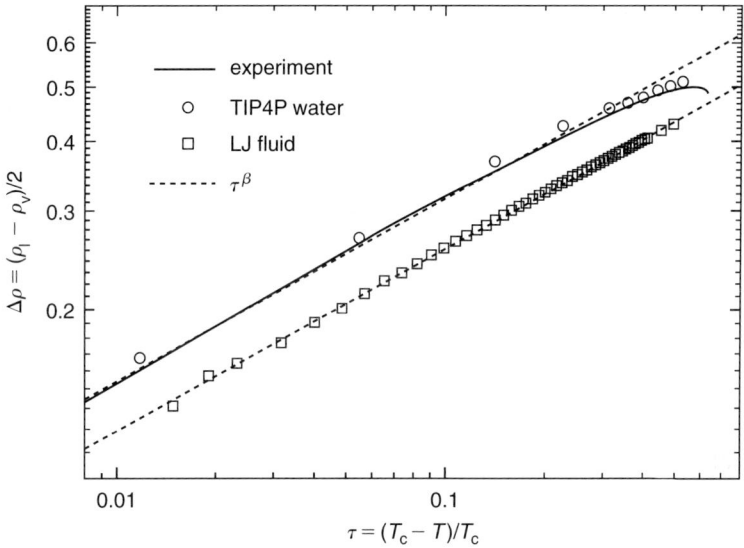

Figure 3: A log–log plot of the order parameter $\Delta\rho$ vs reduced temperature τ for real water [3] (thick solid line), TIP4P water model [28], and for LJ fluids [29]. Scaling equation (1) with $\beta = 0.326$ [12] is shown by straight dashed lines.

change in thermodynamic properties. In Fig. 4, we show the location of minimum of speed of sound [22] and approximate location of the maxima of isobaric and isochoric heat capacities [3]. There is one more important line emanating from the critical point, the line of percolation transitions of physical clusters [23]. This line separates two distinct states of supercritical water: there is an infinite hydrogen-bonded water network at higher density side, and only finite water clusters may be found at the lower density side. The percolation transition of supercritical model water, whose critical temperature is close to the one of real water [24, 25], is shown by a star. The expected line of the percolation transitions of water is shown schematically by thin solid line in Fig. 4. Obviously, more water density is required for the infinite network at higher temperatures, as hydrogen bonds break upon heating. Note that an infinite water network always exists in a liquid phase of water [26, 27]. At subcritical

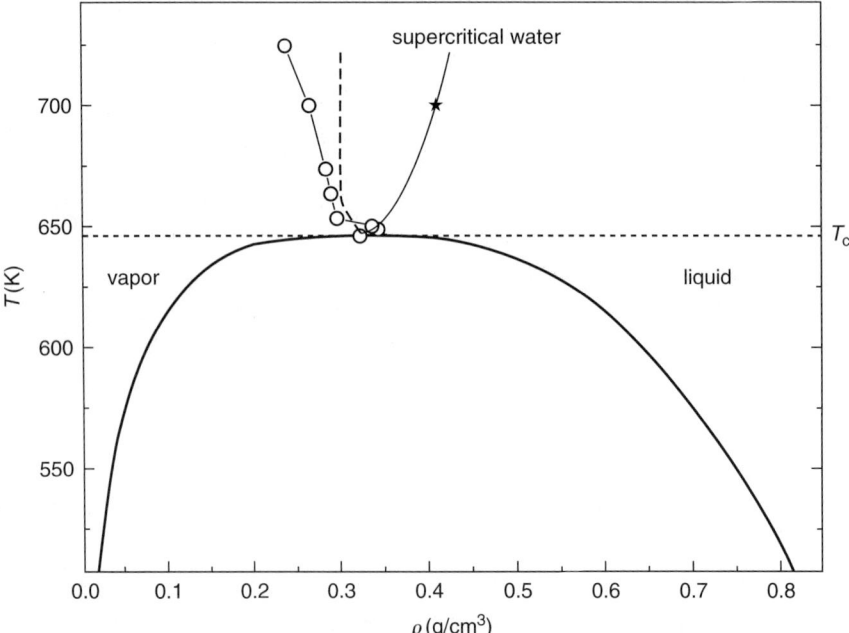

Figure 4: Liquid–vapor coexistence curve of water (thick solid line) and various specific lines emanating from the critical point: maximum heat capacity C_P [3] (dashed line), minimum speed of sound [22] (circles), and percolation transition of water clusters [24] (star).

temperatures, the break of an infinite water network in liquid water may be expected deeply inside the metastable region, i.e. at strongly negative pressures only.

Moving away from the liquid–vapor critical point, role of the density fluctuations diminishes, whereas the role of the molecular structure of fluids and local ordering of molecules become more and more important. Deviation from the asymptotic critical behavior of $\Delta\rho$ seen in Fig. 3 is a manifestation of this tendency. For water, deviations from the universal critical behavior start much closer to the critical point than for LJ fluid. This indicates stronger effect of fluid structure in the case of water. Moreover, due to this effect, the temperature dependence of the order parameter changes even qualitatively: below about 277.1 K (+4°C) [3], the densities of the liquid and vapor phases start to approach each other upon cooling due to the occurrence of the liquid density maximum. This is the most famous example of the specific properties of liquid water. Interestingly, this liquid density maximum is followed by a liquid density minimum in supercooled region. This means that at very low temperatures, liquid water behaves normally, i.e. it becomes more dense upon cooling. The density minimum of liquid water was first found in computer simulations of various water models [6, 30–34] (see Fig. 1) and later was observed in experiments with supercooled confined water [35]. Many other properties of liquid water also show specific behavior upon cooling (see [36–38] for reviews). For example, isobaric heat capacity shows rapid growth upon cooling (Fig. 2), which may be treated as a divergence at $T \approx 228$ K [39, 40]. Isothermal compressibility shows rather similar nonmonotonous temperature dependence [41, 42]. To understand the origin of these specific properties, we consider local ordering of molecules in liquid water at various temperatures.

The structure of condensed water phases is determined by the ability of water molecule to form four highly directional hydrogen bonds with the nearest neighbors. Accordingly, the tetrahedral arrangement of the four nearest neighbors is an important structural element of liquid water. In such arrangement, four nearest neighbors form almost perfect tetrahedron with a water molecule considered in its center. Tetrahedral ordering may be characterized by the tetrahedricity parameter [43], which measures deviation of the tetrahedron from ideal one and allows distinguishing of the tetrahedrally ordered water molecules [6, 10, 44, 45].

Temperature dependences of the fraction of water molecules with tetrahedral arrangement of the nearest four neighbors in liquid water calculated at the liquid–vapor coexistence curve for two water models are shown in Fig. 5. There are less than 10% of such water molecules at the liquid–vapor critical point. Upon cooling, more water molecules gain tetrahedral ordering and their fraction achieves ~50% at ambient conditions. At some temperatures below the freezing temperature, fraction of tetrahedrally ordered water molecules in supercooled liquid water shows a rapid or even a stepwise increase.

Another important structural characteristic of the local order is a coordination number NN, the number of the neighbors in the first coordination shell. There is a clear correlation between NN and the degree of the tetrahedral ordering. The fraction of water molecules with NN = 4, showing tetrahedral ordering, exceeds 90% at low temperatures. This fraction is essentially lower for water molecules having more neighbors,

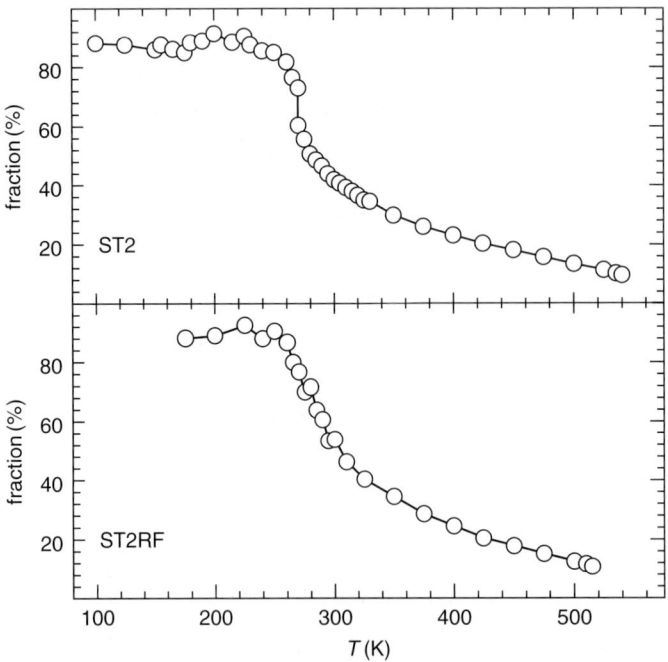

Figure 5: Fraction of water molecules with a tetrahedral arrangement of the nearest neighbors in liquid water along the liquid–vapor coexistence curve of two water models (ST2 and ST2RF) [10].

and it decreases from about 70 to 20% when NN increases from 5 to 9 [44]. The temperature dependences of the fractions of water molecules with different coordination numbers NN in liquid water along the liquid–vapor coexistence curve are shown in Fig. 6 [10, 46]. Molecules with less than four neighbors are important at high temperatures only. There is a specific temperature evolution of water molecules with NN = 4. Starting approximately from the temperature of the liquid density maximum, fraction of such molecules increases in a drastic way from about 20 to 70%, which causes increase in the fraction of tetrahedrally ordered water molecules (see Fig. 5). In parallel, the fraction of molecules with NN > 5 decreases symmetrically. Interestingly, fraction of water molecules with NN = 5 remains high (20 to 30%) in extremely wide temperature range: from the liquid–vapor critical point to the temperatures below the glassy transition. This can indicate a peculiar stability of the five-coordinated water molecules with respect to other kinds of local ordering, or these molecules may be considered intermediate structure between two four-coordinated states. In fact, water molecules with NN = 4 and 5 may be considered two most representative elements of liquid water.

Existence of the various kinds of local ordering in liquid water gives rise to the idea that water is a *mixture* of two or more components. The first mixture model of liquid water was proposed by H. Whiting in 1884 [47]. To explain the density maximum of liquid water, he assumes that in a liquid state there are some particles corresponding to the solid

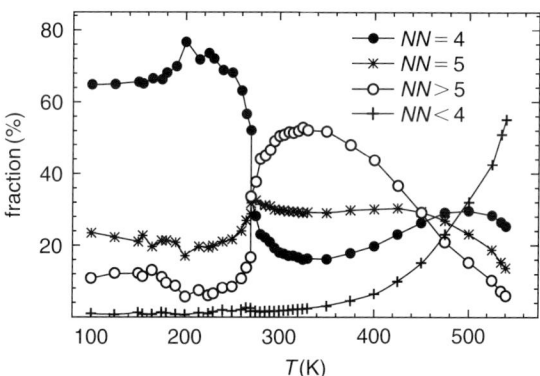

Figure 6: Fraction of water molecules with NN nearest neighbors in the first coordination shell along the liquid–vapor coexistence curve of ST2 water model [10].

state and the fraction of these "solid particles" increases upon cooling. This fraction is about 20% at boiling temperature, increases to about 30% at freezing temperature, and is 50 to 70% in "melting ice", which we may consider a supercooled liquid water. A solid particle represents a four-coordinated water molecule, which is the main structural element of the hexagonal ice. As we can see from Fig. 6, the modern computer simulations nicely confirm the predictions from 19th century. The mixture model of water was further developed in 1892 by W. C. Roentgen, who stated, "In my view an explanation of the mysterious properties of water can be found in the hypothesis that liquid water consists of a mixture of two types of differently constructed molecules, one type being ice-like" [48]. Using the mixture model, W. C. Roentgen has explained the shift of the liquid denisty maximum of water to lower temperature upon increasing pressure, the minimum compressibility of water, lower viscocity of water at higher pressures, and some other water anomalies. Numerous further modifications of the mixture model of water can be found in the literature [27, 49–60]. Exeprimental observations of at least three quite different forms of amorphous ices (see Ref. [61] for review) indirectly support the mixture models.

If liquid water is a mixture of some components, it is natural to expect that at some conditions they may undergo liquid–liquid phase transition, similar to the one in the binary liquid mixtures. Contrary to the mixtures of chemically different compounds, concentrations of "components" in liquid water cannot be imposed independently on temperature and pressure. Besides, the universality class of the liquid–liquid critical points of one-component isotropic fluids may differ from the universality class of Ising model [6]. However, many other features should be similar in both cases. Even when the liquid–liquid transition is unachievable experimentally due to crystallization or due to other processes, its critical point may have a strong "distant" effect on the properties of liquid water at ambient conditions. In a two-component binary mixture, effect of both the liquid–vapor and the liquid–liquid critical points on fluid properties should be taken into account [62]. The liquid–liquid critical point may be distant in terms of temperature, pressure, and also "external field", which may be varied by addition of impurities or by small variation in molecular structure (for example, by deuteration) [63, 64]. For example, mixture of 3-methylpyridine with heavy water possesses a closed-loop immiscibility gap [63, 65]. This gap diminishes in a continuous way by substitution of heavy water by normal water, and at about 83% weight fraction of normal water in heavy water, it shrinks completely. Mixture of 3-methylpyridine with normal water does not show immiscibility, but fluctuations of concentration strongly enhance when the system approaches the temperature/concentration region, where immiscibility gap appears upon deuteration [65]. Therefore, for understanding of the properties of water in fluid phases, it is necessary to take into account all possible critical points, even if they are located rather far from the thermodynamic region of interest or even if they are exhibited by unrealistic "modified" water only. In the latter case, there are no liquid–liquid critical points ("singularity-free scenario" [66–69]), but anomalous properties of liquid water may be attributed to the distant effect of the hidden liquid–liquid transition, which occurs in water with modified molecular parameters.

Experimental discovery of the phase transition from low-density amorphous (LDA) ice to high-density amorphous (HDA) ice [70] was the first evidence that indicated the possibility of the liquid–liquid transitions in supercooled water. Further experiments with amorphous and crystalline ices [71–76] have confirmed apparent first-order character of the transition between two amorphous glassy phases. The phase transition between two liquid phases of supercooled water was detected for the first time in simulations of the ST2RF water model [77]. Later, the liquid–liquid transition(s) was detected in simulation studies of more than six rather different water models [6, 10, 31, 33, 34, 78–87]. Besides, it was found that some simple systems with spherically symmetric pair potential also exhibit liquid–liquid transition, accompanied by water-like anomalies [88–92]. For the most of the studied water models, there are more than one liquid–liquid phase transition. At first, multiple liquid–liquid transitions of supercooled water were found in simulations [31], and later this finding was confirmed by experimental observation of two phase transitions between amorphous ices: between LDA and HDA and between HDA and very-high-density amorphous ices [76]. The character of the transitions between amorphous ices is still not well understood. In particular, HDA ice is heterogeneous on a mesoscopic scale [93–95]. This may be an intrinsic property of this phase, may be related to the experimental problems in obtaining a homogeneous phase, or may reflect existence of one more (unknown) phase. Finally, this may originate from the character of the liquid–liquid phase transition in isotropic one-component fluids. For example, the possibility to change the local order continuously introduces unavoidable disorder, which could vary with the thermodynamic conditions. The critical behavior of such a system could belong to the universality class of the random-field Ising models. Enhancement of disorder with increasing temperature could lead to a rounding of the phase transition and the disappearance of a true critical point. In this case, the two coexisting liquid phases are not infinite, and the two-phase state appears as a domain structure, seen as heterogeneites on a mesoscopic scale.

The number and location of the liquid–liquid phase transitions of real water and their possible critical points in $P - T - \rho$ coordinates are not clear yet. It is natural to assume that each crystalline phase being melted should yield a corresponding liquid with a short-range order, which is reminiscent to the crystalline phase [96, 97]. Five crystalline water phases may be melted directly to a liquid. Therefore, up to five phases of liquid or amorphous water can be expected. However, only the pairs of these phases that are essentially different in a local order may not be miscible at some thermodynamic conditions, giving rise to a liquid–liquid phase transition. Knowledge of the number of the liquid–liquid transitions is important for the correct developing of the equation of state of water at low temperatures, which accounts for available experimental data on liquid water and may be extrapolated to experimentally unachievable supercooled region. When only one liquid–liquid transition is imposed [98, 99], the liquid–liquid critical point is located at positive pressures (the values of about 2.3 kbar [98] and 0.27 kbar [99] were reported). Multiple liquid–liquid transitions are seen in most simulation studies of various water models. Therefore, the scenario with at least two liquid–liquid transitions seems to be the most realistic for real water. Two or three liquid–liquid transitions are usually seen in simulations, and the typical schematic phase diagrams of water are shown in Fig. 7.

The first (lowest density) liquid–liquid transition of water should have the strongest effect on liquid water at ambient conditions, as it is located close to the liquid–vapor coexistence curve. This is a transition between a "normal" water with a density of about 1 g/cm^3 and low-density water phase, enriched with tetrahedrally ordered four-coordinated water

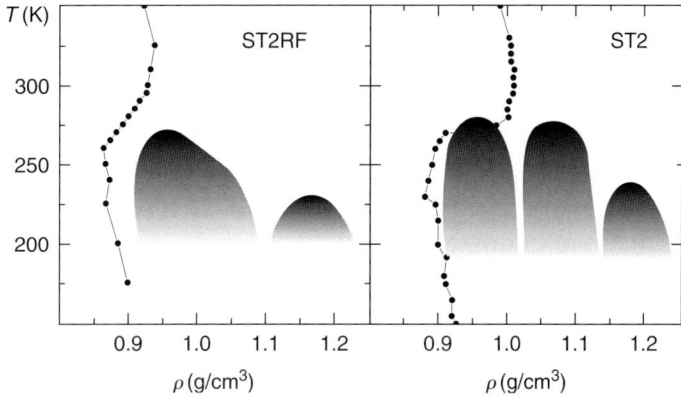

Figure 7: Schematic locations of the liquid–liquid coexistence regions of model water (shadow areas) with respect to the liquid branch of the liquid–vapor coexistence curve (solid circles) for two water models [6].

molecules. The liquid–liquid transition should be accompanied by the percolation transition of the species, representing the minor phase [100], as the phase transition and percolation transition are closely related [23]. Indeed, percolation transition of the four-coordinated water molecules occurs very close to the high-density branch of the first liquid–liquid coexistence curve [45]. This observation corroborates the old idea of H. E. Stanley [101], who has attributed water anomalies to the percolation transition of the four-coordinated water molecules. The critical point of the first liquid–liquid transition in real water may be at positive pressure (like in ST2RF model, left panel in Fig. 7) or at negative pressure (like in ST2 model, right panel in Fig. 7). In both cases, the liquid–liquid critical point has a distant effect on the properties of liquid water at zero pressure. In the former case, liquid water upon cooling along the liquid–vapor coexistence approaches and crosses a Widom line [34, 91] or other specific lines emanating from the critical point. In the latter case, there is a triple point where vapor coexists with two liquid phases of different densities. The liquid–liquid–vapor triple point appears as a break of the liquid branch of the liquid–vapor coexistence curve at some temperature (see solid circles in Fig. 1).

Liquid water shows anomalous behavior not only upon cooling along the liquid–vapor coexistence but also in a subcritical one-phase region when water is being exposed to pressure higher than the saturated vapor pressure. Various dynamic properties of water, such as a self-diffusion coefficient D, shear viscosity η, reorientational correlation time τ_2, monotonically decreases or increases with pressure at high temperatures, and such behavior is observed in other (normal) fluids [3]. However, anomalous increase of D, $1/\eta$, and $1/\tau_2$ upon pressurization was observed at temperatures below about 300 K [102–106]. As a result, D, $1/\eta$, and $1/\tau_2$ pass through a maximum, and the residence time passes through a minimum at $P \approx 1$ to 2 kbar. The observed slowing down of water dynamics with approaching liquid–vapor coexistence curve at low temperatures should be attributed to the effect of the first liquid–liquid transition, as it is accompanied by the sharp increase in the fraction of the four-coordinated water molecules, whose dynamics is slow.

Pressure dependence of the high-frequency sound velocity, determined by inelastic X-ray scattering at $T = 297$ K, changes qualitatively at water density $\rho \approx 1.12$ g/cm^3, which corresponds to the pressure $P \approx 3.7$ kbar [107]. Similar results were obtained from Brillouin scattering measurements, where sound velocity changes behavior at $\rho \approx 1.10$ g/cm^3 ($P \approx 2.9$ kbar), when $T = 297$ K; at $\rho \approx 1.07$ g/cm^3 ($P \approx 2.1$ kbar), when $T = 316$ K; and at $\rho \approx 1.04$ g/cm^3 ($P \approx 1.9$ kbar), when $T = 353$ K [108]. Discontinuity in the pressure dependence is also seen in the Raman spectra of liquid water at ambient temperature, when $\rho \approx 1.13$ g/cm^3 [109]. Clearly, these anomalies cannot originate from the first (lowest density) liquid–liquid transition of water. The thermodynamic points listed above are located close to the line of percolation transition of tetrahedrally ordered water molecules (with any coordination number), which is related to the liquid–liquid phase transition located at higher densities [45]. Hence, it is reasonable to attribute these anomalies to the distant effect of the second or third liquid–liquid critical points of water. The Widom lines or other specific lines emanating from these critical points should approach the liquid–vapor coexistence curve with increasing temperature and may affect the properties of a saturated liquid water (see Ref. [110] for an example).

Summarizing, thermophysical properties of water at high temperatures are similar to those of other fluids and may be described in a universal way, which is based on the theory of the critical behavior. Far away from the liquid–vapor critical point, the details of the fluid structure and intermolecular interactions become progressively more important.

Close to the ambient conditions and in supercooled region (below the melting point), various properties of liquid water show anomalous behavior, which originates from a different local ordering of water at various thermodynamic conditions. Existence of the different kinds of local ordering of water molecules provides not only a variety of the crystalline ices but also a several quite different amorphous (liquid) phases. The distant (or hidden) critical points of the phase transition(s) between the liquid phases of supercooled water seem to be responsible for anomalous water properties at ambient temperatures. Not only water but also other substances (silicon, silica, phosphorus, carbon, triphenyl phosphite, etc.) exhibit liquid–liquid phase transitions, which may be a general phenomenon for fluids. If these transitions are close to the thermodynamic region, where a liquid phase is stable (in the case of water and other substances listed above), their effect on the liquid properties is strong. Otherwise, their effect may be negligible, as in the case of an LJ fluid, where the variety of local ordering is not expected.

2 Surface transitions of water

When a fluid is in contact with some boundary, its density and all other properties become local. This noticeably complicate the description and prediction of the fluid properties compared with a bulk case as they depend on the distance to the boundary. Besides, the presence of a boundary induces wetting/drying phenomena, related to the preferential adsorption of one of the phase. Finally, it causes the appearance of additional (surface) phase transitions, which are characterized by their own critical points. As a result, the phase behavior of a fluid near a surface is much more complicated compared with a bulk case. Additional complications appear due to the fact that the surface phase behavior strongly depends on the strength and range of the fluid–wall interaction. In Section 2, we briefly describe surface transitions of fluids occurring at the bulk liquid–vapor coexistence (wetting and drying) and out-of-the-bulk phase transitions (layering and prewetting). We characterize the surface transitions of water and construct the surface phase diagram of water.

2.1 Surface transitions of fluids

When the system in the two-phase state (liquid–vapor or liquid–liquid coexistence) is placed in contact with a solid surface (or other inert spectator phase), one of the fluid phase is usually preferably attracted to the solid. At the microscopic level, this appears as an enhancement of the density (concentration) of preferable component toward the surface. At the macroscopic level, this preferential attraction is characterized by a contact angle between a solid boundary and a liquid–vapor interface. Zero value of a contact angle corresponds to a complete wetting situation when vapor phase is teared away from the surface and macroscopically thick wetting film intrudes between the surface and the saturated vapor. When the surface–fluid interaction is weak compared with the fluid–fluid interaction, a complete drying situation with a contact angle 180° may occur. In the case of a complete drying, liquid phase is not in direct contact with the surface but separated with a macroscopically thick drying layer. When the contact angle is not equal to 0° or 180°, both coexisting phases remain in contact with the surface. It has been known for about 200 years that the contact angle is determined by a competition between three interface free energies, namely the surface tension between wall and vapor, the surface tension between wall and liquid, and the interface tension between liquid and vapor [111]. Since these properties depend on temperature and liquid–vapor surface tension vanishes when approaching the liquid–vapor critical temperature, a complete wetting or complete drying situation seems to be unavoidable above some particular temperature, which is called a temperature of a wetting or drying transition [112].

A wetting or a drying transition occurs at some temperature in one of the coexisting phases upon heating a system along the bulk coexistence curve. The temperature and character of this transition depend on a balance between interparticle interaction and surface properties. A general theory of the surface transitions was developed based on the lattice systems. A surface phase diagram for Ising magnet in the presence of short-range surface field h_1 is drawn schematically in Fig. 8 (left panel) [113]. When the magnitude of the surface field h_1, acting on the spins in the first surface layer (analogue of a short-range fluid–wall interaction in fluid systems), is rather large, a first-order wetting transition occurs at T_w (line of these transitions is shown by thick solid line). This implies discontinuous, stepwise growth of an infinitely large wetting film near the surface at T_w. Weakening of the surface attraction ($|h_1| \to 0$) shifts the wetting temperature to higher values. Closer to the critical point, the wetting transition is expected to be of a second order, and it is called a critical wetting (dashed line in Fig. 8). In this case, the thickness L_0 of the wetting film adsorbed on the wall diverges to infinity in a continuous fashion with approaching T_w. In simulation studies of 3D Ising lattices, only critical wetting transitions were obtained in the systems, where spin coupling in the surface layer does not exceed the bulk value [114–116]. In the absence of a surface field ($h_1 = 0$), there is no wetting transition.

First-order wetting transition assumes the existence of a prewetting transition, which is located out of the bulk coexistence (dotted line in the left panel of Fig. 8). This is a first-order phase transition between the thin and thick films, adsorbed at the wall. The prewetting transition meets the liquid–vapor coexistence curve at T_w, which in fact is a triple point (solid circle), where three phases coexist. At the prewetting critical point

Surface transitions of water

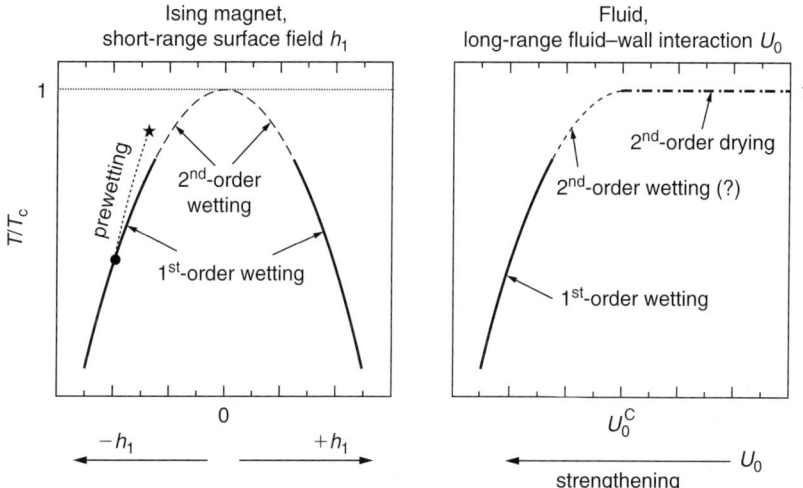

Figure 8: Left panel: schematic surface phase diagram of the Ising model [113] showing the change of the wetting transition temperature with the surface field h_1 acting on spins in the surface layer. There is no wetting, when $h_1 = 0$. Right panel: schematic surface phase diagram of a fluid in the case of a long-range fluid–surface interaction of a strength U_0 [117, 118]. There is no wetting or drying, when $U_0 = U_0^C$. Solid lines: the first-order wetting transitions; dashed lines: the second-order wetting transitions; dotted line: an example of a prewetting transition; solid circle: a triple point of the prewetting and bulk transitions; star: the critical temperature of the prewetting transition.

(star in Fig. 8), the difference between the thin and thick films disappears. The surface phase diagram of the Ising model is completely symmetrical with the respect to zero surface field ($h_1 = 0$) due to the complete symmetry of the coexisting phases in the Ising model. This assumes the existence of a predrying transition, which meets the coexistence curve in the drying temperature at some $h_1 > 0$ (it is not shown in Fig. 8 for simplicity). Prewetting and predrying transitions are equivalent in Ising model but have important distinctions in fluid systems, which possess intrinsic asymmetry of the coexisting phases.

A formation of a wetting layer with increasing temperature is extremely sensitive to the fluid–wall interaction potential, character of the wetting transition, and fluctuation regime [119, 120]. In 3D systems with long-range fluid–wall interaction, the formation of a wetting layer may be described by the mean-field theory. In the particular case of the critical wetting, the thickness L of a wetting layer should diverge logarithmically when approaching the wetting temperature within mean-field approximation [120, 121]. If a fluid–fluid interaction is also of long range, the thickness of a wetting layer shows strong power-law divergence approaching the temperature of the critical wetting transition [122, 123].

Competition between short-range and long-range fluid–wall forces may prevent the wetting transition if these two potentials favor different phases. In this case, the interface between the wetting layer and non-wetting phase will be attracted to the surface, and infinite wetting layer may appear at the bulk critical point only [124]. Below T_c, there is a partial wetting layer, whose thickness L may be large but remains finite at $T < T_c$ and diverges as a bulk correlation length upon approaching the critical point [124]. This case is also relevant for the drying transition, which occurs in a liquid phase near the wall, interacting with fluid molecules via a long-range potential. When the short-range part of this potential favors drying, its long-range tail ultimately attracts liquid–vapor interface to the wall. As a result, the drying transition in the case of long-range fluid–wall interactions occurs at the critical point only [117, 118, 124, 125]. This means that the thickness of a drying layer should diverge approaching T_c roughly as the bulk correlation length and an infinite drying layer never appears. Accordingly, there is no first-order drying transition or predrying transition. The surface phase diagram of fluids, expected for the case of a long-range fluid–surface interaction, is depicted schematically in the right panel of Fig. 8. The wetting or drying transition is absent at some particular strength U_0^C of a fluid–wall interaction [126, 127]. Note that we consider only *attractive* fluid–wall potentials, and the value U_0^C corresponds to some degree of attraction of fluid molecules to the wall. If this attraction is not strong enough ($U_0 > U_0^C$), a drying transition occurs at T_c only (dot-dashed line in the right panel of Fig. 8). In fact, a drying transition never occurs in these cases. However, this transition may strongly affect the density profiles of a liquid phase at subcritical temperatures (see Section 2.3, where this effect is described in detail). The diagram, shown in the right panel of Fig. 8, should be relevant to real fluid systems, including water near various natural surfaces, as the long-range fluid–solid dispersion forces are unavoidable.

Experimental studies of fluids near *solid* surfaces indicate that the temperature of the first-order wetting transition T_w, where vapor, adsorbed film, and liquid coexist, varies from about $0.38\,T_c$ for helium on cesium [128] to $0.89\,T_c$ for mercury on sapphire [129] (where T_c is a critical temperature of the bulk liquid–vapor transition of the respective fluid). Wetting transition of a fluid at the inert *liquid* surface was detected rather close to the critical temperature, that is T_w/T_c is about 0.99, 0.96, and 0.90, for methanol on nonane, decane, and undecane surfaces, respectively [130]. The character of a wetting transition depends on a fluid and on the surface properties, and various kinds of wetting transitions were observed experimentally. The critical wetting transition was observed for methanol at nonane surface and pentane on water surface, whereas methanol at cyclohexane and undecane surfaces shows the first-order wetting transition [131]. A sequential wetting, when the first-order wetting transition is followed by the critical wetting, was observed for pentane at water [132] and for hexane at salty water (brin) [133]. More data on the experimental studies of the wetting transitions can be found in [134, 135]. A drying transition of fluids, in fact, was never observed experimentally [136, 137] (see also Section 2.3 for studies of water).

Coexistence curve of the prewetting transition ends at the prewetting critical point (star in Fig. 8), which is expected to be two dimensional [138, 139]. Accordingly, the critical temperature of the prewetting transition should be lower than the bulk critical temperature. Strengthening of the fluid–wall interaction causes decrease in the wetting temperature T_w, which may be even lower than the bulk freezing temperature. There are several experimental observations of the prewetting phase transition in one-component fluids (see [134] for more details). The critical temperature T_c^{pw} of the prewetting transitions observed for helium on cesium [128] and for mercury on sapphire [129] was found to be $0.48\,T_c$ and $0.99\,T_c$, respectively. Note that in the case of mercury on molybdenum, the prewetting critical temperature ($T_c^{pw} = 1.05\,T_c$) even exceeds the bulk critical temperature [140].

Density profiles in the wetting phase (liquid near a strongly attractive surface) and in the drying phase (vapor near a weakly attractive surface) are not affected by the surface transitions. These profiles reflect the competition between the missing neighbor effect and the fluid–wall interaction and may be described in the framework of the theory of the surface critical behavior (see Section 3). In particular, a gradual density adsorption or a density depletion decays exponentially toward the bulk density, and this decay is determined by the bulk correlation length ξ. Density profile in a phase, which undergoes a wetting or a drying transition, is qualitatively different, and in general case, it consists of three portions. In a vapor phase undergoing a wetting transition, a wetting layer is bounded by the liquid–vapor interface from one side and by the liquid–solid interface from another side. Accordingly, profile of a liquid phase undergoing a drying transition consists of a vapor–solid interface, drying layer, and liquid–vapor interface. The density profile of a liquid phase near a weakly attractive solid surface is shown schematically in the left panel of Fig. 9. The thickness L of a drying layer is controlled by the fluid–wall interaction and by the thermodynamic state (temperature, pressure, chemical potential) of a bulk liquid. L may diverge strongly (as a power law) or weakly (logarithmically) when approaching the drying temperature [127]. The sharpness of a liquid–drying layer interface depends on the bulk correlation lengths in a liquid (ξ_l) and in a vapor (ξ_v) phase. In general, this *intrinsic* interface may be rounded due to the

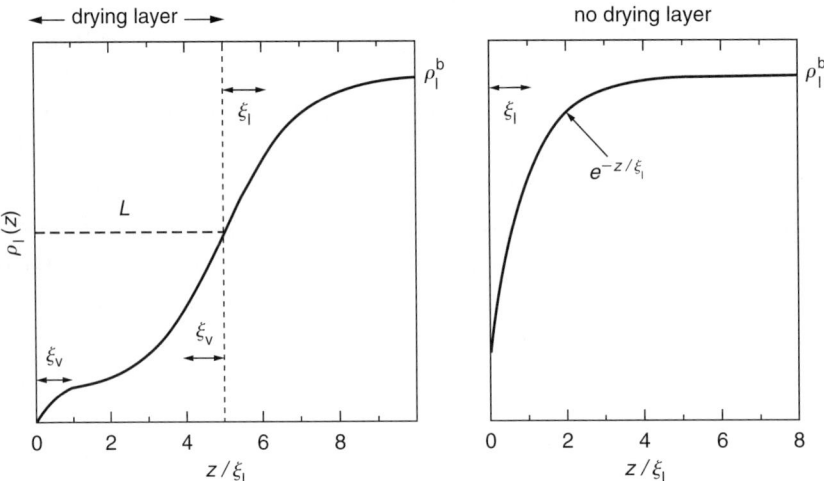

Figure 9: Left panel: density profile of a liquid phase with a drying layer of a thickness L near a weakly attractive surface. The thickness of an interface between the drying layer and solid surface and the thickness of a liquid–vapor interface are controlled by the bulk correlation lengths ξ_l and ξ_v in respective fluid phases. Right panel: a drying layer is completely bound to the wall, two interfaces merge together, giving gradual density depletion, controlled by ξ_l.

fluctuations of the interface position with respect to the wall (capillary waves) [119]. Capillary waves at the liquid–vapor interface near the wall with a long-range fluid–wall interaction are suppressed, and the interface has an intrinsic width. The interface between a drying layer and a solid interface should follow the laws of the surface critical behavior when the thickness of a drying layer $L \gg \xi$. In particular, density depletion of a vapor is governed by the correlation length ξ_v of a bulk vapor (Fig. 9).

When the thickness L of a drying layer is small, three portions of the density profiles, shown in the left panel of Fig. 9, may overlap and affect each other. At L small enough, the interface between a liquid and a drying layer is completely bound to the wall (Fig. 9, right panel). Under such circumstances, the liquid density profiles are determined by the laws of the surface critical behavior and may be described by the exponential equation (see Section 3). The shift of the chemical potential or pressure relatively to the bulk coexistence strongly affects the thickness of a wetting (or drying) layer. In particular, this layer may be strongly suppressed when fluid is confined in pore [127]. In small pores, a drying layer may remain completely bound to the pore wall up to the capillary critical point [141].

The relation between the density profile, which is a microscopic or mesoscopic property, and the contact angle, which is a macroscopic parameter, is not very clear for partial wetting and partial drying situations. Moreover, even for the case of complete wetting, the density profile of a liquid film may be depleted near the surface [142–144], which from the first look seems to be incompatible with a zero contact angle. The degree of the depletion of a liquid density, seen in the situation of a partial wetting (contact angle is less than 90°), does not correlate with the value of a contact angle [145, 146]. Occurrence of two sequential wetting transitions assumes that for the first of these transitions the contact angle is nonzero [147]. For a strongly attractive surface, one or several adsorbed layers of molecules, whose structure and behavior are very different from rest of the fluid, may appear [148, 149]. These layers are identical in both coexisting phases and may be called the *dead layers*. The thickness of dead layers is determined mainly by chemical structure of fluid and solid. Presence of the dead layers complicates studies of the wetting transition in experiments, where density profiles may be studied only in one phase. Such complicated profiles of wetting layer are indeed observed in experiments with binary liquid mixtures [134, 149]. The density profiles of one-component fluids near weakly attractive surfaces are free from this complication, as dead layers of voids cannot exist, but dead layers are possible near strongly attractive surfaces (see Section 2.2).

At some particular strength of the fluid–wall attraction, the prewetting transition is replaced by a sequence of layering transitions. The first layering transition is a 2D condensation of about one monolayer of fluid molecules at the solid surface. The second and subsequent layering transitions correspond to the condensation of a fluid layer on the surface of mono- or multilayer film. Layering transitions are the first-order phase transitions, which occur out-of-the-bulk coexistence at notably undersaturated vapor pressures. The effective dimensionality of the layering transitions is determined by the width of the monolayer film and by the degree of localization of molecules near the surface. Their critical points and asymptotic critical behavior belong to the universality class of the 2D Ising model. The layering transitions were studied experimentally for fluids adsorbed at highly homogeneous and planar crystalline surfaces of graphite, lamellar halides, metal oxides, etc. In the adsorption isotherm, a layering transition appears as a sharp vertical step, providing about monolayer coverage of the surface. Such kinds of behavior was reported for numerous fluid systems at various surfaces (see [28] for review of experimental data), and up to 17 subsequent layering transitions were observed in some cases [150]. The critical temperatures of the first layering transitions were observed below the temperature of the bulk triple point for noble gases, molecular hydrogen, molecular nitrogen, methane, and methyl chloride, and above this temperature for ethylene, ethane, propane, molecular oxygen, and water. With increasing layer number, its critical temperature may increase or decrease, approaching the roughening temperature, which is below the freezing temperature and indicates disappearance of the sharp solid–vapor interface. Two subsequent layering transitions could merge together at low temperatures in one transition, which corresponds to the simultaneous condensation of two layers. Besides, freezing or some structural changes of the condensed layers could also take place during formation of the multilayer film.

The critical temperature T_c^1 of the first layering transition of fluids is typically about 0.30 to 0.55 of the bulk critical temperature T_c. In particular, it depends strongly on the dimensional incompatibility between the adsorbate molecules and substrate [151]. For example, T_c^1/T_c is about 0.40 for LJ fluid near smooth strongly-attractive surface [152]. Some of the experimentally measured liquid–vapor coexistence curves of the layering transitions [153–156] were described by a scaling equation (1), and the critical exponent β of the order parameter was estimated. The values of β obtained from the fits vary from about 0.10 to 0.20 in reasonable agreement with $\beta = 0.125$ expected for 2D critical behavior.

The sequence of layering transitions was obtained for lattice-gas model by various theoretical and simulation methods. For strong surface potentials, the critical temperature T_c^1 of the first layering transition is close to the critical temperature of the 2D system, and it slightly increases with layer number, approaching the roughening temperature. With the weakening of a substrate potential, the critical temperature of the first layering transition increases, and condensation of two or more subsequent layers could occur simultaneously. For yet weaker substrate potentials, the prewetting transition (i.e., condensation of a film of a several molecular layer width) appears in the lattice-gas model instead of the sequence of layering transitions. The surface heterogeneity causes decrease in T_c^1 and may result in the disappearance of the first-order layering transition [157]. Discrete nature of the lattice models yields an infinite sequence of layering transitions. In continuum models, the layering transitions are promoted by the density oscillations near the wall. As these oscillations decay rather quickly, only finite sequence of the layering transitions could be expected for fluids. Layering and prewetting transitions of various model fluids were found using density functional theories and by computer simulations (more details can be found in [28]).

2.2 Layering, prewetting, and wetting transitions of water near hydrophilic surfaces

Adsorption of water from the air on hydrophilic surfaces occurs in various natural processes on the earth. Certain amount of water vapor is always present in the air. About 25 g of water per 1 kg of air corresponds to the 100% relative humidity at ambient conditions. This corresponds to the dew point, where condensation of water vapor into a liquid occurs in a bulk. At these conditions, which exist locally and temporarily on the earth, saturated water vapor coexists with a liquid water, and the volumes of the coexisting phases are determined by the total amount of water in the considered subsystem. Accordingly, different areas of a solid surface, exposed to the air, will be in direct contact with a water vapor or with a liquid water. Above the temperature of a wetting transition, surface should be covered by a macroscopic liquid film in a vapor phase and therefore the whole surface should be in fact in a direct contact with a liquid water only. At lower humidities, only vapor phase is stable and water molecules may adsorb from the vapor phase onto the solid hydrophilic surface. Adsorption of water may be complicated by complete or partial dissociation of water molecules on the surface, which results first in the appearance of the surface hydroxyl groups [158]. For example, water molecules dissociate due to the adsorption on the most of the metallic surfaces, and degree of dissociation depends on temperature and on the surface structure. In fact, these chemical reactions should be considered a modification of the surface. We consider the molecular adsorption of water molecules, which does not include chemical reactions on the surface.

With increasing humidity, growth of the amount of water adsorbed may occur in a continuous way or via the surface phase transitions, such as layering and prewetting, described in Section 2.1. Obviously, the presence of water clusters, water layer(s), or macroscopic water film on the surface essentially modifies the system properties. To predict water behavior near various surfaces, it is, therefore, important to analyze in a systematic way all possible scenarios of water adsorption and to relate them with the thermodynamic conditions and with the properties of a surface. Analysis of the surface phase transitions of water at hydrophilic surfaces (this section) and at hydrophobic surfaces (Section 2.3) will be finalized by constructing the surface phase diagram of water in Section 2.4.

In the adsorption isotherm, surface phase transition (layering or prewetting) should appear as a sharp vertical step at some pressure of a water vapor below the saturated value (Fig. 10). At this particular pressure, two water phases coexist on the surface, and relative fraction of these phases depends on the average surface coverage. Experimental studies of water adsorption on various surfaces give information about the occurrence of the surface phase transitions. In some cases, the corresponding step in the

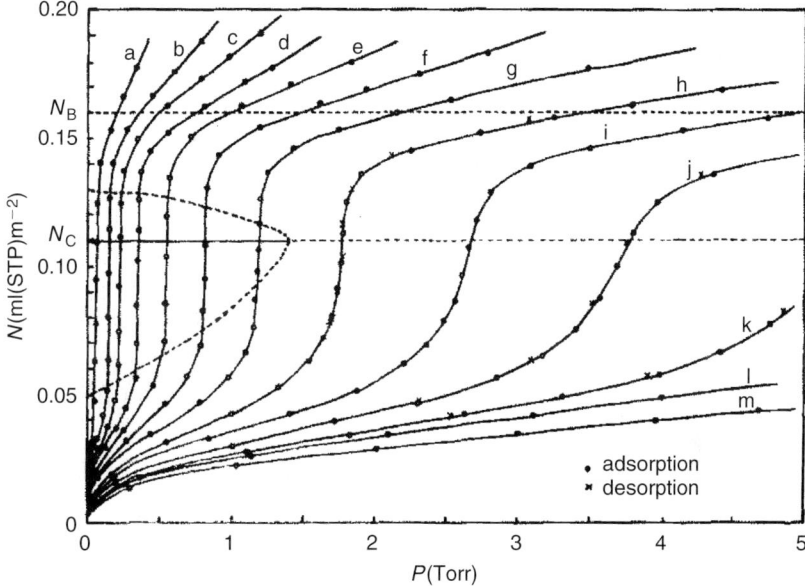

Figure 10: Adsorption isotherms of water on the hydroxylated surface of Cr_2O_3 at 268.9 K (a), 278.4 K (b), 283.2 K (c), 288.2 K (d), 293.3 K (f), 302.7 K (g), 308.0 K (h), 313.3 K (i), 318.2 K (j), 323.3 K (k), 328.3 K (l), and 333.2 K (m). (Reproduced from [159] with permission.)

adsorption isotherm is almost vertical, which strongly supports the first-order character of the transition. In other cases, there is no vertical step in the adsorption isotherm, but sigmoid-like dependence of water density on the vapor pressure, which saturates at about monolayer coverage, is seen. Smearing out of the vertical step in the adsorption isotherm may reflect the limitations of the available experimental techniques. Experimental studies of the phase transitions require long equilibration of a system at fixed temperature and pressure in the close vicinity of the transition, which is accompanied by the strong mass redistribution in a system.

Apart from the technical limitations of the experimental techniques used in the studies of the phase transitions, there are several physical reasons that cause smearing out of the phase transition. First, due to the occurrence of the metastable states, the transition during adsorption and desorption may occur at different pressures. Such hysteresis indicates the first-order character of the phase transition but strongly complicates its localization. Chemical or structural heterogeneity of the surface introduces element of disorder in the system. Due to this disorder, condensation of water layer/film occurs within some interval of vapor pressure, and the step in the adsorption isotherm becomes nonvertical. Finally, above the critical temperature of the surface phase transition, the step in the adsorption isotherm becomes nonvertical intrinsically. There is no phase transition, in this case, and formation of a condensed water layer/film at the hydrophilic surface with increasing pressure occurs in a continuous way. However, not very far from the critical temperature, the corresponding stepwise increase of adsorption in some pressure interval is still pronounced. Even when the surface phase transition is smeared out due to the surface heterogeneity or when it disappears in supercritical conditions, formation of water layer/film at hydrophilic surface is a process, which drastically affects all system properties. Continuous formation of a water layer/film may be characterized by the analysis of water clustering. In particular, first appearance of the condensed water layer/film should be attributed to the percolation transition of water, reflecting formation of an infinite hydrogen-bonded water network. The percolation transition of water at hydrophilic surfaces, which is intrinsically related to the surface phase transition, will be considered in detail below, in Sections 5 and 7. Note that formation of a condensed water layer/film via the first-order phase transition or continuously can occur at hydrophilic surfaces only.

For the strongly hydrophilic surfaces, we may expect existence of the first layering transition, which is characterized by the coexistence of a quasi-2D water vapor with quasi-2D liquid water (or with highly ordered solid quasi-2D phase at low temperatures). Only small clusters of adsorbed water molecules form a quasi-2D vapor, whereas a quasi-2D liquid phase is a dense water monolayer adsorbed on the surface. Note that both these quasi-2D water phases coexist with a bulk water vapor being at pressure lower than the bulk coexistence pressure $P_0(T)$. Adsorption isotherms, showing stepwise increase in the density of adsorbed water with saturation at about monolayer coverage, were reported for some rather homogeneous surfaces. Such kind of behavior was observed mainly for water adsorption on the surfaces of alkali halide crystals (NaCl [165, 166], NaF [162], CaF_2 [163, 167], SrF_2 [164, 168–171]), on the hydroxylated surfaces of some metal oxides (ZnO [159, 171, 172], SnO_2 [159, 173, 174], Cr_2O_3 [159, 171, 175–178],

BeO [179], FeOOH [160]), and on the crystal surface of MgO [180, 181]. At ambient temperature, the condensed 2D water was found to be liquid-like for the layering transition of water on the surfaces of NaF [162], NaCl [182], SrF$_2$, and ZnO [183], whereas it is solid-like for Cr$_2$O$_3$ [159, 178, 183]. At low temperatures, condensed water phase is a 2D ice with some particular crystalline structure (water on the surface of NaCl at $T = 140$ to 150 K [165, 166] and on the surface of MgO at $T = 180$ to 220 K [180]).

In experimental studies, the critical temperature T_c^1 of the layering transition may be estimated from the analysis of the slope of the stepwise increase of adsorption to about monolayer coverage at various temperatures. Below T_c^1, this slope should be infinite, whereas it is finite above T_c^1. This idealized picture cannot be realized in experiment, as even below T_c^1 the step in the adsorption isotherm is nonvertical due to the reasons, described above. However, the temperature at which this slope increases abruptly may be attributed to T_c^1. Using this analysis, the critical temperature of the first layering transition of water on the hydroxylated surface of Cr$_2$O$_3$ was estimated as $T_c^1 \approx 305$ K [159]. This critical temperature is about 0.48 T_c, where T_c is a liquid–vapor critical temperature of a bulk water. The step in the adsorption isotherms of water at the surfaces of ZnO and SnO$_2$ remains nonvertical at ambient temperatures [159]. Extrapolation to lower temperatures allows to expect the step to be vertical at $T < 236$ K for ZnO or even at lower temperature for SnO$_2$. The step in the adsorption isotherm of water on the surface of NaF is almost vertical up to 308 K, which indicates the occurrence of T_c^1 at higher temperatures [162]. There were no more attempts to estimate the critical temperature of the layering transition of water experimentally.

Layering transition of water occurs when the the pressure of the bulk vapor is noticeably below the saturated value. In Fig. 11, the layering transition of water at the hydroxylated surface of Cr$_2$O$_3$ is shown in the pressure–temperature plane with respect to the liquid–vapor bulk transition. The extension of this transition to higher temperature (shown by dotted line) corresponds to the inflection point of the adsorption isotherm, i.e to the line of the maximal compressibility. For other metal oxides, the critical temperature of the layering transition is unknown, and the dotted lines (Fig. 11) indicate pressure at the inflection point of various isotherms. These lines may correspond to the layering transition or to

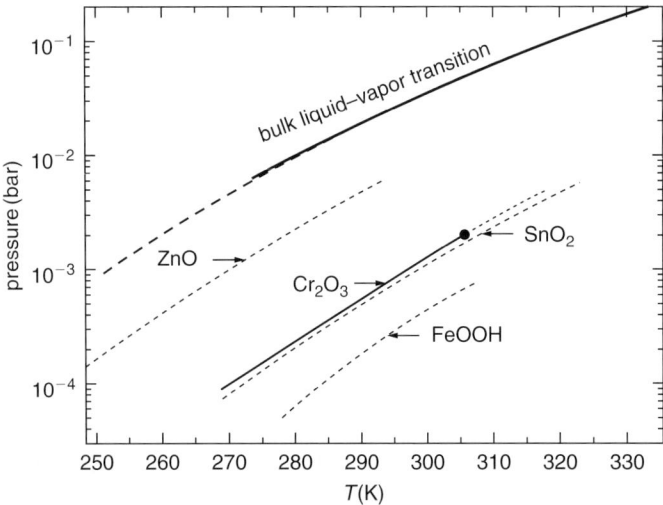

Figure 11: Layering transition of water at the hydroxylated surfaces of metal oxides in the pressure–temperature plane [159, 160]. Solid line with a solid circle: layering transition of water on the surface of Cr_2O_3 and its critical point. Dotted lines: pressures at the inflection points of the adsorption isotherms, which may correspond to the layering transition or to its extension in the supercritical region. Liquid–vapor bulk transition of water is shown by solid line [3], and its extension to supercooled region by Antoine equation [161] is shown by dashed line.

its extension in supercritical region. The experimental data for the bulk liquid–vapor transition of water are shown by a solid line. At low temperatures, these data may be adequately described by Antoine equation $\log(P) = A - B/(T + C)$ [161], and its extrapolation into supercooled region below 273 K is shown by a dashed line. The layering transitions of water on the surface of Cr_2O_3 and SnO_2 occur when the pressure P of the bulk water vapor is 0.02 to 0.04 of its saturated value P_0. The pressure of the layering transition is noticeably lower in the case of FeOOH. In the case of ZnO surface, the layering transition of water occurs much closer to the bulk condensation, at $P \approx 0.20\ P_0$.

Water molecules do not dissociate upon adsorption on the crystal surface of MgO at low temperatures, which allows to study molecular adsorption of water on the nonhydroxylated surface of a metal oxide. Condensation of a 2D gas into a 2D solid layer was observed in the temperature interval from 185 to 221 K at extremely low pressures

($\approx 10^{-11}$ bar) [180]. This is in accord with the water adsorption on the surface of other nonhydroxylated metal oxides. Before the first cycle of water adsorption, the surfaces of Cr_2O_3 and ZnO have no hydroxyl groups, and condensation of the first water layer occurs at very low pressures [171, 172]. So, at strongly hydrophilic nonhydroxylated surfaces of metal oxides, the layering transition occurs at $P < 10^{-4} P_0$. On the surfaces of alkali halide crystals, layering transition of water occurs approximately within the same range of a vapor pressure, as in the case of hydroxylated surfaces of metal oxides (Fig. 12). In the case of NaF [162], this pressure is rather high ($P \approx 0.20$ to 0.30 of P_0), whereas in the case of NaCl [182, 184], CaF_2, and SrF_2, it is by about one order of the magnitude lower.

Differences in the pressure of the layering transitions should be attributed first to the different strength of the water–surface interaction. This strength should correlate with the isosteric heat of adsorption q at

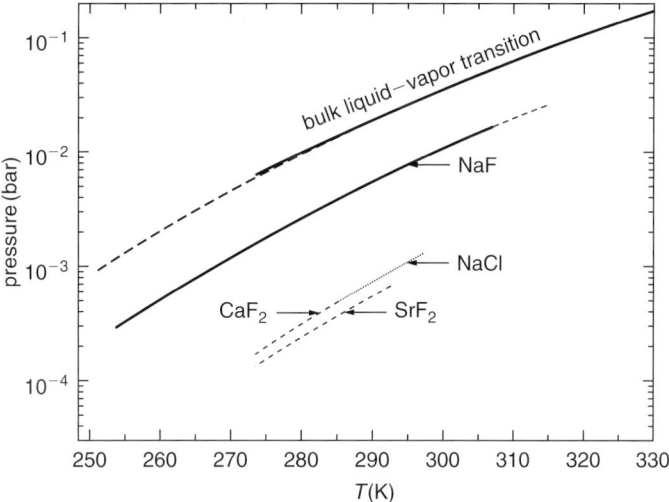

Figure 12: Layering transition of water on the surfaces of alkali halide crystals in the pressure–temperature plane [162–164]. Solid line shows layering transition of water on the surface of NaF. Dotted lines: pressures at the inflection points of the adsorption isotherms, which may correspond to the layering transition or to its extension in the supercritical region. Liquid–vapor bulk transition of water is shown by solid line [3], and its extension to supercooled region by Antoine equation [161] is shown by dashed line.

the coverage, which corresponds to the step-like increase of adsorption. q may be estimated from the slope of the Clausius-Clapeyron line ($\log(P)$ vs $1/T$). For water adsorption on the surface of Cr_2O_3, $q \approx 14.5$ to 15.5 kcal/mol [159, 171]. The comparable value of q (about 15 kcal/mol) was reported for water at FeOOH surface [160]. In the case of the SnO_2 surface, the value of q is slightly lower (≈ 13.5 kcal/mol [159]), whereas it is essentially lower in the case of ZnO surface ($q \approx 11.5$ to 12.0 kcal/mol [159, 171, 185]). Heat of adsorption at about monolayer coverage is much higher for nonhydroxylated surfaces of metal oxides: about 20.5 kcal/mol for the crystal surface of MgO [180] and about 33 kcal/mol for the Cr_2O_3 and ZnO surfaces [171, 185]. So, there is qualitative correspondence between the strength of the water–surface interaction (estimated by q in the vicinity of the layering transition) and the pressure P of the layering transition: transition occurs at lower pressures at more hydrophilic surfaces. Similar relationship is valid for the layering transition of water on the surfaces of alkali halide crystals. The lowest value of q (of about 12.0 kcal/mol [162]) was reported for NaF surface, whereas it is noticeably higher for NaCl surface (14.0 kcal/mol [166] to 15.5 kcal/mol [165]), for Ca_2F surface (about 13.5 kcal/mol [163, 167]), and for SrF_2 surface (13.5 to 15.5 kcal/mol [164, 168, 169, 171]).

Layering transition of water can be also studied by computer simulations. Similar to experiment, it appears as a vertical step in adsorption isotherm. Monte Carlo simulations in the Gibbs ensemble [186, 187] make it possible to equilibrate directly two coexisting phases. This allows to avoid metastable states and to locate accurately the true (equilibrium) phase transition between two stable states, as well as the densities of the coexisting phases. Simulations in the Gibbs ensemble were used to find the layering transition of water at smooth surface, interacting with water via (9-3) LJ potential [32]. The strength of this potential, characterized by its well depth U_0, was varied to determine the water–wall interactions, which enable appearance of the layering transition. The coexistence curves, corresponding to the first layering transition of water at smooth hydrophilic surface of cylindrical pores of various sizes with $U_0 = -4.62$ kcal/mol, are shown in terms of surface number density (ρ^*) in Fig. 13. Both coexisting phases are quasi-2D phases with water molecules localized in the vicinity of the surface. It is clearly seen from the density profiles of the coexisting phases, shown in Fig. 14.

Surface transitions of water

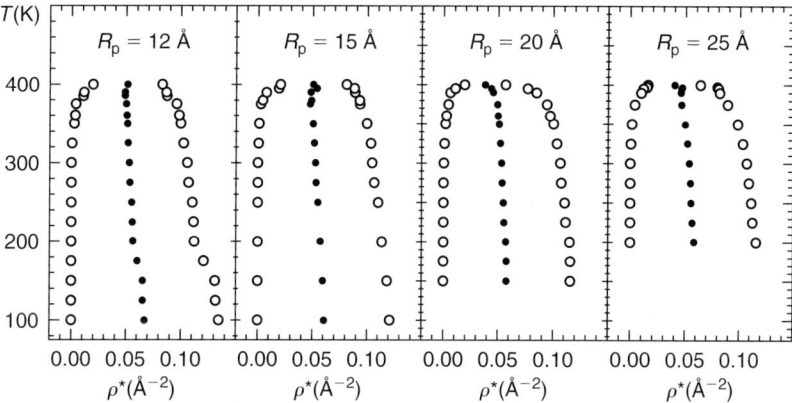

Figure 13: Coexistence curves corresponding to the first layering transition of water in cylindrical pores of various radii R_p and with smooth hydrophilic surface ($U_0 = -4.62$ kcal/mol). Open circles: densities of the coexisting phases. Closed circles: diameter of the coexistence curve.

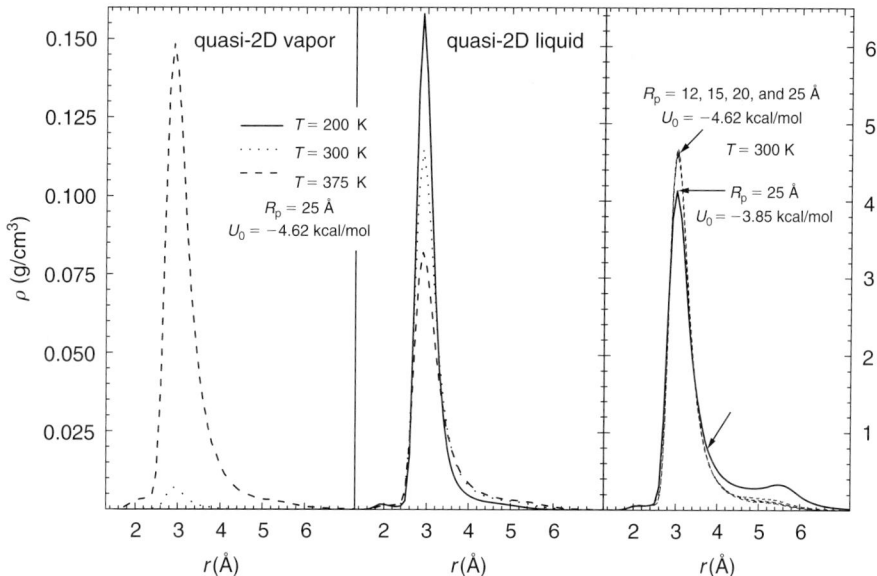

Figure 14: Density profiles of the quasi-2D water phases near hydrophilic wall of the cylindrical pores. Left and middle panels: coexisting quasi-2D vapor and quasi-2D liquid (note the different scales on the ordinates of these two panels). Right panel: quasi-2D liquid water in various pores at $T = 300$ K.

The maximum of the density profiles in both phases coincides with the location of the well depth of the (9-3) LJ water–surface potential, which is at 3 Å from the surface. In the quasi-2D liquid phase, one water molecule occupies about 10 Å^2 of a surface at low temperatures. This value is approximately equal to the projection of the volume occupied by a water molecule in a bulk liquid water with $\rho = 1\,\text{g}/\text{cm}^3$ onto the surface. Arrangement of water molecules in the quasi-2D liquid phase is shown in Fig. 15. At supercooled temperatures, the surface is covered by a dense water layer, which practically does not contain holes. Upon heating from 200 to 375 K, the density of this layer decreases by about 30%, it becomes slightly less localized (see middle panel in Fig. 14), and the holes in the layer appear. However, an infinite hydrogen-bonded water network is always present in a quasi-2D liquid water (see Section 5 for further discussion on the percolation of hydration water).

All four coexistence curves, shown in Fig. 13, are very similar, which indicates a weak sensitivity of the first layering transition to the pore size. Besides, the layering transitions in the slit-like pore and in the cylindrical pore with the same strength of the water–surface interaction are also quite similar (Fig. 16, left panel). The critical temperature of the layering transition of water is just by a few degrees lower in case of the slit-like pore. The degree of the localization of water molecules near the surface is also determined solely by the value of U_0. The density profiles of a quasi-2D

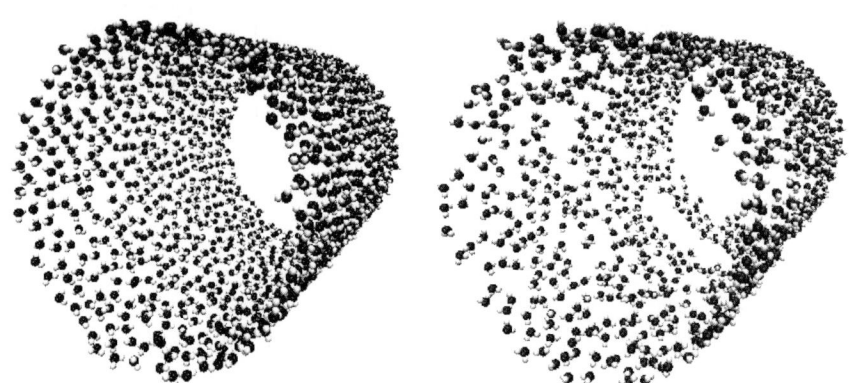

Figure 15: Arrangement of molecules in the quasi-2D liquid water on the inner surface of a cylindrical pore with $R_\text{p} = 25$ Å at $T = 200$ K (left) and at $T = 375$ K (right).

Surface transitions of water

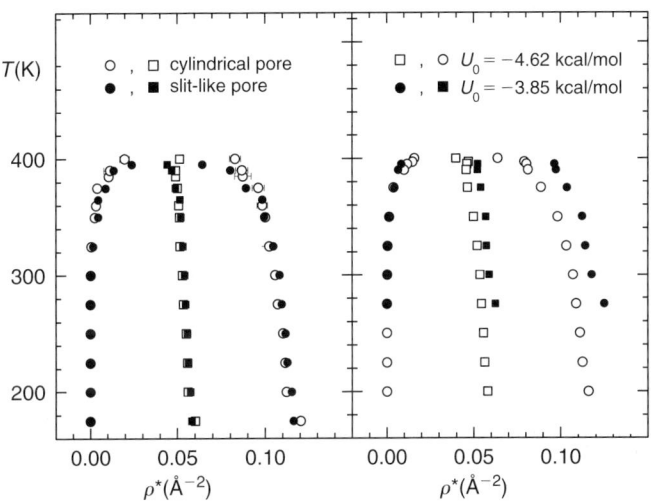

Figure 16: Coexistence curves corresponding to the first layering transition of water in pores with smooth hydrophilic surface. Left panel: cylindrical pore ($R_p = 12$ Å) and slit-like pore ($H_p = 24$ Å) with the same water–wall potential ($U_0 = -4.62$ kcal/mol). Right panel: two cylindrical pores with $R_p = 25$ Å and different water–wall potentials, indicated in the figure.

liquid water in the pores of various radii practically coincide (see dashed lines in the right panel of Fig. 14). So, the first layering transition is not sensitive to the pore size and shape and is determined by the strength of the water–wall interaction. With the weakening of the water–surface interaction from $U_0 = -4.62$ to $U_0 = -3.85$ kcal/mol, the critical temperature does not change practically, whereas the surface density of the quasi-2D liquid slightly increases (Fig. 16, right panel). This simply reflects thickening of the surface layer due to the appearance of a small fraction of water molecules in the second layer (see right panel of Fig. 14). Upon further weakening of the water–surface interaction, the layering transition disappears (see below). Therefore, the temperature $T_c^1 \approx 0.69 T_c$ is the highest possible critical temperature for the layering transition of water. Note that this estimation is valid in the case of the smooth surfaces, when adsorbed water molecules can freely rotate.

With the strengthening of the water–surface interaction, the critical temperature of the layering transition starts to decrease. When the water–surface potential U_0 changes from -4.62 to -7.70 kcal/mol, T_c^1 drops from 400 to 360 K, whereas the surface density of a water monolayer

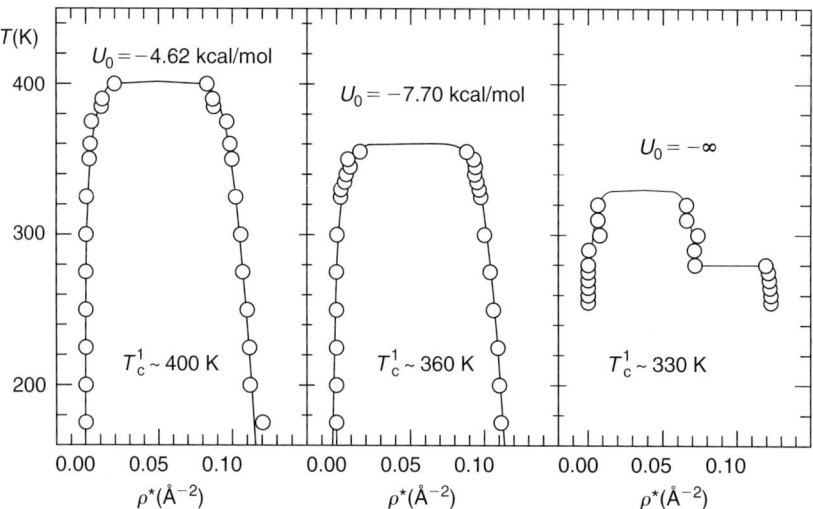

Figure 17: Coexistence curves corresponding to the first layering transition of water in the cylindrical pores with $R_p = 12$ Å and different water–wall potentials (left and middle panels). Coexistence curve of 2D water with all oxygens in one plane ($U_0 = -\infty$, right panel). Solid lines show fits of the coexistence curve to equations (4) and (5). Horizontal line in right panel indicates liquid–solid transition of 2D water.

and the shape of the coexistence curve do not change (Fig. 17). A similar decrease in the critical temperature of the first layering transition with the strengthening of a fluid–wall interaction was observed for the lattice-gas model [188, 189] and for a LJ fluid [190]. It reflects an improving two dimensionality of the system due to the stronger localization of molecules in a plane parallel to the pore wall. Further strengthening of the water–surface potential up to the limit $U_0 \rightarrow -\infty$ causes localization of all water oxygens, but water rotations remain free. In this limiting case, the critical temperature of the layering transition is about 330 K, that is $\approx 0.57 T_c$ for the considered water models. The surface density of a 2D water is about 0.07 Å^{-2}, which is noticeably lower than the value 0.10 Å^{-2} for the quasi-2D water for other studied system with a finite surface attraction. At $T \approx 280$ K, 2D water layer freezes into 2D ice with a surface density of about 0.12 Å^{-2}. Structure of liquid and solid surface water layers is considered in Section 5.2.

The shape of the coexistence curve is determined by the temperature dependences of the order parameter $\Delta \rho$ and of the diameter ρ_d.

Surface transitions of water

The order parameter of the layering transition $\Delta\rho = (\rho_2^* - \rho_1^*)/2$ measures the dissimilarity between the coexisting phases and is equal to zero at $T \geq T_c^1$. Below T_c^1, $\Delta\rho$ should follow the universal scaling law:

$$\Delta\rho = \left(\rho_2^* - \rho_1^*\right)/2 = B\left(\frac{T_c^1 - T}{T_c^1}\right)^\beta = B\tau^\beta, \quad (4)$$

where β is the critical exponent of the order parameter, which is expected to be 0.125, as the layering transition should belong to the universality class of the 2D Ising model. The temperature dependence of the order parameter $\Delta\rho$ of some layering transitions is shown in Fig. 18 as a function of the reduced temperature τ in a double-logarithmic scale. The data shown for two cylindrical pores with different water–surface interaction and for one slit-like pore well agree with the law $\Delta\rho \sim \tau^{0.125}$, expected for 2D systems. The amplitude B is determined by the water–water interaction and therefore should be universal for the same structure of a water monolayer. Indeed, $B = 0.0624 \pm 0.0004\ \text{Å}^{-2}$ for the layering transition of water in five pores with $U_0 = -4.62$ kcal/mol and in one pore with $U_0 = -7.70$ kcal/mol. Higher value of B of about $0.07\ \text{Å}^{-2}$ is seen for

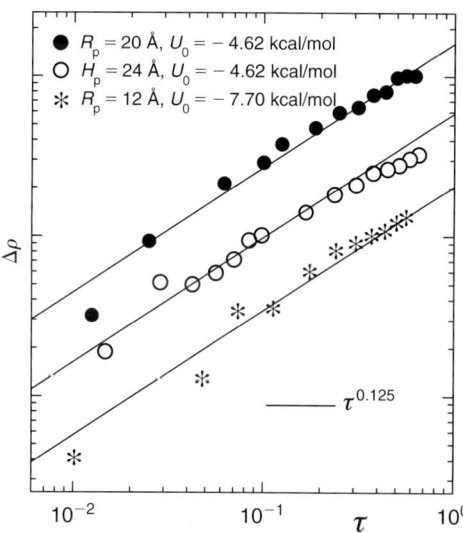

Figure 18: Order parameter $\Delta\rho$ of the first layering transition of water as a function of the reduced temperature $\tau = 1 - T/T_c^1$ in various hydrophilic pores. The lines indicate the power law expected for 2D systems.

water in less hydrophilic pore with $U_0 = -3.85$ kcal/mol. This trend reflects the thickening of a water layer (see the right panel of Fig. 14, where these layers are compared). Amplitude B, which characterizes extension of the coexistence curve in a density range, is noticeably lower for 2D water (compare the width of the coexistence curves, shown in Fig. 17). This should reflect effective weakening of the hydrogen-bonded water network due to the location of all oxygens in one plane.

The diameters of the layering transitions decrease with increasing temperature (Figs. 13 and 16). The temperature dependence of ρ_d can be fitted to the regular equation:

$$\rho_d = \left(\rho_2^* + \rho_1^*\right)/2 = \rho_c^* \left(1 + A_1 \tau + A_2 \tau^2\right), \qquad (5)$$

where ρ_c^* is a critical surface density and A_1 and A_2 are coefficients. The diameters of the layering transitions almost linearly depend on temperature (A_2 is close to zero), but it may show some anomalous trend with approaching T_c^1. To estimate more accurately the critical density of the first layering transition of water, we make a master plot, where the temperature is rescaled by T_c^1 and the surface density is rescaled by ρ_c^* (see Fig. 19). The coexistence curves and diameters of the layering transitions in five different pores with $U_0 = -4.62$ kcal/mol coincide well when the critical density is about 0.044 Å^{-2}. Note the negative anomaly of the diameter with approaching T_c^1, whose temperature behavior is consistent with $\sim \tau^{2\beta} = \tau^{0.25}$ [191].

So, the critical temperature of the first layering transition of water on smooth surface varies within interval $0.57\, T_c \leq T_c^1 \leq 0.69\, T_c$. The restriction on water rotations, as well as the heterogeneity of the surface [157], should decrease T_c^1. Therefore, the critical temperatures of the first layering transitions for real surfaces are expected to be lower. In the case of a realistic silica surface, the first layering transition of water is clearly seen in simulations at $T = 300$ K [192]. There is only one experimentally determined critical temperature of the layering transition of water: $T_c^1 = 305$ K $= 0.48\, T_c$ for water on the hydroxylated surface of Cr_2O_3 [159]. The low experimental value of T_c^1 is in accord with the restriction of water rotations due to the hydrogen bonding with the surface hydroxyl groups. The values of T_c^1 for the first layering transition of water are comparable with the critical temperatures of the layering transition of other fluids, which is typically about $0.30\,T_c$ to $0.55\,T_c$ and strongly depends on the dimensional incompatibility between the adsorbate molecules and substrate [151]. For comparison, T_c^1/T_c is about 0.40 for LJ fluid [152].

Most importantly, T_c^1 for water is essentially higher than the bulk freezing temperature. The similar behavior is observed for ethylene [155, 193, 194], ethane [195], propane [196], molecular oxygen [197], and alcohols [198]. The freezing temperature of 2D water is about $0.49\,T_c$ (Fig. 16), whereas the bulk freezing temperature for the same (TIP4P) water model is about $0.41\,T_c$ [199]. So, there is a rather wide temperature interval where the layering transition of water is a phase transition between two fluid phases. Available experimental estimations of the critical temperature of the layering transition for water give the values $T_c^1 \geq 0.48\,T_c$ [159, 162], which is also above the bulk freezing temperature. Lower experimental values of T_c^1 could be explained by the restrictions on the rotation of water molecules and by the roughness of the surfaces in experiments. Rotational motion of molecules adsorbed on the surface effectively decreases the 2D character in of the system and should cause an increase in T_c^1. This could explain why the critical temperatures of the layering transition of spherically symmetrical molecules (as noble gases, for example) are below the freezing temperature, whereas for strongly asymmetrical molecules (ethylene, ethane, propane, water, and alcohols), the situation is opposite (see Ref. [28] for more details).

The first-order phase transition, corresponding to the condensation of a water layer on various hydrophilic surfaces, as well as their critical points, may be expected at ambient and biologically relevant temperatures. So, strong density fluctuations of hydration water at hydrophilic surfaces should occur in many typical thermodynamic conditions at the earth's surface. In particular, this may be important for the functioning of living organisms, as biosurfaces are covered by a layer of hydration water. Besides, the phase transition is directly related to the percolation transition, which indicates the formation of an infinite and homogeneous hydrogen-bonded network in the case of water (see Sections 5 and 7 for more details). Of course, on real surfaces (including biosurfaces), the layering transition may be smeared out due to the structured character of the surfaces and due to their chemical heterogeneity, but these factors cannot shift the critical temperature noticeably. The transition from the quasi-2D gas-like state, containing small water clusters only, to the quasi-2D state

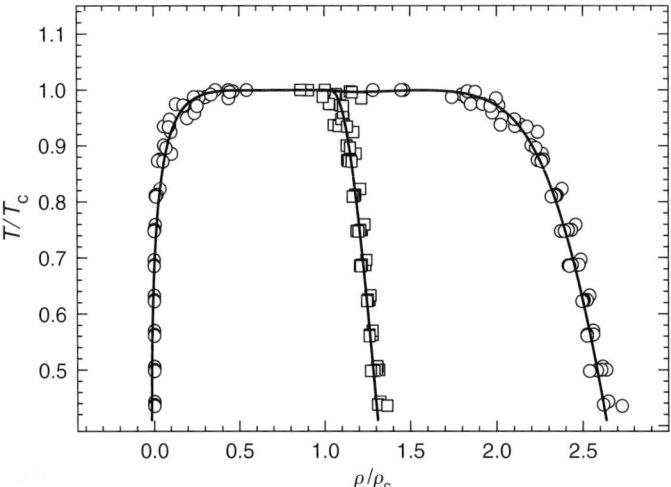

Figure 19: Master plot of the layering transitions of water in five pores with water–wall interaction strength $U_0 = -4.62$ kcal/mol (four cylindrical pores with $R_p = 12$ to 25 Å and one slit pore with $H_p = 24$ Å). The densities of the coexisting phases (circles) and the diameters (squares) are rescaled by the critical density ($\rho_c^* = 0.044$ Å^{-2}). The temperature is rescaled by the critical temperature. The solid lines represent fits of the average coexistence curve and of the average diameter to equations (4) and (5) with the critical exponent $\beta = 0.125$ and amplitude $B = 1.55\rho_c^*$.

of a condensed water monolayer should occur when the average surface hydrophilicity exceeds some critical level. In accordance with the estimations given above, this transition may be expected when the average interaction between one water molecule and the surface is stronger than about -4 kcal/mol, i.e. it becomes comparable with the typical energy of the water–water hydrogen bonds.

For many fluids, not only the first but many more (up to 17 [150]) subsequent layering transitions are observed. In general, the layering critical temperature with increasing layer number approaches the so-called roughening temperature T_r, which is below the freezing temperature. The surface of a solid phase is flat below T_r and rough above this temperature. So, we may expect that the critical temperature T_c^2 of the second and subsequent layering transitions of water should be lower than T_c^1. The second layering transition of water, i.e. quasi-2D condensation of water molecules on the surface already covered by water monolayer,

Surface transitions of water

was never observed in experiment but was detected in the simulations of water phase diagram in some hydrophilic pores [32]. In Fig. 20, the arrangement of water molecules in two coexisting phases (monolayer and bilayer) is shown in a cylindrical pore. Upon the second layering transition, water density in the first layer remains intact. In fact, this transition is a quasi-2D condensation of water molecules in the second layer.

In Fig. 21, the coexistence curves corresponding to the first and second layering transitions are shown for various pores. As expected,

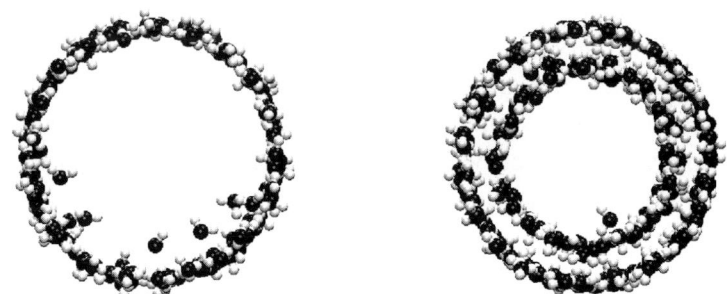

Figure 20: Arrangement of water molecules in two coexisting phases, corresponding to the second layering transition of water in cylindrical pore with $R_p = 12$ Å and $U_0 = -7.70$ kcal/mol (view along the pore axis). Left: water monolayer. Right: water bilayer.

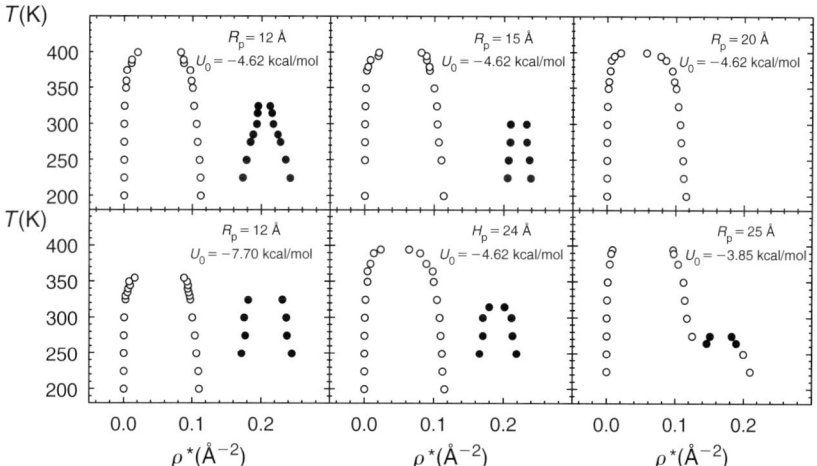

Figure 21: Coexistence curves corresponding to the first (open symbols) and second (closed symbols) layering transitions of water in various pores.

the critical temperature T_c^2 of the second layering transition is lower than T_c^1. In particular, in the pores with $U_0 = -4.62$ kcal/mol, T_c^2 is about 300 to 325 K, whereas $T_c^1 \approx 400$ K. With strengthening of the water–surface interaction to $U_0 = -7.70$ kcal/mol, T_c^1 shifts down by about 40°, but T_c^2 remains unaffected. When the surface hydrophilicity weakens and approaches the limiting value, which may provide layering transition(s), T_c^2 drastically falls down. Interestingly, the density interval corresponding to the second layering transition becomes wider for more hydrophilic surface. With increasing the size of the cylindrical pore with $U_0 = -4.62$ kcal/mol, the second layering transition of water first shrinks in the density range and finally disappears. Accordingly, on a flat surface ($R_p \to \infty$), this transition should be absent. In relatively narrow cylindrical pores, the second layering transition is stabilized, presumably, by the concave surface curvature. The critical temperature of the second layering transition depends on the structure of the first adsorbed water layer rather than on the water–surface interaction.

The critical temperature of the second layering transition is always lower than the critical temperature of the first one, and it was found between $0.48\, T_c$ and $0.59\, T_c$. Similar behavior was found by density functional calculations for a strongly associative LJ fluid in pores [201–203]. Even near strongly hydrophilic surfaces, there are only two noticeable density oscillations of liquid water when it is in equilibrium with a saturated vapor. Water properties (e.g., orientational ordering) in the third and subsequent layers are close to the bulk ones [205–208]. Therefore, the third and subsequent layering transitions of water should not be expected. When the water–surface interaction weakens, the critical temperature of the second layering transition drops down by about 50°, and the triple point, where 2D gas coexists with water monolayer and with water bilayer, may be seen (lower right panel in Fig. 21).

When the water–surface interaction weakens further, the layering transitions disappear and the appearance of the prewetting transition may be expected. Prewetting transition is a first-order phase transition, which occurs (similarly to the layering transitions) at some undersaturated vapor pressure and indicates condensation of a water film on the surface (see Section 2.1). The line of the prewetting transition meets the bulk liquid–vapor transition in a triple point, where bulk vapor and bulk liquid coexist with a water film. The temperature of this triple point is a temperature of the wetting transition. The prewetting transition and the wetting transition of water were never observed experimentally. Absence of the experimental observations of the prewetting transition of water is not surprising, as such studies are complicated by the proximity of the coexisting pressure to the bulk liquid–vapor coexistence and it was reported just for a few fluids (see Section 2.1). The wetting transition upon heating is unavoidable for some range of the fluid–surface interaction and therefore wetting transition of water must be observable at some hydrophilic surfaces. However, as we show below, this interaction range is rather narrow. The strongly hydrophilic surfaces considered above, which are characterized by the occurrence of the layering transition(s), are covered by water film already at the bulk melting temperature. For less hydrophilic surfaces, the wetting transition should occur above the melting temperature, but the temperature T_w may quickly increase with the weakening of the water–surface interaction. In parallel, the wetting transition may change from the first order to second order. As a result, the prewetting transition disappears, and the second-order wetting transition, characterized by the continuous growth of a wetting water film, cannot be easily detected.

Variation in the water–surface interaction can be easily done in simulations of the phase diagram of water in pores. Pore phase diagrams clearly distinguish the surface phase transitions and the bulk phase transition, which appears as a capillary condensation. The collection of the phase diagrams of water in pores with the different surface hydrophilicity is shown in Fig. 22. In the pore with $U_0 = -4.62$ kcal/mol, there is a first layering transition of water and no second one. So, below 400 K, condensation of the first water layer upon increasing pressure occurs discontinuously, but formation of the second layer occurs in continuous fashion. Above 400 K, both layers form continuously. Capillary condensation, that is liquid–vapor phase transition in pore, takes place in the inner part of the pore, i.e. on the surface covered already by a water film. The thickness of the water film at the liquid–vapor coexistence is just two water layers. It is not clear whether in the limit of a semiinfinite system this film turns into macroscopic or it still contains just two layers. In any case, the presence of this water film in a vapor phase is a signature of complete wetting. Clearly, the temperature of the first-order wetting transition of water at this surface may be in deeply supercooled region.

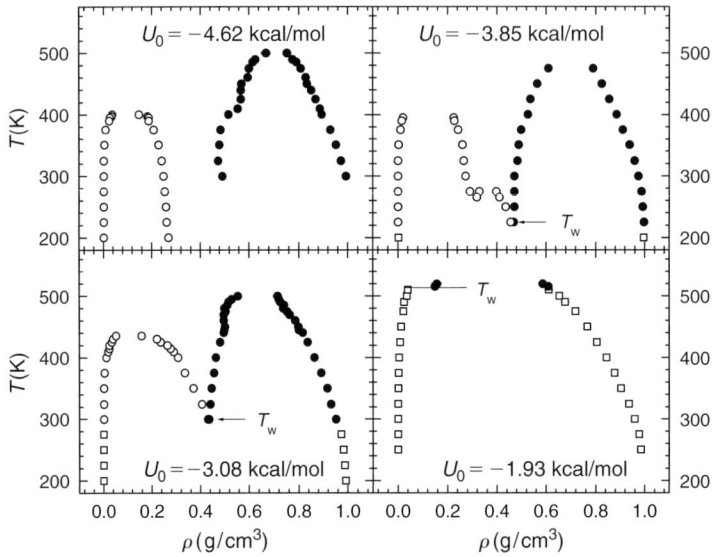

Figure 22: Coexistence curves of water in cylindrical pores with $R_p = 25$ Å and different strength of the water–surface interaction. Surface phase transitions (layering and prewetting) are shown by open circles. Coexistence between vapor and liquid phase is shown by open squares. Coexistence between the water film and the liquid water is shown by closed circles.

Weakening the water–surface interaction from -4.62 to -3.85 kcal/mol noticeably changes the water phase diagram, and the prewetting transition along with two layering transitions may be seen. The triple point, where water vapor coexists with liquid water and water film, indicates the temperature of the first-order wetting transition (T_w) in supercooled region. Similar triple point was observed in computer simulations of lattice-gas model [139, 209]. When the water–surface interaction weakens further to -3.08 kcal/mol, only two phase transitions occur: prewetting transition and capillary condensation. The temperature of the first-order wetting transition appears above the melting temperature at about 300 K. The arrangement of water molecules in three phases (vapor, film, and liquid), coexisting at 300 K, is shown in Fig. 23. The water film formed in a vapor phase upon increasing pressure prior to the capillary condensation contains just two water layers, as in the previous two cases. The density profiles of the coexisting phases are shown in Fig. 24. Two density profiles are shown for the water film at $T = 400$ K: less dense film coexists with a vapor, whereas a denser film coexists with a liquid.

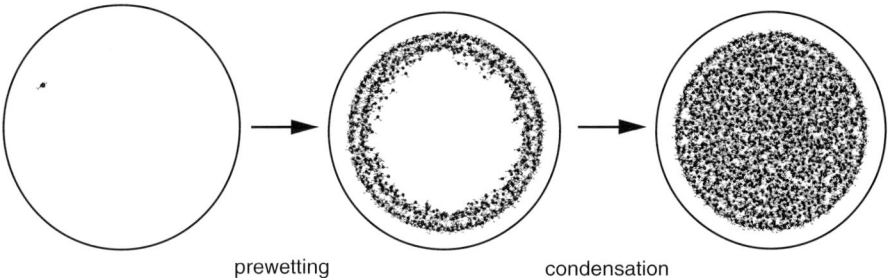

Figure 23: Arrangement of water molecules in three water phases in cylindrical pore with $R_p = 25$ Å and $U_0 = -3.08$ kcal/mol, which coexist at 300 K. In all cases, molecules in the pore of 50 Å length are shown.

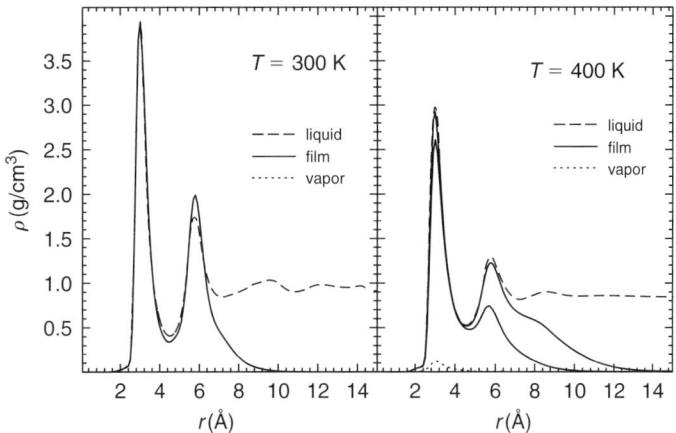

Figure 24: Density profiles of the coexisting water phases, corresponding to the prewetting transition near the hydrophilic surface of the cylindrical pore with $R_p = 25$ Å and $U_0 = -3.08$ kcal/mol.

In latter case, two water layers adjusted to the surface are identical in both coexisting phases, so they remain intact upon capillary condensation. This means that condensation of water occurs in a pore with a wall already covered by two "dead" water layers.

The temperature of the wetting transition is sensitive to the used water model and to the pore geometry. The phase diagram of ST2 water in slit-like pore of 24 Å width and with $U_0 = -4.62$ kcal/mol is shown in Fig. 25. The temperature of the wetting transition is by about 40° higher than in the case of TIP4P water in cylindrical pore with the same water–surface interaction. In parallel, the critical temperature of the prewetting

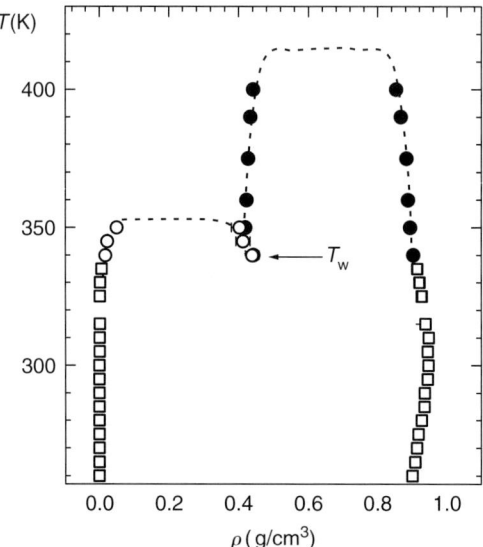

Figure 25: Coexistence curves of water in a slit-like pore with $H_p = 24$ Å and $U_0 = -3.08$ kcal/mol. Prewetting transition is shown by open circles. Coexistence between vapor and liquid phase is shown by open squares. Coexistence between the water film and the liquid water is shown by closed circles.

transition is by about 80° lower. Accordingly, the prewetting transition extends in the temperature range of about 150° for TIP4P water and just 15° for ST2 water. This difference should be attributed to the difference in the bulk phase diagrams of these water models. The temperature interval between the liquid density maximum and the liquid–vapor critical point is ≈330° for TIP4P water and ≈250° for ST2 water (it is about 375° for real water).

Coexistence curves of the prewetting transitions of water in various pores should be compared not in terms of the average density in the pore (as in the Fig. 22) but in terms of a surface density ρ^*. Such comparison is shown in Fig. 26, where the surface density ρ^* in cylindrical pore is calculated with the respect to the cylindrical surface between the first and the second water layers. All three coexistence curves show quite similar extension in the surface density up to 0.20 Å^{-2}, which is about two times the corresponding density interval for the first layering transition. When the critical temperature T_c^{pw} of the prewetting transition increases, the critical density shifts to lower values. The critical point of the prewetting

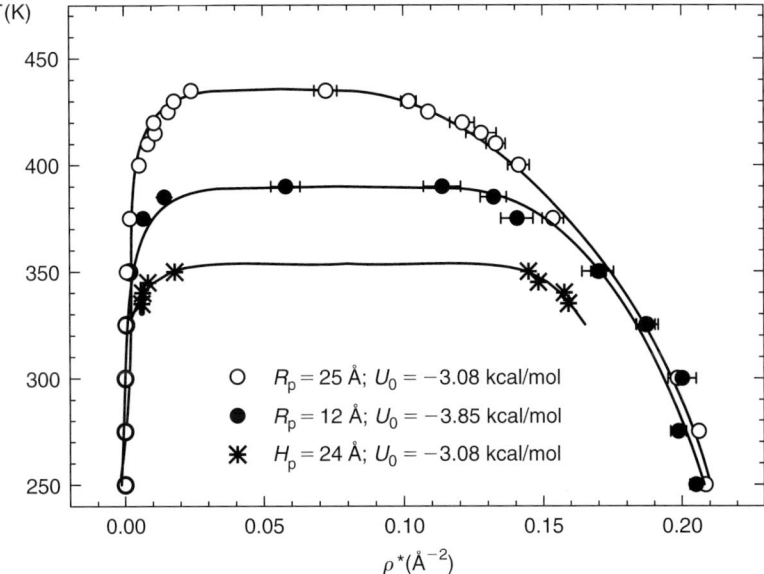

Figure 26: Coexistence curves of water corresponding to the prewetting transition in various pores in terms of a surface density (symbols). The fits of the coexistence curves to the equations (4) and (5) are shown by lines.

transition belongs to the universality class of 2D Ising system. The fits of the coexistence curves of the prewetting transition with a scaling law (4) for the order parameter and with a regular equation (5) for diameter are shown by lines in Fig. 26. For the coexistence curve with the lowest critical temperature, order parameter can be satisfactorily fitted with 2D critical exponent $\beta = 0.125$. However, when the prewetting extends to higher temperature, the effective value of the exponent β starts to increase: $\beta \approx 0.18$ when $T_c^{pw} = 390$ K, and $\beta \approx 0.28$ when $T_c^{pw} = 435$ K. These changes are caused by rather quick depletion of the prewetting film upon heating. At low temperatures, the prewetting film is formed by two completed water layers, but its thickness approaches one layer at high temperatures. Accordingly, the critical density of the prewetting transition moves toward one of the first layering transition when T_c^{pw} is high. The depletion of the prewetting film upon heating is a major factor that determines the shape of the prewetting coexistence curve in a wide temperature range.

Drastic changes in the phase diagram of water occur when the water–surface interaction weakens by just 1 kcal/mol (from −3.08 to

−1.93 kcal/mol). The prewetting transition disappears; instead a step in the vapor branch of the coexistence curve is seen between 510 and 515 K (Fig. 22). This step may indicate the second-order wetting transition, which is expected when the temperature of the wetting transition approaches the bulk liquid–vapor critical temperature [113]. Analysis of the density profiles of the vapor phase at various temperature shows essential increase in the adsorbed water layer when crossing the temperature of the wetting transition (Fig. 27). Increase in temperature of just 5° makes the density of water layer several times higher. However, upon condensation above the temperature of the wetting transition, density of water layer increases noticeably (dashed curve in Fig. 27), whereas identical water films near the surface are expected in two coexisting phases in case of wetting transition. It is not clear whether this effect may be

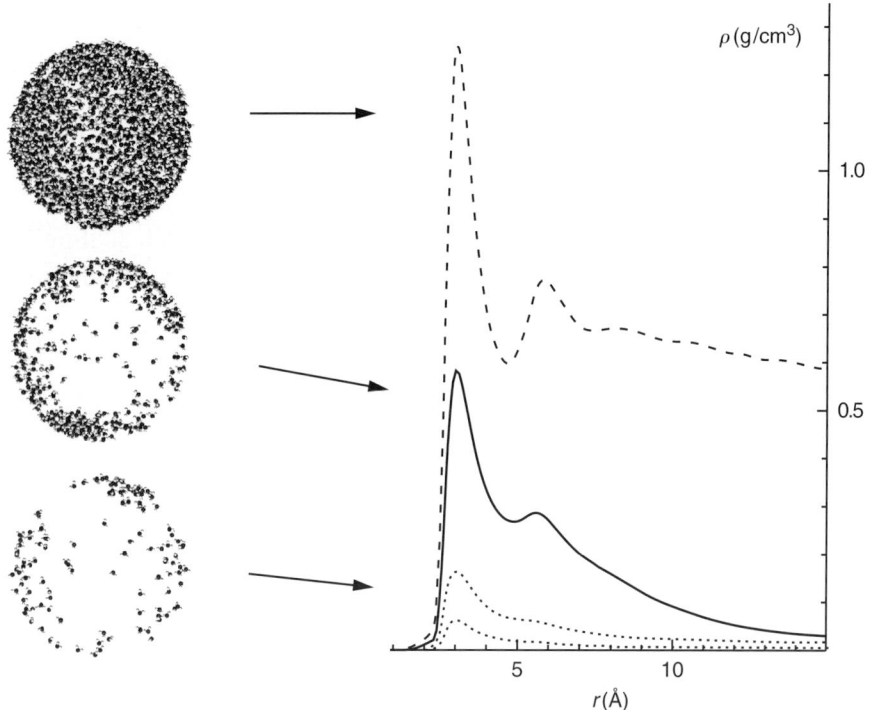

Figure 27: Density profiles of water in a cylindrical pore with $R_p = 25$ Å and $U_0 = -1.93$ kcal/mol. Solid and dashed lines correspond to the vapor and liquid phases, coexisting at $T = 515$ K. Two dotted lines correspond to the vapor phase at $T = 450$ K and $T = 510$ K.

attributed to the shift of the liquid–vapor transition in confinement. Note that qualitatively different behavior of water condensation is observed at more hydrophilic surfaces even upon confinement in narrower pores: two water layers are identical in both coexisting phases (see Fig. 24).

In simulations, the thickness of a wetting layer above the wetting temperature does not exceed two molecular layers, i.e. it remains microscopic. There are two possible explanations for this behavior. First, a wetting layer may be suppressed by confinement, which causes the shift of the chemical potential relatively to the bulk liquid–vapor coexistence [120]. If so, a macroscopic wetting layer may be expected in a semiinfinite system, and thicker wetting layer should be detected in simulations of water in very large pores, where the liquid–vapor coexistence is closer to the bulk one. Such scenario seems to be relevant for moderately hydrophilic surfaces, which become wet at high temperatures only. Alternatively, the wetting transition at T_w may be a condensation of a water bilayer even in a semiinfinite system. This expectation is supported by a numerous evidences on the specific structure of exactly two water layers adsorbed at hydrophilic surfaces. Such wetting transition may be caused by the short-range water–surface interaction and results in the essential modification of the surface. Water molecules, adsorbing on such modified surface, interact with a solid surface via a long-range part of the water–surface potential. Presence of the water bilayer effectively suppresses a missing neighbor effect for water molecules in a third layer. So, an attractive tail of a long-range surface-water interaction represents overall effect of surface and water bilayer on rest of the water molecules. Therefore, the temperature of the wetting transition of water on the surface of a highly structured water bilayer should be expected not very far from the critical point. A wetting film is expected to grow continuously when approaching the temperature of the second wetting transition, which is a temperature of the critical wetting transition governed by the long-range fluid–wall forces [210, 211]. Taking into account that the wetting transition of water with formation of a thick liquid film was never observed experimentally, the latter scenario with two sequential wetting transitions seems to be more realistic. This scenario seems to be consistent with the growth of the wetting film in a pore with strongly attractive walls. The density profiles of the phases at the liquid–vapor coexistence in a cylindrical pore with $R_p = 25$ Å and $U_0 = -4.62$ kcal/mol are shown in Fig. 28.

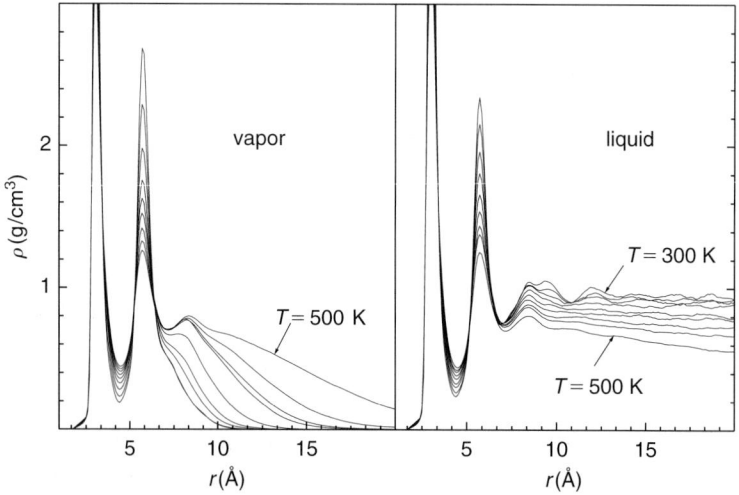

Figure 28: Density profiles of water in a cylindrical pore with $R_p = 25$ Å and $U_0 = -4.62$ kcal/mol at $T = 300, 325, 350, 375, 400, 425, 450, 475$, and 500 K.

Two water layers near the surface are identical in two phases in a wide temperature range from 300 to 500 K. Water adsorption on these two "dead" layers enhances upon heating (see right panel in Fig. 28). This may be considered a formation of wetting layer with approaching temperature of the wetting transition, which is close to the bulk critical temperature.

2.3 Drying transition of water near hydrophobic surfaces

The interfaces between liquid water and hydrophobic surfaces are abundant on the earth and involved in various phenomena, such as attraction between extended hydrophobic surfaces [212], slipping flow of liquid water near a hydrophobic surface [213], etc. Knowledge of the density distribution of a liquid water near hydrophobic surfaces is important for the understanding of the mechanisms and driving forces of these phenomena. For liquid water near hydrophobic surfaces, we may expect density depletion due to two main reasons. *First*, absence of fluid molecules from the side of a solid surface causes worsening of the potential energy of the fluid molecules near the surface. The effect of a missing neighbor is universal for any system of interacting particles near a boundary, including fluids near solid surfaces. This effect is governed by the bulk correlation length and correlates with the average number of the nearest neighbors in a bulk fluid. When the missing neighbor effect is not compensated by an attractive fluid–surface interaction, fluid density is depleted near the surface. Density depletion near weakly attractive surfaces may be seen for a fluid in a wide range of thermodynamic states, including a supercritical region and a vapor phase. As the effect of a missing neighbor is governed by the bulk correlation length, which diverges with approaching liquid–vapor critical temperature, it should be considered in the framework of the surface critical behavior of fluids (see section 3). The *second* factor, which may cause density depletion of a liquid near a weakly attractive surface, is a drying layer. In this section, we consider the effect of a drying layer on the liquid density profiles and an interplay of this effect with the effect of confinement in pores.

Drying transition may occur in a liquid phase upon heating along the liquid–vapor coexistence curve (see Section 2.1). This transition has drastic effect on the liquid–solid interface: above the temperature T_d of a drying transition, the liquid is separated from the solid surface by a macroscopic vapor layer. However, even below T_d and out of the liquid–vapor equilibrium, distant effect of the drying transition may noticeably affect the liquid density profile. Therefore, it is important to know the temperature of the drying transition of water and its sensitivity to the water–surface interaction. This allows description of the density profiles of liquid water near hydrophobic surfaces at various thermodynamic conditions.

Nanobubbles seen in a liquid water near some hydrophobic surfaces [145, 214–221] may indicate the formation of a water vapor layer. However, the amount and size of these nanobubbles decrease essentially in degassed water [145] and increase due to the contact with air [216, 217]. These facts indicate that the nanobubbles originate from the gases dissolved in water and do not represent a water vapor layer expected above the temperature of a drying transition. Clearly, the presence of a dissolved gases may strongly affect the properties of the interface between water and hydrophobic surface. For example, the dissolved gases may be adsorbed at the hydrophobic surface without producing nanobubbles, and this adsorption enhances slipping of a liquid water over a hydrophobic surface [222].

Various experimental methods were used to study density depletion of a liquid water near hydrophobic surfaces. In some experimental studies (ellipsometry [223, 224] and neutron reflectivity [225]), density depletion was not found [225]. In many other studies (neutron reflectivity [145, 226–228], X-ray reflectivity [229–231], ellipsometry [232], thermal conductivity [233], liquid water intrusion in hydrophobic pores [234]), noticeable depletion of the liquid water density near various hydrophobic surfaces was detected. Density depletion was found sensitive to the presence of dissolved gases [227], but no such sensitivity was observed in other studies [230, 231]. The available experimental methods do not give explicit profiles of a liquid water density but rather allow estimation of a density deficit near the surface. For simplicity, this deficit may be attributed to the homogeneous vapor film (the so-called slab approximation), whose thickness D can be used as a measure of a liquid density depletion. If the realistic liquid density profile is close to a sigmoid-like function, D approximately corresponds to a distance between the inflection point of this function and the surface.

In experimental studies, the effective thickness D of a depletion layer is about 6 Å for silica surface hydrophobized with the monolayer of trimethylsilil groups [234]. For water near a octadecyltriethoxysiloxance monolayer, D is between 0.8 and 4.0 Å [230]. Near octadecyltrichlorosilane monolayer, various values of D were reported: 1.1 Å [231], <2.5 Å [233], and 5.2 Å [227]. The thickness of a depletion layer of liquid water near a film of polystyrene was estimated as 5 to 10 Å by ellipsometry studies [232] and as 1.3 to 3.0 Å by neutron reflectivity experiments [145]. These estimations of D clearly evidence the absence of a *vapor* layer at hydrophobic surfaces. Even if one attributes liquid density depletion to the vapor layer only, the thickness of this layer is just two to three molecular diameters, which is comparable to or less than the width of the liquid–vapor interface.

So, a stable macroscopic vapor layer between a liquid and a solid or its formation via a surface phase transition (drying transition) was never observed experimentally for water and for any other fluid. This well agrees with the theoretical expectations [117, 118, 124] that a drying transition is suppressed by the long-range fluid–surface interaction up to the bulk liquid–vapor critical temperature T_c. Even for the extremely weak fluid–surface interface (liquid neon on cesium), formation of a macroscopic vapor layer was not observed even close to the liquid–vapor critical temperature T_c [136]. As the long-range dispersion forces between fluid molecules and a surface are unavoidable, we may expect the drying transition of water (and of other fluids) at T_c. Accordingly, the formation of a thin vapor layer outside the liquid–vapor coexistence via a predrying transition is not possible. However, the formation of a thin vapor layer upon heating along the liquid–vapor coexistence via a partial drying transition cannot be excluded at some levels of the surface hydrophobicity [117, 118].

The available experimental techniques give very rough estimation of the liquid density profiles near the surfaces. Currently, quantitative description of the fluid density profiles by experimental methods is problematic, and computer simulation is the main "experiment" that provides such information. In simulations, the properties of a fluid near a surface can be studied in the pore geometry only because a semiinfinite system can not be simulated. Liquid–vapor coexistence of fluids confined in pores occurs at different chemical potential compared with the bulk coexistence, and the shift of the phase transition depends on the pore size, shape, and fluid–wall interaction. Confinement may strongly affect fluid density profiles in narrow pores (see Section 4.2), but insight into the surface transitions may be obtained by the increasing pore size, with some meaningful extrapolation on semiinfinite system.

When considering the shift of the temperature T_w of the wetting transition upon strengthening the water–surface interaction (Section 2.2), T_w was found at about $0.88T_c$, when $U_0 = -1.93$ kcal/mol. Further weakening of the water–surface interaction should shift T_w to higher temperatures and $T_w = T_c$ at some particular value of U_0. In Fig. 29, the density profiles of liquid and vapor water phases in cylindrical pores of various strengths of the water–surface interaction are compared at $T = 520$ K, which is not very far from the critical temperature $T_c = 580$ K of the water model used. When $U_0 = -1.54$ kcal/mol, water density increases toward the surface both in the liquid and in the vapor phases. In a liquid phase, this effect can only enhance upon heating (see Section 3 below). So, for this interaction, a wetting transition may be expected very close to T_c. For more hydrophobic surfaces with $U_0 \geq -1.16$ kcal/mol and $U_0 \geq -0.77$ kcal/mol, liquid density depletes near the surface, whereas vapor density enhances. At higher temperatures, liquid and vapor density

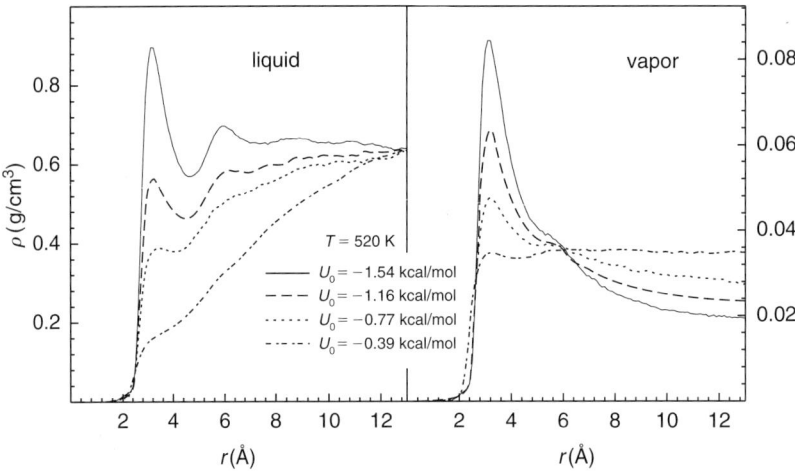

Figure 29: Density profiles of liquid water in cylindrical pores ($R_p = 25$ Å) with smooth surfaces of various strengths U_0 of the water–surface interaction.

profiles approach each other and become identical at $T = T_c$. If the fluid density profile at the critical point is close to horizontal one (do not show adsorption or depletion), the surface is neutral and does not prefer liquid or vapor phase. From the analysis of the density profiles, shown in Fig. 29, we may expect approximately horizontal density profiles for water at the critical point when $U_0 \approx -1.0$ kcal/mol. Only for this unique level of the surface hydrophilicity/hydrophobicity, neither wetting nor drying transition occurs in the system and we cannot expect distortion of the liquid density profiles due to the developing drying layer or distortion of the vapor density profiles due to the developing wetting layer. At all other water–surface interactions, these distortions may noticeably affect fluid density profiles.

Near a hard wall, liquid–vapor coexistence occurs above the temperature of a drying transition. This situation is unrealistic, as the long-range interactions between fluid molecules and solid surface typical of real systems are absent near a hard wall. However, it is useful to use this model surface as a reference one to study the effect of the weak attraction on the density profiles. In Fig. 30, density profiles of liquid water near a hard wall and near a weakly attractive wall with $U_0 = -0.39$ kcal/mol are compared at various temperatures. Even in the case of a hard wall, a drying layer cannot be detected at $T = 300$ K, despite the fact that the

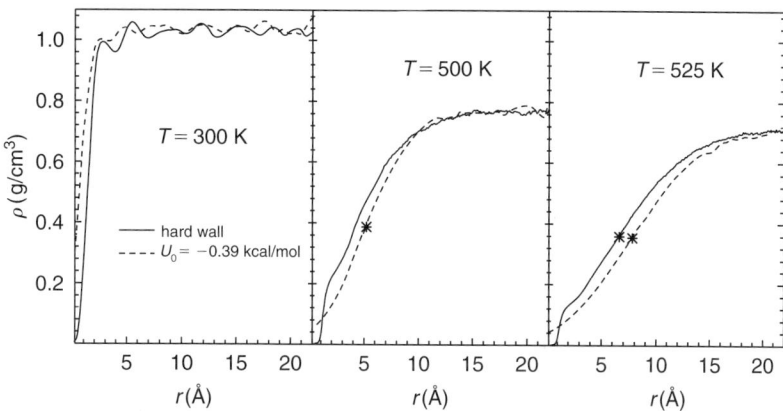

Figure 30: Density profiles of liquid water in cylindrical pores ($R_p = 25$ Å) with hard wall and with weakly attractive wall.

system is definitely above the temperature T_d of a drying transition, and a macroscopic vapor layer should be present near hard wall in semiinfinite liquid coexisting with a vapor. Above T_d, the width L of a drying layer is proportional to the bulk correlation length ξ. Therefore, a drying layer should grow upon heating. Indeed, at higher temperatures, liquid density profiles near a hard and a weakly attractive walls (middle and right panels in Fig. 30) evidence the formation of drying layer. When a vapor layer between a surface and a liquid is macroscopic, a liquid–vapor interface is located at some distance from the surface, and this distance noticeably exceeds the width of the liquid–vapor interface. The density profiles in the region of the liquid–drying layer interface may be described by the interfacial equation

$$\rho_l(z,\tau) = \frac{\rho_l^0(\tau) - \rho_v^0(\tau)}{2} \tanh\frac{z - L}{2\xi} + \frac{\rho_l^0(\tau) + \rho_v^0(\tau)}{2}, \quad (6)$$

where L is a location of the inflection point of the interface with respect to the surface, ξ is a correlation length, ρ_l^0 and ρ_v^0 are the densities of the coexisting bulk liquid and vapor phases. A complete density profile of liquid water near a hydrophobic surface includes also a vapor–solid interface (see Section 2.1). A macroscopic vapor layer is suppressed when the system is out of the bulk liquid–vapor coexistence (for example, due to confinement) or when the temperature is below T_d. In such cases, a liquid–vapor interface is attached to the solid surface and a macroscopic vapor layer is absent. However, a microscopic drying layer strongly affects a liquid density profile, which still can be adequately described by the equation (6). Due to the proximity of the interface to the surface, ρ_v^0 is not a saturated vapor density, but a fitting parameter, which goes to zero with decreasing L.

The location of the inflection points of the interfacial-like density profiles of a liquid water is indicated by asterisks in Fig. 30. The thickness L of a drying layer is about 5.2 Å at $T = 500$ K and about 7.8 Å at $T = 525$ K. When a very weak attractive potential with $U_0 = -0.39$ kcal/mol is applied, inflection point can still be detected and L shrinks from 7.8 to 6.6 Å at $T = 525$ K. At $T = 500$ K, inflection point at the density profile is not seen and liquid density decays exponentially toward the surface. In this case, the drying layer is absent, and the liquid density depletion is determined solely by the missing neighbor effect (see Section 3 for more details).

Analysis of the liquid density profiles in the pores of various sizes can give information about drying layer in a semiinfinite system. Liquid density profiles at $T = 520$ K in various cylindrical and slit-like pores with the same weakly attractive walls are compared in Fig. 31. In slit-like pores, a drying layer is absent even in the wide pore with $H_p = 50$ Å. In cylindrical pores, a drying layer is absent in a pore with $R_p = 15$ Å, but

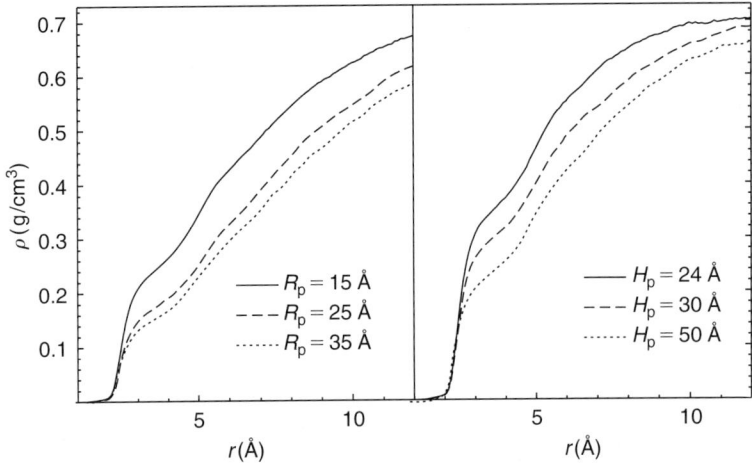

Figure 31: Density profiles of liquid water in cylindrical pores (left panel) and slit-like pores (right panel) with $U_0 = -0.39$ kcal/mol at $T = 520$ K.

it is seen in the pores with $R_p = 25$ and 35 Å. So, a drying layer should be expected near considered surface in a semiinfinite system. Such analysis is necessary in order to estimate a drying layer at each temperature. Clearly, the presence and thickness of a drying layer in a semiinfinite system should depend on the water–surface interaction.

It is important to know, how appearance of a drying layer and its thickness L depend on temperature, pore size and U_0. Available simulation data for water do not allow reliable estimations of the effect of these factors on a drying layer. However, the important knowledge may be furnished from the data for a LJ fluid obtained in much larger pores. In Fig. 32, dependence of the thickness L of a drying layer for LJ liquid near a weakly attractive wall is shown as a function of a reverse pore size H_p. This dependence is close to linear and allows estimation of L in the limit $H_p \to \infty$: $L \approx 2.7\sigma$. The dependence presented was obtained at $T = 0.93T_c$ and for the fluid–wall potential with a well depth U_0 of about 70% of that for the fluid–fluid pair potential. The case presented in Fig. 32 for water is different ($T = 0.90T_c$ and U_0 is just about 10% of a typical pair water–water hydrogen bond), but the rough estimations can be done. Solid circles in Fig. 32 represent L in the liquid water phase

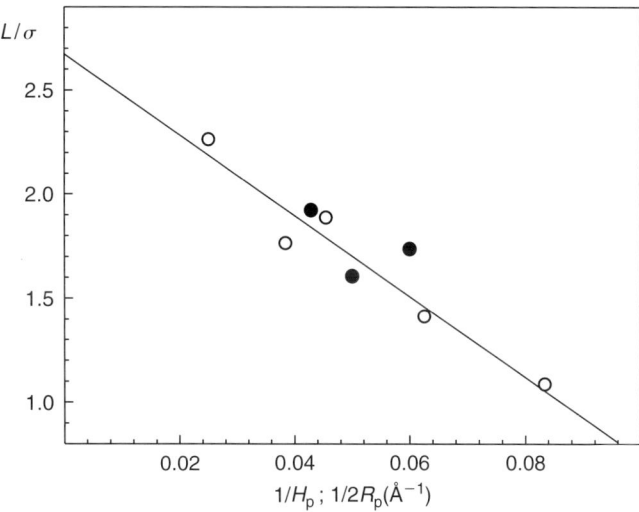

Figure 32: Thickness L of a drying layer as a function of the reverse system size: LJ liquid confined in slit-like pores of width H_p (open circles) and liquid water confined in cylindrical pores of radius R_p (closed circles).

in cylindrical pores with $R_p = 25$, 30, and 35 Å. The pore size and the layer width are normalized by the diameter of water molecule, which is about 3 Å. Extrapolation to semiinfinite system gives the drying layer in liquid water of about 8 Å thick. So, even near the strongly hydrophobic (paraffin-like) surface, a drying layer is strongly attached to the surface at high temperatures.

Temperature dependence of the thickness of a drying layer of LJ fluid near two different weakly attractive walls is shown in Fig. 33 [127]. The reliable estimations of L from the liquid density profiles using equation (6) can be done when L exceeds about 1.5 to 2σ. Near a weakly attractive surface with a well depth of a fluid–wall potential of about 20% of a fluid–fluid one, the thickness L of a drying layer increases with temperature as a correlation length $\xi \sim \tau^{-0.63}$. When the fluid–wall interaction is three times stronger, L decreases and its temperature dependence becomes logarithmic: $L \sim \ln\tau$. So, it seems that the thickness of a drying layer does not increase with temperature in terms of the correlation length, even near strongly hydrophobic surfaces. This means that the effect of drying layer on liquid water profiles near a paraffin-like surface may be notable only in the close proximity of the critical point when the correlation length

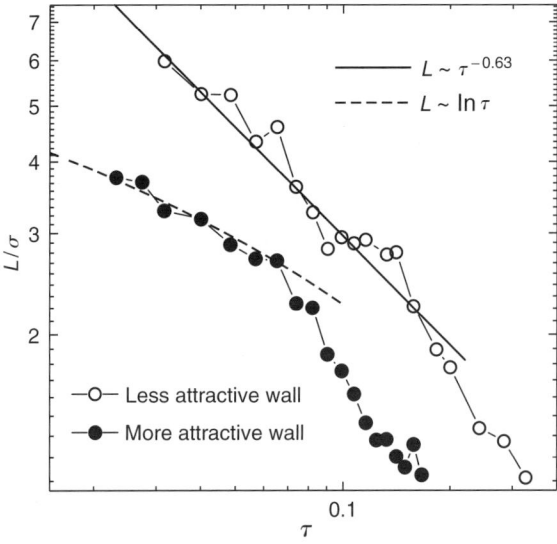

Figure 33: Temperature dependence of the thickness L of a drying layer near two weakly attractive walls ($\tau = 1 - T/T_c$).

diverges at $T \to T_c$. At ambient and modarate temperatures, a drying layer cannot be defined, as it "enters" the first density oscillation caused by the water–surface potential. In these cases, liquid density depletion can be described solely by the missing neighbor effect (see Section 3).

Obviously, when the surface hydrophilicity increases, the drying layer collapses quickly (see left panel in Fig. 29). So, manifestations of a drying layer and, accordingly, an interfacial-like profile of liquid water, are expected to be very rare. Notable drying layer may occur at extremely high temperatures (more than 500 K for liquid water near paraffin-like surface). Appearance of an interfacial-like profile of liquid water cannot be excluded near superhydrophobic surface, which shows a contact angle higher than 150° at ambient temperatures [235]. Finally, a drying layer may be important near the surfaces, which exhibit short-range repulsion of water molecules. We are not aware of the existence of such surfaces in nature, but water shows a first-order predrying transition near the liquid branch of the liquid–vapor coexistence curve in simulations (see Fig. 34) [205]. Thus, for the vast majority of the hydrophobic surfaces

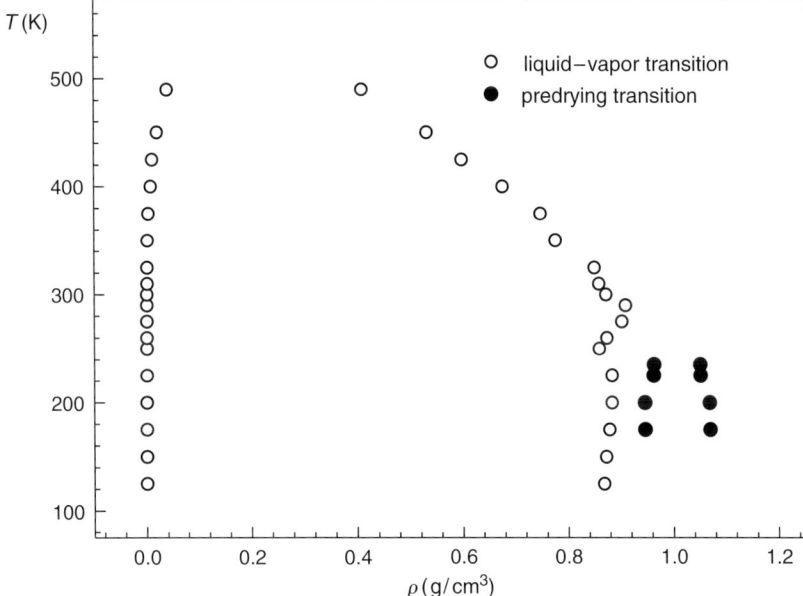

Figure 34: Phase diagram of water in the cylindrical pore with a repulsive step of +0.2 kcal/mol height.

and in a wide temperature range, a drying transition should not affect liquid density profiles noticeably. When the effective thickness of a drying layer is about one to two molecular diameter, liquid–drying layer interface and drying layer–solid interface merge resulting in a gradual density depletion due to missing neighbor effect (see Fig. 9). In such a situation, there is no inflection point of the liquid density profiles, which is characteristic of the interfacial-like profiles (equation (6)), and a drying layer cannot be defined. Note that the concave curvature of the surface (for example, in cylindrical pores) makes the effect of a drying layer (as well as other surface effects) more important, whereas the effect of the convex surface is opposite.

Finally, we would like to note some confusion in literature when a possible drying transition of water near hydrophobic surfaces is considered. First, a well-known phenomenon of a capillary evaporation of a fluid in a pore (see Section 4.3) was mistakenly mixed [236–245] with a drying transition, which may occur in a semiinfinite system. As a result, the words "drying transition," "drying," "capillary drying," and "dewetting transition" were used to describe liquid–vapor transition of confined fluid instead of the physically correct term "capillary evaporation." Second, an absence of a drying transition in the presence of a long-range fluid–wall interactions is not well recognized. Therefore, an interface between a liquid water and a hard wall (or with a vapor) is sometimes used as a close analogue of an interface between liquid water and hydrophobic surface [237, 246–248]. However, the difference between the two cases is drastic: being in contact with liquid phase, a hard wall is always dry, whereas a weakly attractive wall is never dry at liquid–vapor coexistence. A "descriptive" use of the word "drying" (or even "dewetting") to characterize a liquid density depletion near a weakly attractive surface [239, 247] is misleading and physically unjustified, as this depletion may occur not only in a liquid fluid phase but also in a vapor phase and in a supercritical fluid, i.e in the thermodynamic states, where no drying transition occurs in all senses (see Section 3.2). The third source of confusion originates from the numerous attempts to present behavior of a liquid water near hydrophobic surface, including a possible drying transition, as some peculiar property of water. However, the drying transitions in water and in LJ liquid are very similar and closely follow general theoretical expectations for fluids. The specificity of water is in a wide abundance of a solid surface, weakly interacting with water.

2.4 Surface phase diagram of water

The analysis of the surface transitions of water near various surfaces, presented in Sections 2.2 and 2.3, enables construction of a surface phase diagram of water. Knowledge of a surface phase diagram allows prediction of the phase state of water, transitions between these states, and density distribution near various solid surfaces. This diagram shows location of the surface phase transitions as a function of a fluid–wall interaction and temperature. In particular, it shows how the temperatures of the wetting and drying transition and the critical temperatures of the layering and prewetting transitions depend on the strength of a fluid–wall interaction. Besides, it indicates the conditions, which provide fluid density depletion or enchancement near the surface. Various regimes of the surface phase behavior are usually presented in terms of temperature vs strength of fluid–wall potential at the bulk liquid–vapor coexistence curve. The surface phase transitions, which occur out of the liquid–vapor coexistence, could be shown as projections on this plane.

Obtaining the surface phase diagram of water or some other fluid from experiment is problematic, as it is not easy to characterize the surface transitions even for one particular strength of a fluid–wall interaction, whereas for the diagram, this strength should be varied continuously. In simulations, the situation is somehow better, as we can use structureless surfaces, and variation of U_0 is not a problem. However, constructing of a surface phase diagram is a difficult task even in this case. First, this requires simulations of the phase transitions (liquid–vapor and surface phase transitions) in a wide temperature range. These simulations are time consuming and require the use of the sophisticated simulation techniques. Besides, it also very difficult to prove the absence of the phase transition(s). Second, simulations are restricted to the pore geometry and therefore extrapolation to semiinfinite system requires simulations of the phase transitions in the pores of various sizes but with the same U_0. Nevertheless, extensive and systematic simulation studies of confined water [28, 30, 32, 205, 207, 208, 249, 250] allow construction of the surface phase diagram of water. This diagram is based mainly on the simulation studies of the TIP4P model of water near a smooth surface interacting with water oxygens via LJ (9-3) potential. When appropriate, the diagram for the model water will be related to the experimental studies of water.

As the long-range interaction between water and solid surface is intrinsic for real interfaces, we may expect that the surface phase diagram of water should be similar to the one shown in the right panel of Fig. 8. It it reasonable to start the surface phase diagram from the specific point corresponding to the strength U_0 of the water–wall interaction, which provides coincidence of the wetting and drying transitions at T_c (see star in Fig. 35). As we discussed in the Section 2.3, this value is about -1.0 kcal/mol for water. For this strength of the water–wall interaction, vapor density profiles always show adsorption, liquid density profiles always show depletion, and at T_c the fluid density profile is close to horizontal. Only for this surface, neither wetting nor drying transition occurs. The same strength U_0 of the water–wall interaction divides regime of the capillary evaporation from the regime of the capillary condensation for

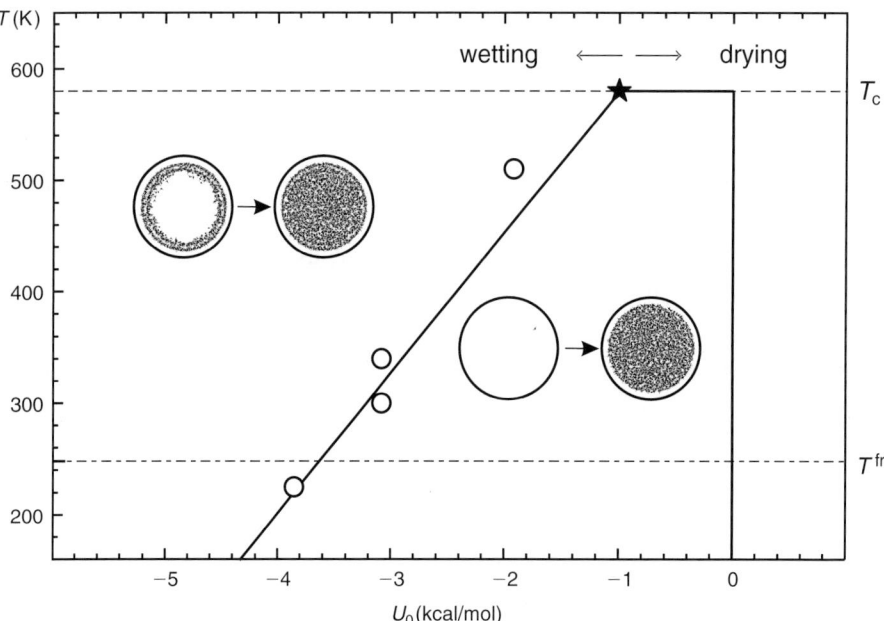

Figure 35: Surface phase diagram of water. Solid lines indicate drying and wetting transitions. Horizontal dashed lines indicate liquid–vapor critical temperature T_c and freezing temperature T^{fr}, respectively. Insets show arrangement of molecules in the coexisting phases of water in cylindrical pores ($R_p = 25$ Å, $T = 300$ K) to the left ($U_0 = -3.08$ kcal/mol) and to the right ($U_0 = -0.77$ kcal/mol) from the inclined line of the wetting transitions.

confined water, which is in equilibrium with a saturated bulk fluid (see Section 4.3 for more details). If the pore walls have $U_0 > -1.0$ kcal/mol, water vapor is a stable phase in the pore (capillary evaporation). When $U_0 < -1.0$ kcal/mol, the pore is filled with a liquid water [208].

The strength of the water–wall interaction with $U_0 = -1.0$ kcal/mol approximately corresponds to the surface whose hydrophobicity is between that of paraffin surface (U_0 is about -0.3 to -0.4 kcal/mol) and carbon surface (U_0 is about -1.5 to -1.7 kcal/mol [251, 252]). More hydrophobic surfaces cover the range of U_0 from -1.0 to 0 kcal/mol. For these surfaces, the temperature of a drying transition is equal to T_c, as any long-range attraction of molecules makes a drying layer miscroscopic (horizontal solid line at $T = T_c$). Even for the strongly hydrophobic surface with $U_0 = -0.39$ kcal/mol, a thickness of a drying layer exceeds molecular width close to T_c only (see Section 2.3). So, for the vast majority of hydrophobic surface, at ambient temperature, a liquid water density depletion is governed by the missing neighbor effect and the drying layer is absent. Hydrophobicity of the surface can be improved by the structuring of the surface, and the contact angle of a liquid water at the superhydrophobic surfaces, produced in such way, achieves the values close to 180° [235]. For these surfaces, a noticeable drying layer with an inflection point of the density profile at the distance of a several molecular diameters from the surface can be expected already at ambient temperature.

The case $U_0 = 0$ corresponds to the hard wall. At this strength of a water–wall interaction, a temperature of a drying transition jumps from T_c to supercooled temperatures. For both the short-range and the long-range repulsive water–wall potentials ($U_0 > 0$), liquid water exists only above the temperature of a drying transition. Behavior of water near the surfaces of this kind is mainly of theoretical interest, as it is difficult to find a surface that does not attract water molecules at least via dispersion forces. However, the surfaces, which repel water molecules, may be based on magnets. Diamagnetic water molecules are repelled by a magnetic field, and this effect causes levitation of water droplets in a strong enough magnetic field [253]. Practical implementation of surfaces, repelling water molecules, will give possibility to obtain a macroscopic vapor layer between a liquid water and a surface, which may have various practical applications. In simulations, a first-order drying transition of water was obtained when a repulsive step of just +0.2 kcal/mol height was added to a hard wall potential (see Section 2.3).

The surface may be considered as hydrophilic when U_0 is lower than -1.0 kcal/mol. For these surfaces, a wetting transition occurs at some temperature. The inclined line of the wetting transitions is close to linear. It starts from the "neutral" wall at T_c (star in Fig. 35) and enters a supercooled region when $U_0 \approx -4.0$ kcal/mol. Two points at $U_0 = -3.08$ kcal/mol were obtained for two different water models (TIP4P and ST2) and in the pores of different geometries. Simulations of SPCE water at carbon-like surfaces of various hydrophilicities show that the contact angle of liquid water is equal to zero when $U_0 < -3.13$ kcal/mol at $T = 300$ K [251, 252]. This estimation well agrees with the line of the wetting transitions, shown in Fig. 35. This line separates two areas in the $T - U_0$ plane, where water condensation on the surface is quite different. To the right of this line, liquid water condenses on the surface, whose coverage by water is noticeably below the monolayer coverage. This behavior is characteristic of all hydrophobic surfaces ($U_0 > -1.0$ kcal/mol) and hydrophilic surfaces below the respective temperature of a wetting transition. To the left of this line, liquid water condenses on the surface, which is already covered by at least two water layers. At not very high temperatures, water molecules in these two layers are highly ordered, whereas starting from the third layer, water structure is close to the structure of a bulk liquid water.

In a wide temperature range, two water layers represent a wetting film at hydrophilic surfaces. Condensation of these two layers with increasing vapor pressure may occurs continuously, via one or two layering transitions or via prewetting transition (see Section 2.2). The dependence of the critical temperatures of the surface phase transitions on the strength of the water–wall interaction is shown in Fig. 36. Closed circles indicate the critical temperatures T_c^{pw} of the prewetting transitions, whereas the open circles correspond to the temperatures of the first-order wetting transition, where three water phases (vapor, film, and liquid) coexist. Accordingly, the vertical lines, connecting the respective open and closed circles, are the lines of the prewetting transitions. Note that open symbols indicate the states at the bulk liquid–vapor coexistence, whereas the prewetting transitions and their critical point occur at lower pressures. The prewetting transitions, shown in Fig. 36, meet the

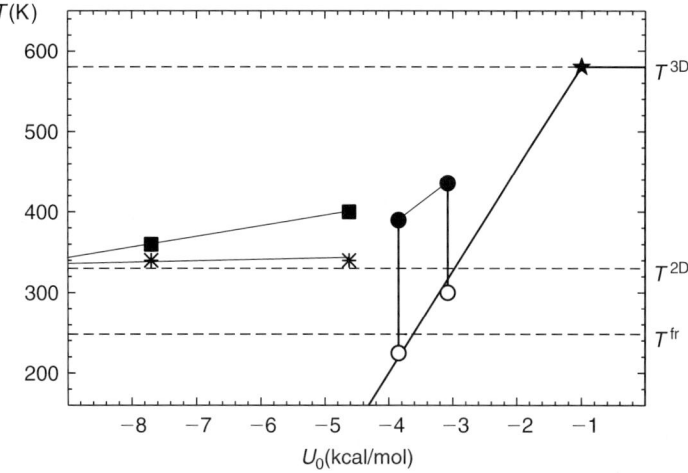

Figure 36: Surface phase diagram of water. Solid inclined line indicates wetting transitions. Horizontal lines indicate liquid–vapor critical temperature T^{3D}, freezing temperature T^{fr} and critical temperature T^{2D} of 2D water. Closed and open circles indicate the critical temperatures of the prewetting transitions and the temperatures of the vapor–film–liquid triple points, respectively. Closed squares indicate the critical temperatures of the first layering transitions, which approach T^{2D}, when $U_0 \to -\infty$. Asterisks indicate the critical temperatures of the second layering transitions.

liquid–vapor coexistence at the temperatures T_w of the first-order wetting transitions.

With weakening of the water–wall interaction, the temperature interval of the prewetting transition shrinks and T_c^{pw} approaches the line of the wetting transitions. There are two possible scenarios for the evolution of the wetting transition with further weakening of U_0. In the first scenario, the line of the critical temperatures of the prewetting transitions meets the line of the wetting transitions below T_c at some U_0 (see Section 2.1). For weaker water–wall interactions, prewetting transition is absent and the wetting transition is of the second order. This scenario is expected for the short-range fluid–wall interactions [113]. For the long-range fluid–wall interactions, the wetting transition may be of the first order only [117, 118]. So, the second scenario assumes that the line of T_c^{pw} meets the line of the wetting transitions at $T = T_c$. Which of these scenarios is valid for water is not clear. Note that this question is important for the systems within rather narrow interval of U_0 and for high temperatures only.

Another unsolved problem is related to the possibility of the sequential wetting of hydrophilic surfaces by water. Specific structure of two water layers, adsorbed on the surface, allows considering their condensation apart from the condensation of thicker water films. In fact, excluding very high temperatures, the line of the wetting transitions shown in Figs. 35 and 36 corresponds to the wetting of a surface by two water layers. This wetting transition may be caused by the short-range fluid–wall interaction, and it may be followed by a second wetting transition with high wetting temperature. The second wetting transition occurring at the "liquid-like" surface formed by two water layers should be expected to be of a second order and governed by long-range fluid–wall forces [210, 211]. Accordingly, a continuous growth of the wetting film on the surface of two water layers with approaching the second wetting temperature is expected. Experiments show a possibility of the wetting transitions at *liquid* surface for other fluids [130] and two sequential wetting transitions in particular [133]. Therefore, scenario with two sequential wetting transitions may be realistic for water, and surface phase diagram of water may be more complicated.

The critical temperatures of the first layering transitions are shown by squares in Fig. 36. With strengthening water–wall interaction, the critical temperature of layering transition approaches T^{2D}, as expected. It is important that the critical temperatures of layering transition on the surfaces of various hydrophilicity and 2D critical temperature T^{2D} are noticeably above the bulk freezing temperature T^{fr}. Therefore, we may expect the effect of 2D critical point on various properties of hydration water at ambient temperatures.

3 Surface critical behavior of water

The presence of a boundary breaks the translational invariance of a bulk system and introduces an anisotropy. As a consequence, all system properties become local, that is dependent on the position of the elementary volume considered relative to the boundary. In the simplest case of a single planar surface, all properties depend on the distance z to the boundary. The surface perturbs the bulk properties of a fluid over some distance from the surface, whereas the system remains undisturbed (bulk-like) far from the surface. The critical behavior of fluids near the surface strongly differs from the bulk behavior [254]. On approaching the bulk critical point, the surface critical behavior intrudes deeply into the bulk, as the range of the surface perturbation is governed by the bulk correlation length [255]. Knowledge of the laws of the surface critical behavior makes it possible to describe the fluid density profiles at various thermodynamic conditions.

The main unavoidable effect of any surface appears in the missing neighbors in the first (surface) layer. In the absence of other factors (nonzero surface field, restructuring of particles near a surface), the average interaction energy of particles in the surface layer decreases in absolute value, and density near the surface becomes less than in the bulk. If the effect of a surface appears as a effect of missing neighbors only, the situation corresponds to the so-called *ordinary* transition [254], which can be realized in Ising lattice simply by setting the surface field to zero. However, this situation is not typical of fluids, as in general, the fluid–wall interaction causes the preferential adsorption of one of the phases. A weakening of the energy of the intermolecular interaction per molecule near a surface due to the missing neighbors and a preferential adsorption/desorption of molecules due to the fluid–wall interactions is a normal situation for fluids, and the corresponding transition is called *normal* [256–258]. The theory of normal transition is developed mainly for the case of an infinitely strong fluid–wall interaction, whereas more realistic cases of moderate or weak fluid–wall interactions are much less studied. In the latter case, we may expect behavior corresponding to the normal transition asymptotically close to T_c only, whereas at lower temperatures, the behavior corresponding to the ordinary transition should dominate. The temperature crossover between these two regimes has not been studied yet. Besides, for any finite surface field, magnetization profiles in Ising lattices are nonmonotonous in supercritical region with the maximum at some distance from the surface determined by the surface field [259–261]. The same effect, expected in two coexisting phases at subcritical temperatures, has not been studied yet. So, the theory of the surface critical behavior in the presence of an arbitrary nonzero surface field is far from being complete (see [262] for more details).

3.1 Surface critical behavior of fluids

Experimental studies of the surface critical behavior of fluids require obtaining an information concerning the density (concentration) profiles near surfaces. Currently, this is a difficult experimental task and, as a rule, only rough estimations can be done. Surface critical behavior of one-component fluids at subcritical temperatures has not been studied yet experimentally, as it is very difficult to measure the fluid density profiles in two coexisting phases. In supercritical region, the excess adsorption should strongly increase with approaching the critical point, and this can be measured in order to test the theoretical predictions. When the bulk critical point is approached upon cooling along isochore ($\tau \to 0$), the excess adsorption should diverge as $\sim \tau^{\nu-\beta} = \sim \tau^{-0.31}$ for an infinitely strong surface field [255]. Some experimental studies of adsorption of one-component fluids in supercritical region are consistent with this expectation [263, 264], whereas in other experiments [265], excess adsorption diverges significantly stronger. Besides, in some cases, the critical adsorption turns to critical desorption very close to T_c [266–268].

For binary mixtures, the experimental results on the critical adsorption are also contradictory. In some cases, the excess adsorption (depletion) diverges upon approaching the critical point or coexistence curve stronger than $\tau^{\beta-\nu}$ [149, 269]. Other experiments indicate that critical adsorption remains strongly undersaturated even very close to T_c [148]. The local order parameter near the surface, which is a difference between the concentrations of the coexisting phases, was found to follow a power law with exponent $\beta_1 \approx 0.8$ [270–272], i.e. in accord with the power law expected for the ordinary transition [254]. However, these experimental studies were not carried out close to the liquid–liquid critical point and above the temperature of a wetting transition. Therefore, the character of the expected crossover from ordinary to normal transition, as well as the relation of crossover temperature and the temperature of the wetting (drying) transition, remains unclear. Nonmonotonous concentration profile near the surface was observed in supercritical binary mixtures [135, 273], which qualitatively agrees with the theoretical expectations [259–261]. However, both theory and experiment do not describe the concentration profiles at subcritical temperatures.

Computer simulation studies of the liquid–vapor coexistence curves in pores yield a powerful tool to study the surface critical behavior of fluids near various surfaces [30]. Surface critical behavior of LJ fluid near weakly attractive (strongly solvophobic) surface, whose interaction with fluid molecules is about 70% of the fluid–fluid interaction, was studied in detail [28, 29, 141, 262, 274]. The liquid–vapor coexistence curve of LJ fluid in slit pore of width $H_p = 12\sigma$ shown in Fig. 37 indicates that the density of a liquid phase is lower than that in the bulk fluid at the same temperature. The origin of this effect is obvious when looking at the density profiles (Fig. 38): depletion of density near the surface, located at $r = 0$. This depletion increases when approaching the critical temperature, and it is clearly seen not only in the liquid but also in the vapor phase at high temperatures. Just near the surface, there is a pronounced density oscillation. It is caused by a quasi-2D localization of molecules in a plane parallel to the surface in the well of the surface–fluid interaction potential. The localization of molecules in this plane causes, in turn, fluid density oscillations, which decay quickly when moving away from the surface. For weakly attractive surfaces, only the first density oscillation remains noticeable in the two phases close to the critical temperature. For strongly attractive surfaces, several density oscillations are seen in the density profile up to the critical temperature. In the general case, the density profile of fluid appears as a gradual (exponential) decay near a weakly attractive or gradual growth near a strongly attractive surfaces, with oscillatory deviations from this gradual behavior caused by preferential localization of molecules. Note that in the absence of a potential well (for example, near a hard wall), density profiles in the coexisting

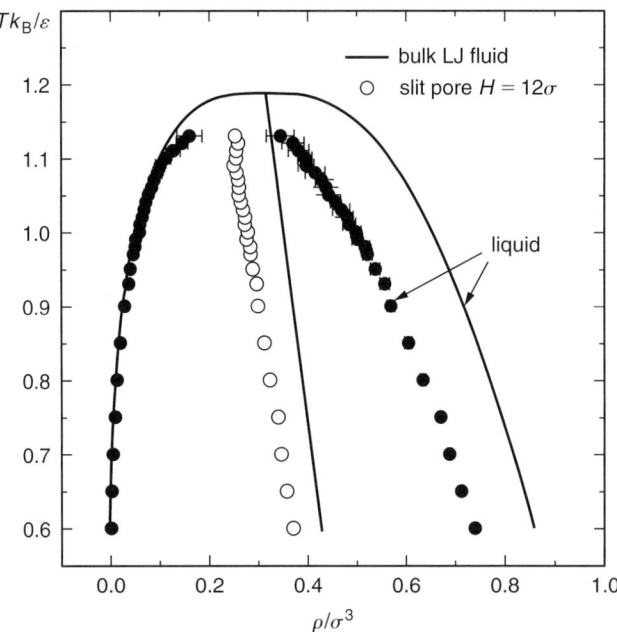

Figure 37: Liquid–vapor coexistence curve of LJ fluid in slit pore of the width $H_p = 12\sigma$ with weakly attractive walls (solid circles) and the coexistence curve diameter (open circles). Coexistence curve and diameter of the bulk LJ are shown by solid lines (data from [29]).

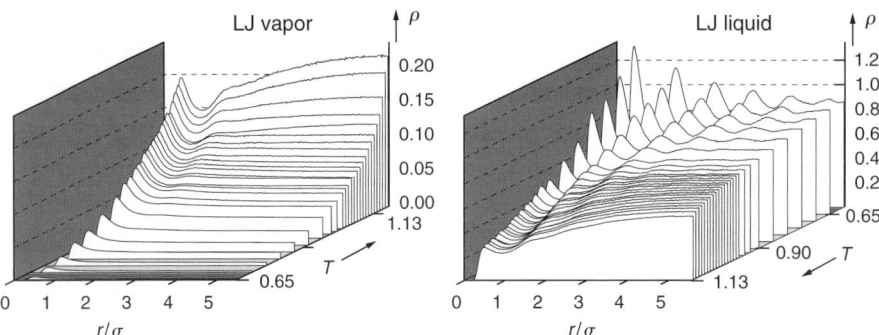

Figure 38: Density profiles of vapor and liquid phases of LJ fluid near weakly attractive surface along the pore coexistence curve ($H_p = 12\sigma$). Surface is located at $r = 0$ (data from [29]).

phases are always gradual and the density oscillations may be achieved in the liquid phase at pressures, which noticeably exceed the pressure of the liquid–vapor coexistence.

In the case of inhomogeneous fluid near the surface, the local order parameter $\Delta\rho(\Delta z, \tau)$ at the distance Δz from the surface may be defined similarly to the order parameter of bulk fluid:

$$\Delta\rho(\Delta z, \tau) = (\rho_l(\Delta z, \tau) - \rho_v(\Delta z, \tau))/2, \qquad (7)$$

where $\tau = 1 - T/T_c$ measures temperature deviation from the bulk critical point. The profiles of the local order parameter near weakly attractive surface show depletion toward the surface at any temperature (see left panel in Fig. 39, for example). As the surface perturbations, such as the missing neighbor effect and effect of the short-range surface field, should decay exponentially with an increasing distance from the surface [254, 255], the order parameter profile is expected to follow the equation [250]:

$$\Delta\rho(\Delta z, \tau) = \Delta\rho_b(\tau)\left[1 - \exp\left(-\frac{\Delta z}{\xi}\right)\right], \qquad (8)$$

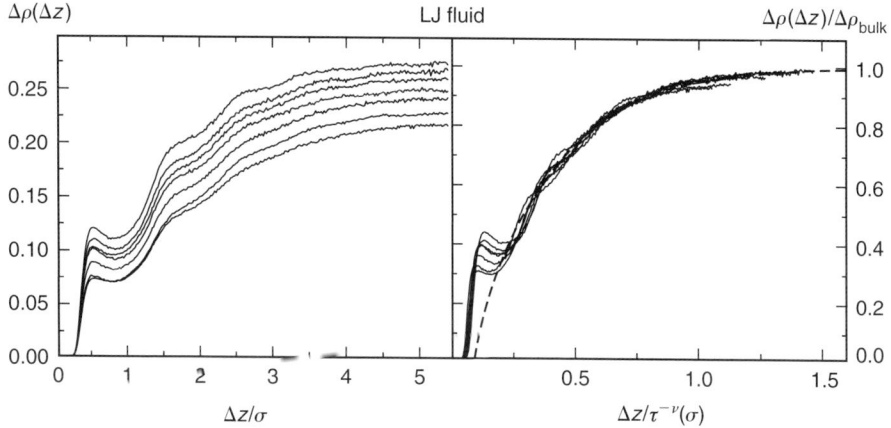

Figure 39: Left panel: profiles of the local order parameter $\Delta\rho(\Delta z)$ of LJ fluid near weakly attractive surface along the pore coexistence curve for temperatures $1.04 \leq T \leq 1.10$ ($H = 12\sigma$). Surface is located at $\Delta z = 0$. Right panel: master curve of the order parameter profiles shown in the left panel. Thick dashed line represents equation (8) with $\xi_0 = 0.30\sigma$ (data from [29]).

which assumes that $\Delta\rho(\Delta z, \tau) = 0$ at $\Delta z = 0$ and $\Delta\rho_b(\tau)$ and ξ are the bulk order parameter and the bulk correlation length, respectively. Equation (8) suggests that order parameter profiles should collapse on a single master curve if $\Delta\rho(\Delta z, \tau)$ is normalized by $\Delta\rho_b(\tau)$, whereas distance Δz is normalized by the bulk correlation length $\xi = \xi_0 \tau^{-\nu}$. An example of such master curve shown in the right panel of Fig. 39 evidences that fluid behavior near the surface is consistent with the prediction of the theory of the surface critical behavior.

Another theoretical prediction concerns the temperature dependence of the layer magnetization [254]. For fluids, this is a temperature dependence of the local order parameter at various fixed distances from the surface. The fluid analogous of the first surface layer in the Ising model is a monolayer of molecules, the equilibrium position of which is close to the minimum of the fluid–wall interaction potential. At low temperatures, the localization of molecules in the surface layer causes, in turn, localization of the molecules in the second and subsequent layers, and several such layers of fluids may be identified with layers in magnet. At high temperatures, the definition of a fluid layer as being of one molecular diameter width has no structural grounds, and at $\Delta z > \sigma$, an arbitrarily thin fluid layer may be used for the analysis. The temperature dependences of the local order parameter $\Delta\rho_i$ in layers of one molecular diameter width are shown in Fig. 40. The behavior of the first layer differs from the behavior of other fluid layers: $\Delta\rho_1$ changes practically linearly with τ over the whole temperature range. In the second and subsequent layers, the behavior of $\Delta\rho_i(\tau)$ follows the bulk critical behavior $\sim \tau^\beta$ with exponent $\beta \approx 0.326$ at low temperatures and crosses over to surface critical behavior with exponent $\beta_1 \approx 0.8$ at higher temperatures. The value $\beta_1 \approx 0.82$ was predicted for the order parameter of the ordinary transition in magnets, which occurs in the absence of the surface field. Thus, surface critical behavior of the order parameter of fluid in a wide temperature range corresponds to the ordinary transition in magnets.

A temperature crossover of the local order parameter is better seen in the right panel of Fig. 40, where data are shifted vertically. The crossover temperatures, estimated as a crossing point of two straight lines in a double-logarithmic scale, are indicated by stars. The crossover temperature depends on the distance to the surface. Moving away from the surface, the crossover from bulk to the surface critical behavior occurs closer to the critical temperature. It was found in [29] that such crossover upon increasing temperature occurs when the distance from the fluid layer to the surface is about two bulk correlation lengths. Note that asymptotically close to the critical point, a critical behavior of normal transition universality class is expected [256–258]. It is remains unclear, however, at what temperature and in what way a crossover from ordinary to normal transition may occur with temperature and whether such a crossover occurs in the whole system simultaneously or whether its intrusion is governed by temperature. This and related problems are reviewed in Ref. [262].

Similar to the case of a bulk fluid, the local density in each of the coexisting phase is a superposition of symmetric and asymmetric contributions:

$$\rho_{l,v} = \rho_d(\Delta z, \tau) \pm \Delta\rho(\Delta z, \tau), \qquad (9)$$

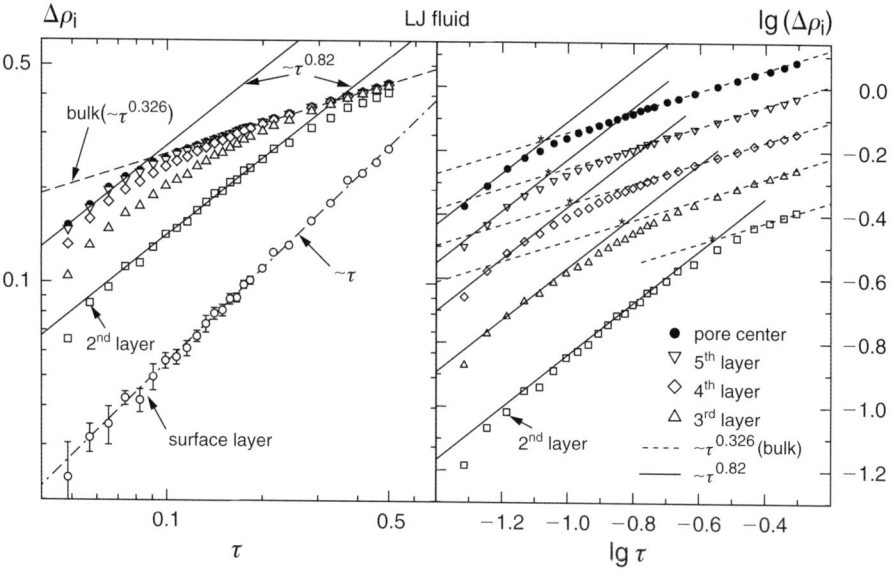

Figure 40: Left panel: temperature dependence of the order parameter $\Delta\rho_i$ of LJ fluid near weakly attractive surface averaged over the ith molecular layer in double-logarithmic scale. The power laws, which represent bulk critical behavior, surface critical behavior, and the behavior in the first surface layer, are shown by dashed, solid, and dot-dashed lines, respectively. Right panel: the same data as in the left panel, but data for the third and subsequent layers are shifted vertically. The temperatures at which a crossover from bulk-like to surface-like behavior occurs in each layer are denoted by stars (data from [29]).

where "+" and "−" relate to the liquid and vapor phase, respectively. The local diameter $\rho_d(\Delta z, \tau)$ reflects not only the intrinsic asymmetry of liquid and vapor, as in the bulk case, but also response on the surface perturbation. Indeed, the effect of missing neighbors decreases density of both liquid and vapor phases near the surface, whereas the surface attraction tends to the increase of density in both phases. The local diameter is identical in the two coexisting phases. With increasing temperature, the profile of the local diameter continuously crosses over to the density profile in the one-phase supercritical region at $T > T_c$. The profiles of the local diameters being normalized by the bulk diameter at the same temperature do not collapse on a single master curve contrary to the profiles of the local order parameter $\Delta\rho$ [262]. It is not clear whether the nonuniversal behavior of the local diameter preserves also in the case of a short-range fluid–surface interaction or it originates from the tail of the long-range potential only.

When the wetting or drying transition is absent or suppressed, the density profile in one of the coexisting phase appears as a result of competition between two opposite trends: density depletion due to the missing neighbor effect and density increase due to the attractive fluid–wall interaction. These effects depend on temperature and on bulk fluid density in different ways. In particular, the missing neighbor effect strongly increases with increasing bulk fluid density. The density profile of a liquid in the case of a weak fluid–surface interaction is dominated by the missing neighbor effect and may be described by exponential equation

$$\rho_l(\Delta z, \tau) = \rho_l^b(\tau) \left[1 - \exp\left(-\frac{\Delta z}{\xi^{ef}}\right)\right], \quad (10)$$

where $\rho_l^b(\tau)$ is a bulk liquid density and ξ^{ef} is an effective correlation length, which in general case differ from the ξ due to the surface attraction. The density profiles of a liquid LJ fluid near weakly attractive wall nicely follow equation (10) at high temperatures (see Fig. 41). Oscillatory deviations are notable in the range of about one molecular diameter near the surface only. At low temperatures, when correlation length is much less than molecular diameter, the density oscillations dominate the liquid density profile and gradual exponential depletion is difficult to

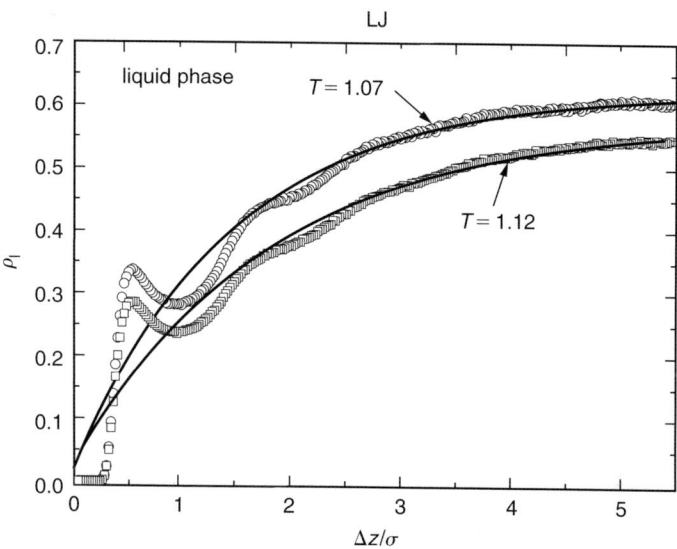

Figure 41: Density profiles of LJ liquid near a weakly attractive wall at two temperatures (symbols) and their description by equation (10) (solid lines).

notice (Fig. 38). Note that the profiles shown in Fig. 41 correspond to the case when the drying layer is suppressed by confinement in relatively narrow pore.

A vapor phase near a weakly attractive surface shows several kinds of density profiles depending on temperature [274]. At low temperatures, when the bulk vapor density ρ_v^b is low, a gradual adsorption of vapor may be described by the exponential equation

$$\rho_v(\Delta z) = \rho_v^b + \left(\rho_s - \rho_v^b\right) \exp\left(-\frac{\Delta z}{\xi^{ef}}\right), \tag{11}$$

where ρ_s ($> \rho_v^b$) is an effective density of fluid near the surface ($\Delta z = 0$). The effective correlation length ξ^{ef} in equation (11) is close to the bulk correlation length ξ when the effect of missing neighbors is negligible. With increasing temperature, the bulk vapor density increases, making the effect of missing neighbors more important, and adsorption of vapor near the surface decreases (see Fig. 42, left panel). There is a specific crossover temperature when two surface effects completely compensate each other, providing a zero excess adsorption of vapor.

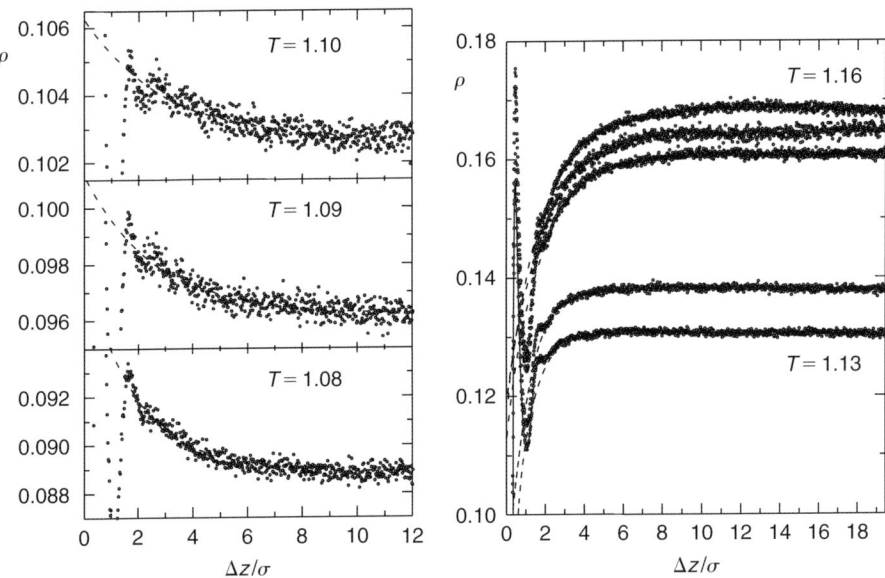

Figure 42: Density profiles of a vapor phase of LJ fluid near a weakly attractive surface at various temperatures. Adsorption of fluids at low temperatures (left panels) and depletion at high temperatures (right panels) may be described by equation (11) with $\rho_s > \rho_b$ and $\rho_s < \rho_b$, respectively (dashed lines) (data from [274]).

Above this temperature, vapor density profiles are dominated by missing neighbor effect and show depletion toward the surface (see Fig. 42, right panel). In this regime, $\rho_v(\Delta z)$ may be again well described by equation (11), with $\rho_s < \rho_v^b$ and with ξ^{ef} close to the bulk correlation length. Note that the exponential behavior may be distorted by the long-range tail of the fluid–wall potential. However, this effect on the density profile is smaller than the effect of a (short range) potential well [141, 274].

3.2 Surface critical behavior of water

Surface critical behavior of water has been studied by computer simulations of the liquid–vapor coexistence curves in pores of various sizes and geometries [28, 205, 250, 262]. The behavior of the local order

Surface critical behavior of water

parameter $\Delta\rho(\Delta z)$ may be analyzed in detail when the surface transitions (wetting or drying) are suppressed and both coexisting phases remain near the surface, at least in some wide temperature interval. A drying transition of water near a hydrophobic surface is efficiently suppressed by a weak but long-range water–surface interaction potential and by the shift of the chemical potential in pore with respect to the value of the bulk phase transition. The liquid–vapor coexistence curves of the TIP4P water model in slit pores of various sizes with hydrophobic walls ($U_0 = -0.39$ kcal/mol) are shown in Fig. 43. Similar to the LJ fluid near weakly attractive wall, the density of liquid water in a hydrophobic pore is essentially smaller than one of the bulk liquid water.

The density profiles of liquid water and water vapor near the hydrophobic surface, calculated in the largest pore studied ($H_p = 30$ Å), are shown in Figs. 44 and 45, respectively. An inspection of the liquid density profiles evidences a gradual decline in the water density toward the surface without indications for the formation of a vapor layer. Under such circumstances, the local order parameter $\Delta\rho(\Delta z, \tau) = (\rho_l(\Delta z, \tau) - \rho_v(\Delta z, \tau))/2$ may be studied in the whole temperature interval of the liquid–vapor coexistence. Some profiles $\Delta\rho(\Delta z, \tau)$ are shown in Fig. 46.

Order parameter profiles of water may be successfully described by the exponential equation (8). Values of $\Delta\rho^b$ obtained from the fits are equal to the values of the order parameter in the pore interior and were found close to the bulk order parameter at the same temperature. The fitting parameter ξ in equation (8) is shown in Fig. 47. It diverges when approaching the bulk critical temperature ($T_c = 581.9$ K for TIP4P water model corresponds to the right vertical axis at the left panel). The character of this divergence is seen better when the dependence ξ vs reduced temperature τ is shown in a double-logarithmic scale (right panel). In the studied temperature range, a power-law behavior $\sim \tau^{-\nu}$ with $\nu = 0.63$ (dashed line) is not observed for $\xi(\tau)$, which may be roughly described as $\sim \tau^{-0.95}$. However, the obtained values of the fitting parameter ξ in a wide temperature interval may be satisfactorily described by the extended scaling equation [250]:

$$\xi = \xi_0 \tau^{-\nu}(1 + a_\xi \tau^{0.52}), \qquad (12)$$

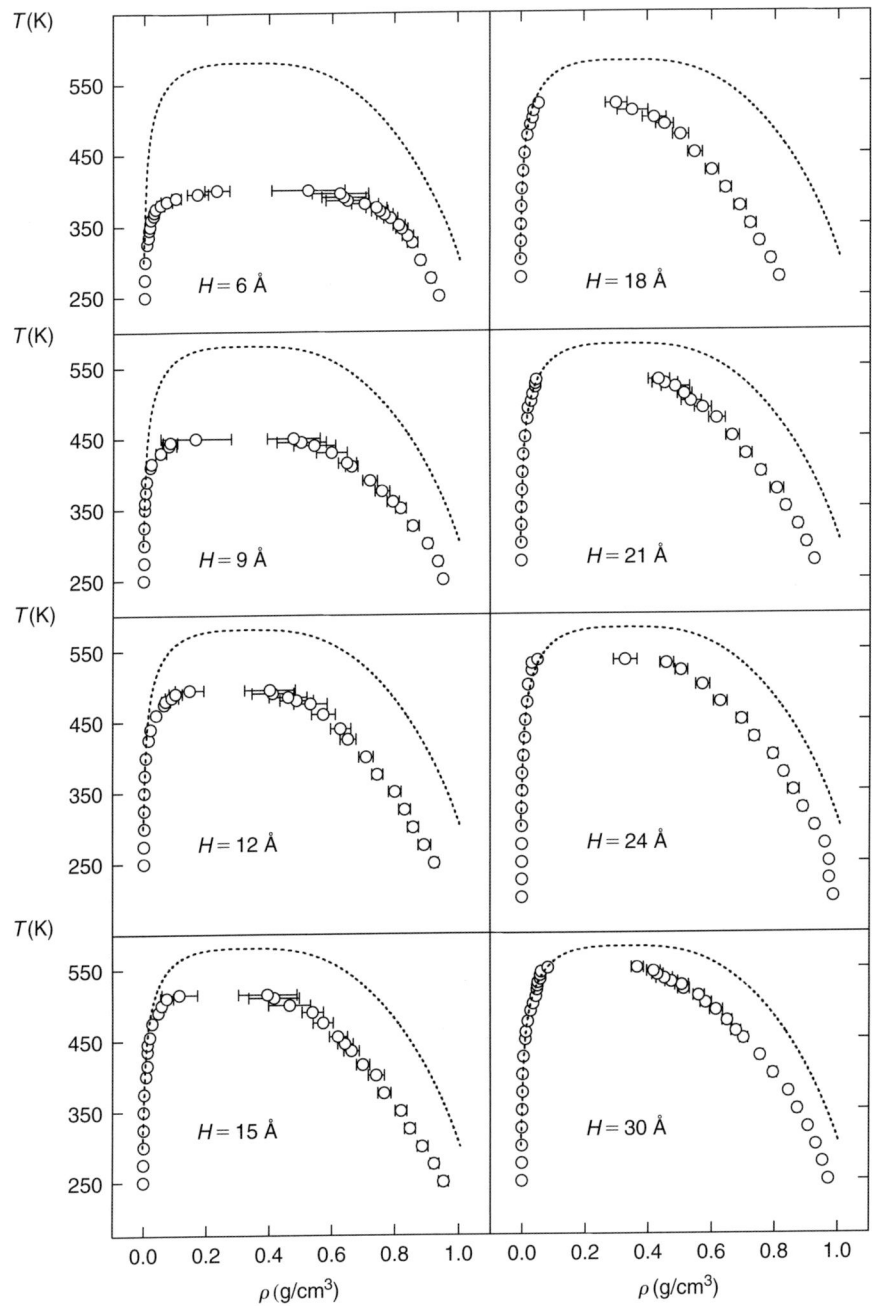

Figure 43: Liquid–vapor coexistence curves of water in slit-like pores with hydrophobic walls ($U_0 = -0.39$ kcal/mol). The bulk coexistence curve is shown by the dotted curve (reprinted, with permission, from [250]).

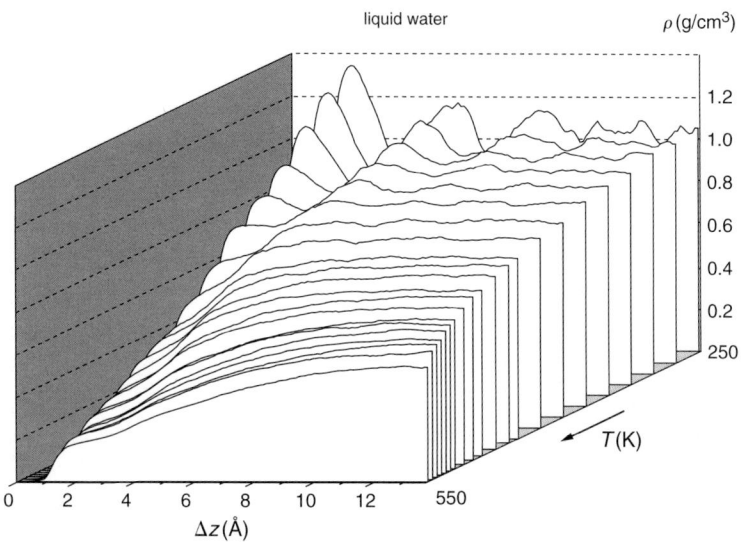

Figure 44: Density profiles of liquid water near hydrophobic surface with $U_0 = -0.39$ kcal/mol along the pore coexistence curve. Surface is located at $\Delta z = 0$ (data from [250]).

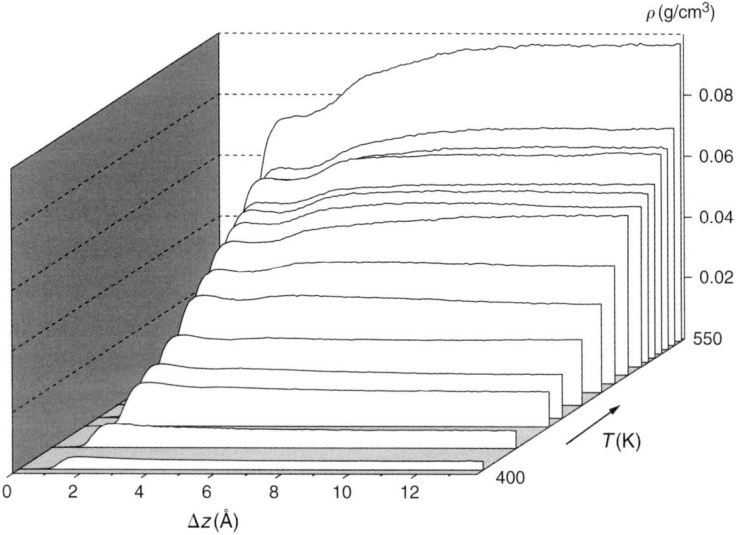

Figure 45: Density profiles of water vapor near hydrophobic surface with $U_0 = -0.39$ kcal/mol along the pore coexistence curve. Surface is located at $\Delta z = 0$ (data from [250]).

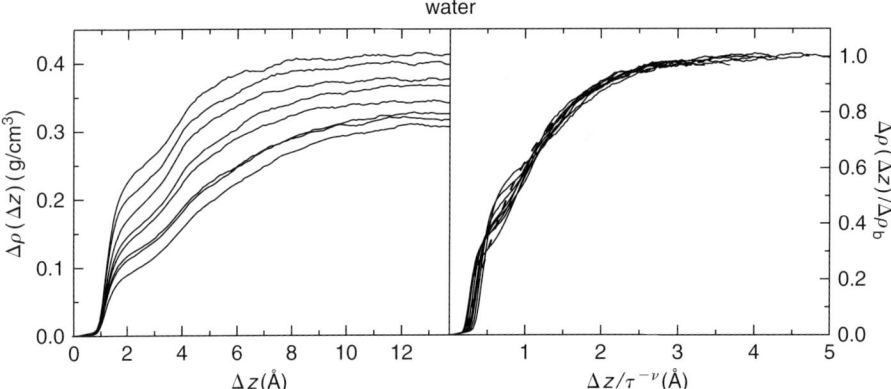

Figure 46: Left panel: profiles of the local order parameter $\Delta\rho(\Delta z)$ of water near hydrophobic surface along the pore coexistence curve ($H_p = 30$ Å) for temperatures $T = 460, 475, 490, 500, 510, 520, 525,$ and 530 K. Right panel: master curve of the order parameter profiles shown in the left panel. Thick dashed line represents equation (8) with $\xi_0 = 0.80$ Å (data from [250, 262]).

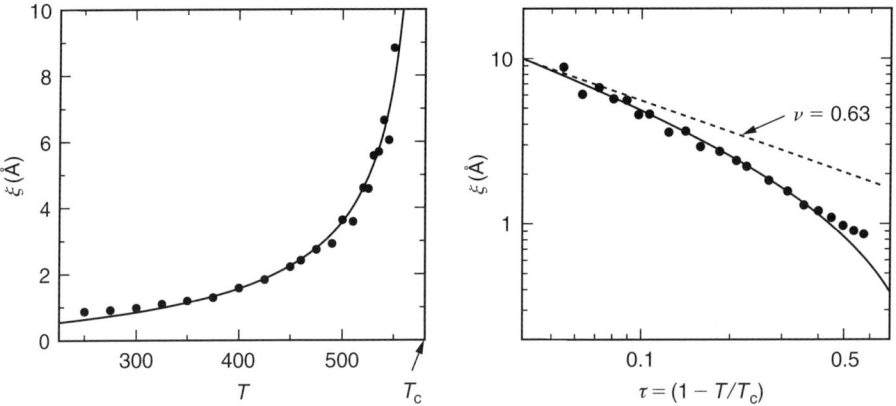

Figure 47: The temperature dependence of the correlation length (ξ) of water along the liquid–vapor coexistence curve obtained from the fits of the order parameter profiles in slit pore of $H_p = 30$ Å to equation (8). Left panel: ξ as function of T in normal scale; right panel: ξ as function of τ in double-logarithmic scale. The asymptotic critical behavior is indicated by dashed line. The fit to equation (12) with $\xi_0 = 1.60$ Å and $a_\xi = 0.97$ is shown by solid lines in both panels (data from [262]).

where term in parentheses is a correction to scaling. A relatively large (close to 1) value of the coefficient a_ξ indicates that asymptotic range, where ξ follows the power law $\sim \tau^{-\nu}$, is rather narrow and it is not achievable in simulations. It seems to be comparable but narrower than the asymptotic range for ξ in LJ fluid, which may be estimated as $\tau < 0.1$ [29]. Note that the results shown in Fig. 47 are a single estimation of the correlation length of water in simulations available so far. The amplitude $\xi_0 \approx 1.60$ Å in equation (12) is about factor of 2 of the experimental value for the correlation length of real water [19]. To clarify the origin of this discrepancy, the correlation length of a bulk model water should be determined by the computer simulations. However, such studies have not been done yet.

To examine the temperature dependence of the local properties of water, the fluid layers should be reasonably defined. The first surface layer may be defined in a natural way based on the first density oscillation, which is clearly visible at the density profiles even at high temperatures (see Figs. 44 and 45). Water molecules located in a shell bounded from one side by the van der Waals contact between water and surface ($\sigma/2 = 1.25$ Å from the surface, where $\Delta z = 0$) and by the first minimum in the liquid density distribution at low temperatures from another side (typically about $\Delta z \approx 3.75$ Å) should be attributed to the surface layer. At low temperatures, the second density oscillation indicates the second layer of water, whose thickness is about 3 Å. This oscillation practically disappears at high temperatures, where density changes gradually at the distances $r > 5$ Å from the surface. In case of a smooth density variation, the layer thickness could be chosen to be arbitrarily small. It is convenient to set the thickness of third and subsequent water layers to 3 Å.

The temperature dependence of the order parameter $\Delta \rho_i$ averaged over ith water layer is shown in the right panel of Fig. 48 in a double-logarithmic scale. The order parameter of "inner" water, calculated for water near the pore center, follows closely the bulk behavior with the critical exponent $\beta \approx 0.326$ (dashed line) up to $\tau = 0.08$ (up to about 50° below T_c). The temperature dependence of the order parameter in the surface layer shows an essentially different behavior: starting from the extremely low temperature $\tau = 0.57$ (close to the bulk freezing

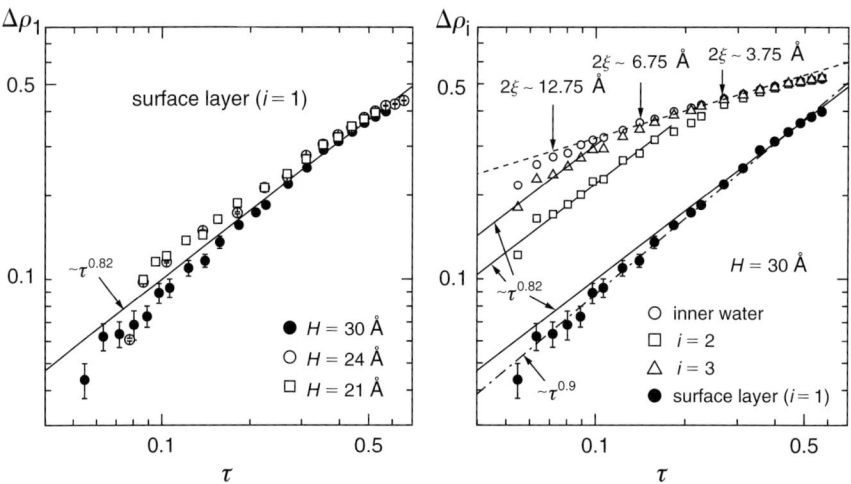

Figure 48: Left: temperature dependence of the order parameter in the surface layer $\Delta\rho_1$ of water confined in several slit-like pores. Right: temperature dependence of the local order parameter $\Delta\rho_i$ in various water layers in the slit-like pore with $H_p = 30$ Å. Arrows indicate the reduced temperatures, when 2ξ achieve the second, third, and fifth water layers (data from [250]).

temperature $\tau = 0.59$ of this model water [199]), $\Delta\rho_1$ follows a scaling law $\sim \tau^{\beta_1}$ with $\beta_1 \approx 0.9$, which is close to but higher than $\beta_1 = 0.82$ expected for the ordinary transition [254]. The second and third water layers show a crossover from bulk-like behavior with $\beta \approx 0.326$ to the surface behavior with $\beta_1 \approx 0.82$ at $\tau = 0.22$ and 0.11 respectively. A lower values of the local order parameter near the surface indicates that the densities of the coexisting phases near the surface are close to each other than those of phases far from the surface or in the bulk. A higher value of the exponent β of the order parameter near the surface indicates, moreover, that liquid and vapor densities approach each other with temperature much faster near the surface than in the bulk. This disordering effect of a surface is very similar in Ising lattices and in LJ fluid (see Section 3.1).

Temperature dependence of the order parameter of water in the surface layer found in the large slit-like pore is not notably sensitive to the pore size. Although in smaller pores the density difference between liquid and vapor phases is slightly larger, the exponent β_1, which determines the slope of the temperature dependence of $\Delta\rho_1$, is close to 0.82 in all pores

(Fig. 48, left panel). Thus, a similar behavior of water in the surface layer may be expected in any hydrophobic confinement.

Surface critical behavior does not disturb notably the order parameter in the second water layer up to about 425 K. Starting from this temperature, the surface perturbation spreads over the surface water layers and intrudes deep into the confined water. Using the values of the correlation length obtained from the fits of the order parameter profile, we notice that $\Delta\rho_i$ deviates from the bulk behavior when 2ξ achieves approximately the outer border Δz_i of the ith layer. The temperatures when 2ξ is equal to the distance of the outer border of the second water layer to the surface ($\Delta z_2 \sim 3.5$ Å), third layer ($\Delta z_3 \sim 6.5$ Å), and fifth layer (outer border of "inner water" $\Delta z_i \sim 12.5$ Å) are indicated by arrows in Fig. 48 (right panel).

The local diameter $\rho_d(\Delta z, \tau) = (\rho_l(\Delta z, \tau) + \rho_v(\Delta z, \tau))/2$ of water decreases toward the hydrophobic surface (an example for one temperature is shown in Fig. 49, left panel). In bulk water, $\rho_d(\tau)$ reflects the intrinsic asymmetry of liquid and vapor phases. Its temperature evolution may be described by a regular temperature dependence in a wide temperature range, although a critical anomaly may be detected close to the critical temperature (see Section 1). Diameter of the coexistence curve of confined water reflects also response on the surface perturbation. Near the hydrophobic surface, the effect of missing neighbors dominates over the surface attraction and water local diameter decays toward the surface. With increasing temperature, the profile of the local diameter continuously crosses over to the density profile in the supercritical region. In particular, $\rho_d(\Delta z)$ close to the critical point looks very similar to the density profile at the critical isochore (an example of the density profile in an one-phase region is shown by dashed line in Fig. 49, left panel).

When the profiles of the local diameters are normalized by the bulk diameter at the same temperature, they do not collapse on a single master curve, as it happens with the profiles of the local order parameter (Fig. 49, right panel). This nonuniversality may be caused by the long-range water–surface potential. As behavior of water near a surface with short-range water–surface interaction is not yet studied, this idea remains speculative. The local diameter ρ_d calculated in the surface layer vanishes upon increasing temperature much faster the bulk diameter (Fig. 50). It is

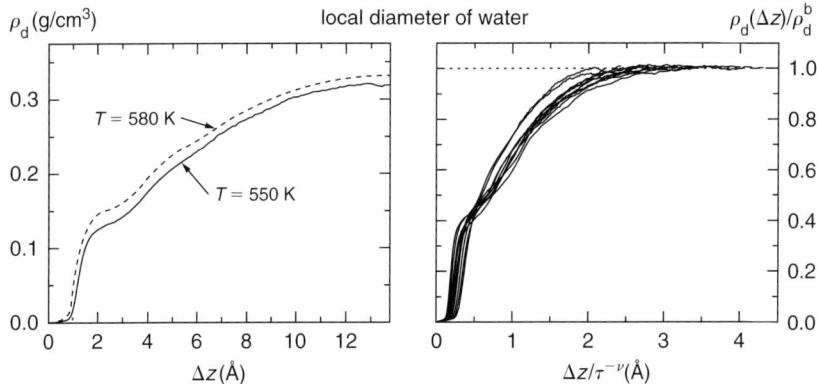

Figure 49: Left: the local diameter $\rho_d(\Delta z)$ at $T = 550$ K (solid line) and water density profile in one-phase region at about the bulk critical temperature $T \approx 580$ K (dashed line). Right: the profiles of local diameters at several temperatures ($T = 490, 500, 510, 520, 525, 530, 535, 540, 545,$ and 550 K) normalized by the respective bulk values as functions of the rescaled distance to the surface in pore with $H_p = 30$ Å (data from [262]).

interesting to note that ρ_d of water in the surface layer is a simple linear function of τ in the whole temperature range of the liquid–vapor coexistence. The water diameter in the pore interior shows notable deviations from the bulk starting from about 50° below the pore critical temperature. The origin of this behavior is unclear and deserves to be studied.

Knowledge of the surface critical behavior of the order parameter and diameter allows description of the local density of liquid water and vapor in a universal way. According to the equation (9), the local densities of water in the surface layer ($\Delta z = 0$) may be written as

$$\rho_{l,v}(0, \tau) = \rho_c(0) \pm B_1 \tau^{\beta_1} + A_1 \tau + \ldots, \tag{13}$$

where "+" should be taken for liquid phase and "−" for vapor. The first temperature-dependent term in equation (13) arises from the order parameter $\Delta \rho$ and its temperature dependence follows the power law with the critical exponent $\beta_1 \approx 0.82$. This contribution describes an increase in a dissimilarity of the liquid and vapor densities near the surface when moving away from T_c. The second (asymmetric) contribution reflects the change of a density due to the water–surface interaction, and it should

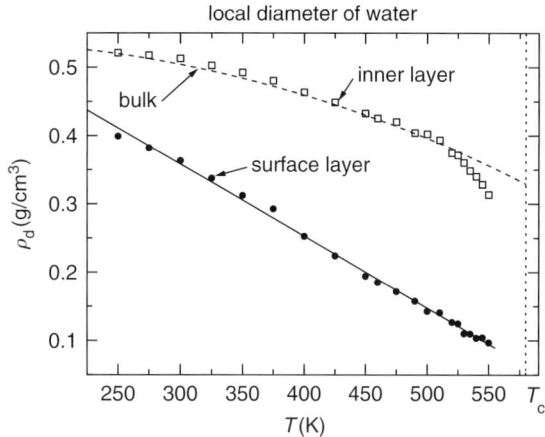

Figure 50: Diameters ρ_d averaged over the first (surface) water layer and near the pore center (inner layer) as functions of temperature T in pore with $H_p = 30$ Å. Solid line: linear fit of ρ_d in the surface layer; dashed line: bulk diameter of model water; vertical dotted line: the bulk critical temperature (data from [250]).

be the same in two coexisting phases. In both phases, the asymmetric contribution contains constant $\rho_c(0)$, which is the local water density in the surface layer at $T = T_c$. The temperature dependence of the asymmetric contribution is dominated by a linear term $A_1\tau$, which is positive for all one-component fluids. This linear term may be just a leading temperature-dependent term of regular contribution to the local diameter. Since the diameter of a bulk fluid contains also some nonanalytic terms (see Section 1), some "singularities" in the surface diameter also can be expected. We may speculate that the bulk term $\sim \tau^{1-\alpha}$ should correspond to the anomaly $\sim \tau^{2-\alpha}$ near the surface, as far as the surface exponent $\alpha_s = \alpha - 1$ in the case of the ordinary transition [275]. Accordingly, the anomaly $\sim \tau^{2\beta}$ in bulk diameter may appear as $\sim \tau^{2\beta_1}$. The terms with powers of $2 - \alpha \approx 1.9$ and $2\beta_1 \approx 1.6$, which represent the possible nonanalytic contributions to the surface diameter, can be hardly distinguished from the regular contributions. Moreover, these powers are much larger than the exponent $\beta_1 \approx 0.8$ of the leading symmetric term in equation (13), which yields a dominant nonanalytic (singular) contribution to the fluid density near the surface. Note, however, that situation may change very close to the critical temperature, where a crossover to the normal transition is expected (see Section 3.1).

We would expect that the amplitude B_1 of the leading singular term in equation (13) should not depend on the water–surface interaction potential, at least in the first approximation. This term arises from the bulk order parameter, whose amplitude B_0 is determined by the water–water interaction only. Therefore, we believe that the water–water interaction gives a major contribution to the amplitude B_1. In contrast, the parameters of the asymmetric terms in equation (13) should strongly depend on the water–surface interaction. In particular, ρ_c in the surface layer is essentially below the bulk critical density, when a weak fluid–wall interaction provides "preferential adsorption of voids," whereas ρ_c may exceed the bulk critical density in the case of a strong water–surface interaction. It is difficult to predict the values of the temperature-dependent terms in the asymmetric contribution, as the surface diameter reflects interplay between the natural asymmetry of liquid and vapor phases, described by the bulk diameter, and preferential adsorption of one of the "component" (molecules or voids).

The temperature behavior of the local water densities near the surface, described by equation (13), intrudes into the bulk with approaching critical temperature. It was found that the surface behavior of the symmetric part (order parameter) spreads over the distance about 2ξ from the surface. Temperature crossover of the asymmetric contribution from bulk to surface behavior needs to be studied. Although both the missing neighbor effect and the effect of the short-range water–surface interaction decay exponentially when moving away from the surface, the effective correlation lengths or/and amplitudes of two effects in general may be different. Approaching the bulk critical temperature, symmetric contribution vanishes, whereas the asymmetric contribution remains finite at $T = T_c$. In this sense, one may speculate that the asymmetric contributions dominate the density profile of water near the critical point.

Near hydrophobic surface, the profile of liquid water shows exponential decay described by equation (10) with the fitting parameter ξ^{ef}, which is close to ξ at high temperatures and lower than ξ at ambient and low temperatures [250]. The liquid density profiles are perfectly exponential at $\Delta z > 3.75$ Å, i.e. beyond the first surface water layer (Fig. 51). When applying equation (10) at low temperatures, the distance Δz should be replaced by $\Delta z - \lambda$, where parameter λ is about 1.5 Å at $T = 400$ K

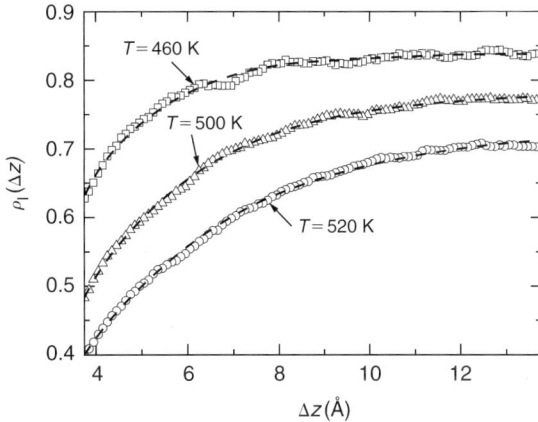

Figure 51: Profiles of liquid water $\rho_l(\Delta z)$ in pore under pressure of saturated vapor at several temperatures (symbols). Fits of the gradual parts of $\rho_l(\Delta z)$ ($\Delta z > 3.75$ Å) to the exponential equation (11) are shown by dashed lines.

and vanishes upon approaching the critical temperature. When surface hydrophilicity increases, the effect of missing neighbor may be effectively compensated and liquid water profile approaches the horizontal line and then crosses over to the gradual increase of water density toward surface. Increase of the surface hydrophilicity results in an increase in localization of water near the surface and, therefore, increase in density oscillations, which may prevent observation (detection) of the gradual trends in the water density profile, especially at low temperatures.

Distribution of the water molecules in vapor phase at low temperature and low density is determined mainly by water–surface interaction. Close to the triple point temperature, water vapor shows adsorption even at the strongly hydrophobic surface. In this regime, the vapor density profiles $\rho_v(\Delta z)$ can be perfectly described by the Boltzmann formula for the density distribution of ideal gas in an external field:

$$\rho_v(\Delta z, \tau) = \rho_v^b \exp\left(\frac{-U_w(z)}{k_B T}\right), \quad (14)$$

where $U_w(\Delta z)$ is the water–surface interaction potential, ρ_v^b is the vapor density far from the surface, and k_B is the Boltzmann constant. The vapor density profile at $T = 300$ K and equation (14) for this temperature are shown in the upper-left panel in Fig. 52. The ideal-gas approach

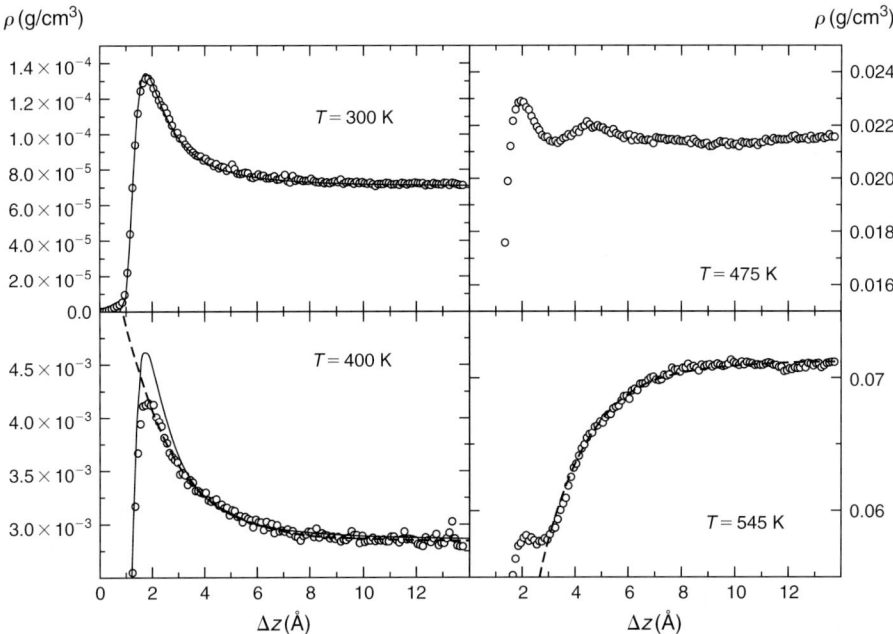

Figure 52: Profiles of water vapor $\rho_v(\Delta z)$ near hydrophobic surface at several temperatures along the pore coexistence curve ($H_p = 30$ Å). Solid lines represents equation (14). Thick dashed lines show the fits to the exponential equation (11) with $\rho_s > \rho_v^b$ and $\xi = 1.88$ Å for $T = 400$ K and with $\rho_s = 0$ and $\xi = 1.80$ Å for $T = 545$ K.

overestimates the adsorption of water vapor on the surface at higher temperature when the density of the saturated vapor exceeds $\sim 10^{-3}$ g/cm^3 (see solid line at the panel $T = 400$ K in Fig. 52). In this regime, the water–water interaction is no more negligible, and a vapor density profile becomes exponential (dashed line in the lower-left panel in Fig. 52).

A further increase in the temperature (density) of the saturated vapor promotes the effect of missing neighbors, and at some thermodynamic state, it may be roughly equal to the effect of surface attraction. The signature of such balance is an almost flat density profile. For the water–surface interaction with a well depth $U_0 = -0.39$ kcal/mol, this happens at $T \approx 475$ K and $\rho_v \approx 0.02$ g/cm^3 (right-upper panel in Fig. 52). At the more hydrophilic surface, the flat density profile may be found at higher temperature. One may expect that at some level of hydrophilicity, the flat density profile of water may appear at the bulk critical point only.

Presumably, this particular strength of water–surface interaction should correspond to the value $U_0 \approx -1$ kcal/mol, which provides an absence of a wetting or drying transition (see Section 2.3). This conjecture has not been tested yet. Increase in the bulk density beyond the value corresponding to the flat density profile results in the density depletion. In case of strongly hydrophobic surface with $U_0 = -0.39$ kcal/mol, vapor shows the depletion of density in a wide temperature range below the critical point. In this regime, $\rho_v(\Delta z)$ may be described by the same equation (10) as profile of liquid water. Such description works perfectly at the distances more than one to two molecular diameters from the surface (lower-right panel in Fig. 52).

So, the water density profiles near the surfaces and their temperature evolution follow the laws of the surface critical behavior, which are universal for fluids and Ising magnets [254]. Nothing peculiar can be found in the surface critical behavior of water in comparison with LJ fluid (see Section 3.1). Many questions concerning the surface critical behavior of fluids and Ising magnets remain open [262] and should be studied. This may provide the possibility to describe the density profiles of water and other fluids analytically in a wide range of thermodynamic conditions near various surfaces.

4 Phase diagram of confined water

4.1 Effect of confinement on the phase transitions

Confinement in pores affects all phase transitions of fluids, including the liquid–solid phase transitions (see Ref. [276, 277] for review) and liquid–vapor phase transitions (see Refs. [28, 278] for review). Below we consider the main theoretical expectations and experimental results concerning the effect of confinement on the liquid–vapor transition. Two typical situations for confined fluids may be distinguished: fluids in open pores and fluids in closed pores. In an open pore, a confined fluid is in equilibrium with a bulk fluid, so it has the same temperature and chemical potential. Being in equilibrium with a bulk fluid, fluid in open pore may exist in a vapor or in a liquid one-phase state, depending on the fluid–wall interaction and pore size. For example, it may be a liquid when the bulk fluid is a vapor (capillary condensation) or it may be a vapor when the bulk fluid is a liquid (capillary evaporation). Only one particular value of the chemical potential of bulk fluid provides a two-phase state of confined fluid. We consider phase transions of water in open pores in Section 4.3.

In closed pores, there is no particle exchange between confined and bulk fluids. Depending on temperature and on the average fluid density in the pore, the confined fluid exists in a one-phase or a two-phase state. When the average density is within the two-phase region, a fluid separates into two coexisting phases. Each phase (liquid or vapor) possesses its own spatial heterogeneity due to the contact with pore walls. Additionally, the coexisting phases are separated by a liquid–vapor interface, which is normal to the pore walls. So, the fluid is extremely inhomogeneous even in pores with ideal geometries (cylindrical or slit-like) and smooth walls. The surface phase transitions, which occur at chemical potential different from that of the liquid–vapor phase transition (layering, prewetting/predrying), appear in confinement as additional two–phase regions, which are marked by their own coexistence curves. In general, there are triple points of two phase transitions, where three fluid phases may exist simultaneously in pores. The surface phase transitions occuring out of the liquid–vapor coexistence are not very sensitive to confinement (see Section 2) but may strongly affect the liquid–vapor phase transition in pores.

Liquid–vapor phase transitions of confined fluids were extensively studied both by experimental and computer simulation methods. In experiments, the phase transitions of confined fluids appear as a rapid change in the mass adsorbed along adsorption isotherms, isochores, and isobars or as heat capacity peaks, maxima in light scattering intensity, etc. (see Refs. [28, 278] for review). A sharp vapor–liquid phase transition was experimentally observed in various porous media: ordered mesoporous silica materials, which contain non-interconnected uniform cylindrical pores with radii R_p from 10 Å to more than 110 Å [279–287], porous glasses that contain interconnected cylindrical pores with pore radii of about 10^2 to 10^3 Å [288–293], silica aerogels with disordered structure and wide distribution of pore sizes from 10^2 to 10^4 Å [294–297], porous carbon [288], carbon nanotubes [298], etc.

It is very difficult to measure the coexistence curves of confined fluid experimentally, as this requires estimation of the densities of the coexisting phases at various temperatures. Therefore, only a few experimental liquid–vapor coexistence curves of fluids in pores were constructed [279, 284, 292, 294–297]. In some experimental studies, the shift of the liquid–vapor critical temperature was estimated without reconstruction of the coexistence curve [281–283, 289]. The measurement of adsorption in pores is usually accompanied by a pronounced adsorption–desorption hysteresis. The hysteresis loop shrinks with increasing temperature and disappears at the so-called hysteresis critical temperature T_{ch}. Hysteresis indicates nonequilibrium phase behavior due to the occurrence of metastable states, which should disappear in equilibrium state, but the time of equilibration may be very long. The microscopic origin of this phenomenon and its relation to the pore structure is still an area of discussion. In disordered porous systems, hysteresis may be observed even without phase transition up to hysteresis critical temperature $T_{ch} > T_c$, if the latter exists [299]. In single uniform pores, T_{ch} is expected to be equal to [300] or below [281–283] the critical temperature. Although a number of experimentally determined values of T_{ch} and a few the so-called hysteresis coexistence curves are available in the literature, hysteresis coexistence curves may give only very approximate information about the liquid–vapor phase transition of fluids in pores.

Phase transitions of confined fluids were extensively studied by various theoretical approaches and by computer simulations (see Refs. [28, 278] for review). The modification of the fluid phase diagrams in confinement was extensively studied theoretically for two main classes of porous media: single pores (slit-like and cylindrical) and disordered porous systems. In a slit-like pore, there are *true* phase transitions that assume coexistence of *infinite* phases. Accordingly, the liquid–vapor critical point is a true critical point, which belongs to the universality class of 2D Ising model. Asymptotically close to the pore critical point, the coexistence curve in slit pore is characterized by the critical exponent of the order parameter $\beta = 0.125$. The crossover from 3D critical behavior at low temperature to the 2D critical behavior near the critical point occurs when the 3D correlation length becomes comparable with the pore width H_p.

In cylindrical pore, the first-order phase transitions are rounded. This rounding decreases exponentially with increasing cross-sectional area of the cylinder [301], leading to rather sharp first-order phase transitions even in narrow pores [300, 302–305]. Theory [306] and computer simulations [32, 249, 303, 304, 307–310] show that phase separation in a cylindrical pore appears as a series of alternating domains of two coexisting phases along the pore axis. The characteristic length of these domains is related to the interfacial tension: it increases exponentially with pore radius R_p and decreases exponentially with temperature [306]. At low temperatures, it could be larger than 10^5 times the pore diameter even in very narrow pores [308, 309]. A fluid confined in an infinite cylindrical pore is close to a 1D system and thus it should not exhibit a true liquid–vapor critical point above zero temperature. However, a "pseudocritical point" could be defined as the temperature when the surface tension between the domains of the two coexisting phases disappears. Above the pseudocritical point the alternating domain structure vanishes and the fluid becomes homogeneous along the pore axis. The densities of the coexisting liquid and vapor domains could accurately be defined in cylindrical pore in a wide temperature range, excluding the vicinity of the pore critical temperature. This means that the coexistence curve and temperature dependence of the order parameter could be studied also for fluid confined in cylindrical pore.

It is not clear how two phases coexist in disordered pores: as alternating domains or as two infinite networks. Disordered porous materials with low porosity are more reminiscent of interconnected cylindrical pores and therefore a domain structure seems to be more probable [299, 311–315]. In highly porous materials, such as highly porous aerogels, infinite networks of two coexisting phases may be assumed. The critical point of fluids in disordered pores is expected to belong to the universality class of the random-field Ising model [316–318].

In large pores, the shift of the first-order phase transition of fluids is described by the Kelvin equation, and in general, it is inversely proportional to the capillary size [319]. In cylindrical pores, the shift of the phase transition is more significant than its rounding [320]. Density functional approaches predict a reduction in the critical temperature ΔT_c in narrow slit and cylindrical pores as

$$\Delta T_c = (T_c - T_{cp}) \sim H_p^{-1}(\sim R_p^{-1}), \tag{15}$$

where T_c and T_{cp} are the critical temperatures of bulk and confined fluid, respectively [307, 321]. However, this approach is not valid close to the critical point, where the correlation length becomes comparable with the pore size. In the asymptotic critical range, scaling theory predicts the following reduction of the critical temperature [322, 323]:

$$\Delta T_c = (T_c - T_{cp}) \sim H_p^{-\theta}, \tag{16}$$

where $\theta = 1/\nu \approx 1.6$. In the framework of the mean-field theory of critical phenomena, T_c is expected to decrease as $\Delta T_c \sim H_p^{-2}$ [323].

The experimental determination of the critical temperature T_{cp} of fluids in pores is a difficult problem. Usually, adsorption measurements are the main way to locate the liquid–vapor phase transition. When approaching the pore critical point, the jump in the adsorption decreases and should disappear. But due to the nonuniform distribution of pore sizes in real porous materials, this jump is smeared out and it is difficult to determine accurately its disappearance. The most accurate results were obtained for fluids in silica aerogels, where the shifts of the critical temperature ΔT_c from $0.002T_c$ to $0.007T_c$ was observed [294–296]. In porous glasses with mean pore radius $R_p = 157$, 121, and 39 Å, the shifts $\Delta T_c = 0.0015T_c$, $0.0029T_c$, and $0.047T_c$, respectively, were obtained [289, 292]. In the ordered cylindrical mesopores with radius $R_p = 17$ and 14.5 Å, shifts $\Delta T_c = 0.019T_c$ and $0.063T_c$, respectively, were obtained for sulfur hexafluoride [279]. Much stronger shift $\Delta T_c = 0.13T_c$ is reported for similar fluid (hexafluoroethane) confined in pores with a radius $R_p = 26$ Å [284]. Even stronger decrease in the critical temperature for a number of fluids is reported in Refs. [281–283]. Shift of the critical temperature decreases with increasing pore size: $\Delta T_c = 0.30T_c$ to $0.35T_c$ ($R_p = 12$ Å), $\Delta T_c = 0.18T_c$ ($R_p = 22$ Å), $\Delta T_c = 0.17T_c$ ($R_p = 30$ and 32 Å), $\Delta T_c = 0.11T_c$ ($R_p = 39$ Å). The obvious discrepancy in the estimated shifts of the liquid–vapor critical temperature in pores is caused mainly by different methods to define a disappearance of the phase transition based on the shape of the adsorption isotherms.

The disappearance of the adsorption–desorption hysteresis with temperature may be used as a very rough estimation of the liquid–vapor coexistence curve in confinement. In Vycor glass, the difference between the bulk critical temperature and the pore hysteresis critical temperature was found to be $\Delta T_{ch} = 0.128T_c$ [288]. In controlled-pore glass with a mean pore radius of 50 Å, ΔT_{ch} is about $0.044T_c$ [291]. The most detailed studies of the disappearance of hysteresis with temperature were reported for cylindrical mesopores with radii R_p from 12 to 110 Å (see [281–283] and [285] for a data collection). The observed values of ΔT_{ch} range from $0.49T_c$ to $0.59T_c$ at $R_p = 12$ Å, from $0.29T_c$ to $0.42T_c$ at $R_p = 21$ Å, and attain $0.033T_c$ at $R_p = 110$ Å [281–283, 285]. The obtained shifts of the hysteresis critical temperature were found approximately proportional to R_p^{-1}.

Not only the critical temperature but also the critical density and the shape of the coexistence curve of fluids may change drastically due to confinement. In aerogels, an increase in the critical density (up to 17% with respect to the bulk value) is accompanied by a strong narrowing of the two-phase region [294, 295]. This narrowing is much stronger at higher temperatures, giving rise to an unusual bottle-like shape of the coexistence curve in a wide temperature range. The shape of the coexistence curve in pore may be described using the dependence of the order parameter $\Delta \rho$ on the reduced temperature deviation $\tau_{pore} = (T_{cp} - T)/T_{cp}$

from the pore critical temperature T_{cp}. The fit of the top of the coexistence curve to a simple scaling law,

$$\Delta\rho^{av} = \left(\rho_l^{av} - \rho_v^{av}\right)/2 = B_0^{ef} \tau_{pore}^{\beta^{ef}}, \qquad (17)$$

where $\rho_{l,v}^{av}$ are the density of liquid and vapor phases averaged over the pore, shows decrease in the amplitude B_0^{ef} by a factor of 2.6 (for nitrogen [296]) and even 14 (for helium [294]) with respect to the bulk value. The effective critical exponent β^{ef} was found to be close to the bulk values for nitrogen ($\beta^{ef} = 0.35$ in the gel and $\beta^{ef} \approx 0.327$ in the bulk [296]) and for helium ($\beta^{ef} = 0.28$ in the gel and $\beta^{ef} = 0.355$ in the bulk [294]). The narrowing of the coexistence region is essentially weaker in the case of neon in an aerogel, where the estimated value of β^{ef} is comparable to or larger than the bulk value [297]. Liquid–vapor coexistence curves reconstructed from the measurements of adsorption of fluids in porous glasses [292] and in cylindrical silica pores [279, 284] show a strong increase in the critical density (up to 100%) and a narrowing of the two-phase region. Fitting of the coexistence curve of fluids in porous glasses to the scaling law (17) yields a value for the critical exponent β^{ef} of about 0.5 [292]. The value of $\beta^{ef} = 0.37$ is reported in [284] for fluid in cylindrical silica pores. Note that available experimental estimates of β^{ef} for liquid–liquid coexistence curves of binary mixtures confined in pores show an increase of β^{ef} to about 0.5 upon decreasing pore size [278]. In simulation studies of argon adsorption in silica glasses, β^{ef} is about 0.5 for the hysteresis coexistence curve [324].

When considering the effect of confinement on the critical temperature, critical density, and the shape of the coexistence curve, it is necessary to take into account density distribution in both coexisting phases. In particular, the critical temperature is predicted to be strongly influenced by the strength of a fluid–substrate interaction [322, 323]. Scaling theory predicts that in a pore of fixed size, the maximum shift of the critical temperature ΔT_c, expected at an infinite fluid–substrate interaction, is 2.07 times larger than the minimum ΔT_c seen at zero fluid–substrate interaction. Mean-field theory predicts a larger value (≈ 2.60) of this ratio [323]. Occurrence of the surface phase transition may change both the critical temperature and the critical density in a drastic way. For example, when a wetting film of some thickness appears in a vapor phase, this effectively reduces the pore size, causes strong increase in the critical density and decrease in the critical temperature. Even when the surface phase transition does not occur, nonhomogeneous density distribution shifts the pore critical density relatively to the bulk value. In the case of an attractive fluid–substrate interaction, ρ_c^{av} averaged over the pore is higher than ρ_c of the bulk fluid, whereas in the central part of the pore, the critical density appears to be below the bulk value [322, 323]. Accordingly, in pores with weakly attractive walls, the average critical density is lower than the bulk value, whereas a higher critical density is seen in the pore interior. Finally, the shape of the pore coexistence curve is affected by the surface critical behavior. This behavior is characterized by the value of the surface critical exponent $\beta_1 \approx 0.8$, which is noticeably higher than the bulk critical exponent $\beta \approx 0.326$. Experimental observations of rather high values of the effective exponent β^{ef} for some system seem to be related to the disordering effect of a surface.

The coexistence curves and properties of confined fluid were extensively studied by computer simulations. Shift of the parameters of the liquid–vapor critical point of fluids in pores was seen in many simulation studies. The most accurate results were obtained by simulations of LJ fluid in the Gibbs ensemble [10, 28–30, 32, 127, 141, 186, 187, 205, 249, 250, 262, 274, 325, 326], but this method is restricted to the pores of simple geometry only. In the narrow slit pore with weakly attractive walls and widths of 6, 7.5, and 10 σ, the liquid–vapor critical point of LJ fluid decreases to $0.889T_c$, $0.919T_c$, and $0.957T_c$, respectively [325, 326]. For comparable fluid–wall interaction, the liquid–vapor critical temperature is about $0.964T_c$ and $0.981T_c$ in the pores with a width $H_p = 12\,\sigma$ and $H_p = 40\,\sigma$, respectively [29]. The dependence of the pore critical temperature on the pore width is shown in Fig. 53. This dependence may be satisfactorily described by equation (15) (solid line) when we take into account that centers of molecules do not enter an interval of about $0.5\,\sigma$ near each wall. The critical temperatures of LJ fluid in the pores with strongly attractive walls are noticeably lower than in pores with weakly attractive walls (compare circles and squares in Fig. 53) [325, 326]. This should be attributed to the effective decrease in the pore width due to the appearance of adsorbed film on the pore walls, which is almost identical in both phases. In this case, dependence of $T_{c,p}$ on H_p may be satisfactorily described by equation (15) (dashed line) if we take into account that

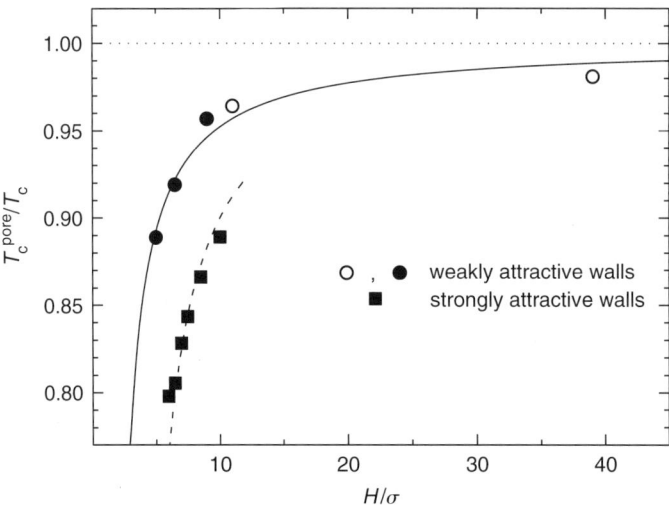

Figure 53: The dependence of the shift of the critical temperature of LJ fluid in slit pores on the pore width H_p. Closed [325, 326] and open [141] circles: pores with weakly attractive walls. Squares: pores with strongly attractive walls. Solid and dashed lines represent equation (15).

the effective width of hydrophilic pore is about 2σ smaller than that of hydrophobic pore.

4.2 Phase transitions of confined water

Various phase transitions of confined water and related phenomena may play an important role in technological and biological processes. First, we consider the effect of confinement on liquid–vapor phase transition of water. Then, freezing and melting transitions of confined water are analyzed. Finally, we discuss how confinement may affect the liquid–liquid phase transitions of supercooled water.

Similar to other fluids, a liquid–vapor phase transition of confined water appears, for example, as a rapid change in the mass adsorbed when the pressure of external bulk water is varied. Numerous examples of the adsorption isotherms of water in various pores, obtained in experiments or in simulations, can be found in literature (some of them we consider below in Section 4.3). However, there are only a few studies where the equilibrium liquid–vapor phase transition of confined water, the corresponding coexistence curve, or the critical parameters were determined. To our knowledge, there is only one experimental estimation of the critical temperature of confined water. Dielectric measurements of water confined in porous Vycor glass with the average pore size of about 70 Å evidence the liquid–vapor critical point of confined water at about 15° below the bulk value [327].

The hysteresis coexistence curves of water in carbon slit pores were estimated from the adsorption isotherms simulated at several temperatures [328]. The hysteresis critical temperature (minimal temperature where hysteresis is not seen in simulations) is about $0.85T_c$, $0.71T_c$, and $0.65T_c$ for water in the pores of the width H_p = 16, 10, and 8 Å, respectively. For cylindrical pores of the radii R_p = 18.5, 8.4, and 6.8 Å, the hysteresis critical temperature is about $0.85T_c$, $0.69T_c$, and $0.58T_c$, respectively. The obtained hysteresis coexistence curves were used to estimate the pore critical temperature from their fits to the simple scaling law, and the values obtained exceed the values of the hysteresis critical temperatures by about 20 to 40° [328]. Note very approximate character of the estimations based on the hysteresis coexistence curves. Liquid–vapor coexistence curve of water in a carbon nanotube of 13.5 Å radius was calculated by constant volume simulations with an average density of water inside the two-phase region [329]. In such simulations, there is an explicit interface between the coexisting phases, and the densities of both phases can be estimated by fitting of this interface by interfacial-like equation. The pore critical temperature is about $0.80T_c$, and the critical density is below the bulk value. In some other simulation studies, true phase transition of water in slit carbon pores of various width (from 6.4 to 16.4 Å) [330] and in cylindrical silica pores (of the radii 5.2 to 14.4 Å) [192] was estimated at one temperature only. This was done by the constant volume simulations of the state points within the two-phase region, where the phase separation does not occur due to the small size of the simulation box. Note that such estimations are very approximate, as the method used gives the so-called finite-size loop of the adsorption isotherm, which has no relation to the Van der Waals loop [331, 332].

The most detailed studies of the liquid–vapor coexistence curve of water [10, 28, 30, 32, 205, 249, 250] were performed by simulations in the Gibbs ensemble [186, 187]. This method provides the direct equilibration between the coexisting phases, but its applicability is limited by the pores with simple geometries and smooth surfaces, as equilibration of pressure requires continuous variations of the volumes of the coexisting phases. Evolution of the liquid–vapor phase transition of water in pore with decreasing pore width was studied systematically for slit-like and cylindrical pores with smooth hydrophobic walls [250]. Interaction between water and pore walls ($U_0 = -0.39$ kcal/mol) roughly corresponds to the interaction of water molecule with paraffin-like surface or with methylated parts of biomolecular surfaces. The liquid–vapor coexistence curves of water obtained in hydrophobic slit-like pores of various width H_p and in cylindrical pores of various radii R_p are shown in Fig. 43 and in Fig. 54, respectively. The liquid–vapor coexistence curves of confined water terminate at lower temperatures in comparison with a bulk case. For slit-like pores, the apparent flattening of the top of the coexistence curves is noticeable when the pore width decreases from $H_p = 30$ Å to $H_p = 6$ Å (Fig. 43). The pore critical temperature T_{cp} may be estimated as an average of two temperatures: the highest temperature, where two-phase coexistence was obtained, and the lowest temperature, where the two phases become identical in the Gibbs ensemble MC simulations (see [32] for the details). The values of T_{cp} obtained in such a way evidence that in the smallest pores ($H_p = 6$ Å), the critical temperature

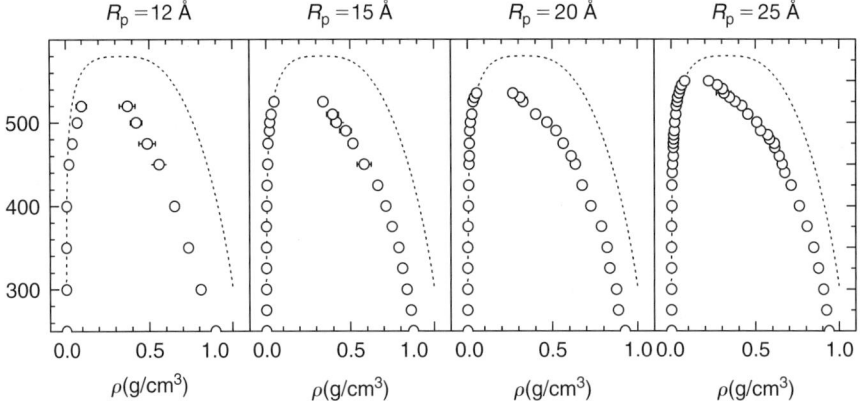

Figure 54: Liquid–vapor coexistence curves of water in cylindrical hydrophobic pores with $U_0 = -0.39$ kcal/mol and various radii. Bulk coexistence curve is shown by dashed line.

Table 1: Pore critical temperature T_{cp} and pore critical density ρ_{cp} of water in slit-like pores of various width H_p with hydrophobic walls ($U_0 = -0.39$ kcal/mol) (data from [250]).

H_p (Å)	T_{cp} (K)	ρ_{cp} (g/cm³)
	Slit-like pores	
6	402.5 ± 2.5	0.358 ± 0.007
9	452.5 ± 2.5	0.294 ± 0.007
12	497.5 ± 2.5	0.262 ± 0.007
15	520.0 ± 5.0	0.246 ± 0.007
18	525.0 ± 5.0	0.235 ± 0.007
21	535.0 ± 5.0	0.241 ± 0.007
24	540.0 ± 5.0	0.229 ± 0.007
30	555.0 ± 5.0	0.227 ± 0.007
bulk 3D water	580 ± 2.5	0.330 ± 0.003
quasi-2D water	330.0 ± 7.5	

is about 180° below the bulk critical temperature T_c (see Table 1), i.e. $T_{cp} \approx 0.69 T_c$. Note that this value is still notably higher than the critical temperature T_{2D} of quasi-2D water with water oxygens located in one plane [32].

The evolution of the pore critical temperature of water in slit-like pores with the pore width is shown in Fig. 55. Note that a thickness of water layer in pore is notably smaller than pore width H_p in narrow pores because the space of about 1.25 Å width near each pore wall is not accessible for water molecules. Therefore, a real thickness of water phases is equal to $H_p - 2.5$ Å. A critical temperature of quasi-2D water, which may be considered as being confined in pore of width $H_p = 5$ Å, and the respective critical temperatures of water in various pores are shown in Fig. 55. To compare the water critical temperature in pores with theoretical equations (15) and (16), ΔT_{cp} is analyzed as a function of ($H_p - 2.5$ Å) in double-logarithmic scale (Fig. 56). When all data points in Fig. 56 being fitted to equation (16), the value of $\theta \approx 0.82$ was obtained [250]. However, the most of the data points may be well fitted by

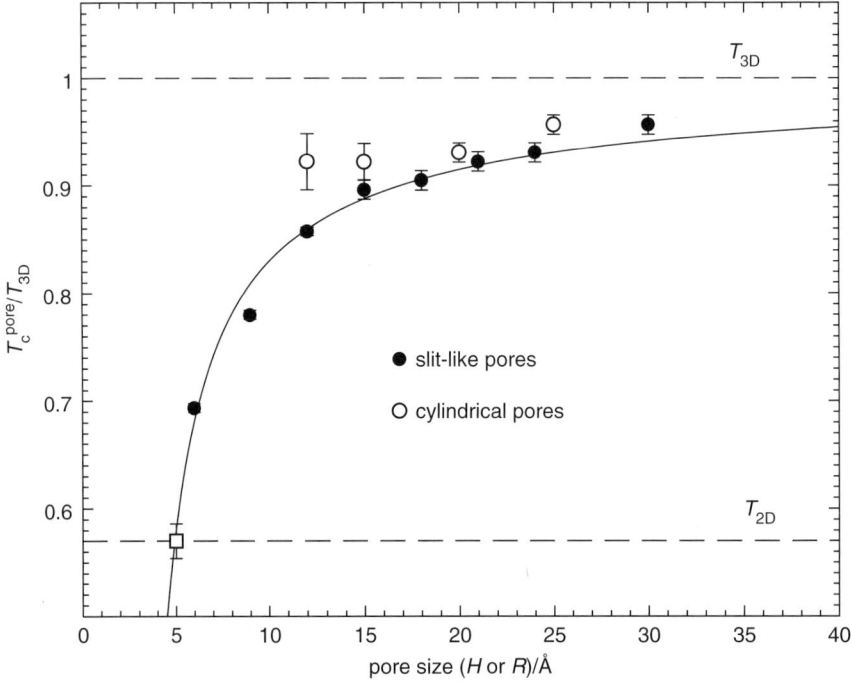

Figure 55: The dependence of the pore critical temperature (T_c^{pore}) on the pore size (H or R). Closed symbols: slit-like pores; open symbols: cylindrical pores. The critical temperatures of bulk water (T_{3D}) and quasi-2D water (T_{2D}) are shown by dashed lines. The square corresponds to the critical temperature of the quasi-2D water, attributed to $H = 5$ Å. The solid line is a fit of equation (16) to the data for slit-like pores (see text for details). Reprinted, with permission, from [250].

a linear law (equation (15)), which is equivalent to $\theta = 1$ in equation (16) (see solid line in Fig. 56). The value ΔT_{cp} in the largest pore studied ($H_p = 30$ Å) shows a trend toward $\theta = 1/0.63$ (see dashed line in Fig. 56). This suggests a crossover between two kinds of behavior at a pore size $H_p = 24$ Å. The latter pore contains a water film of roughly eight molecular diameters width at low temperatures.

The observed evolution of the pore critical temperature with the size of the slit-like pore is in general agreement with theoretical predictions. To make a comparison with simulations for the 3D Ising model in pores, one should express the results obtained for fluids in terms of layers. In particular, the shift of the critical temperature of water in the pores,

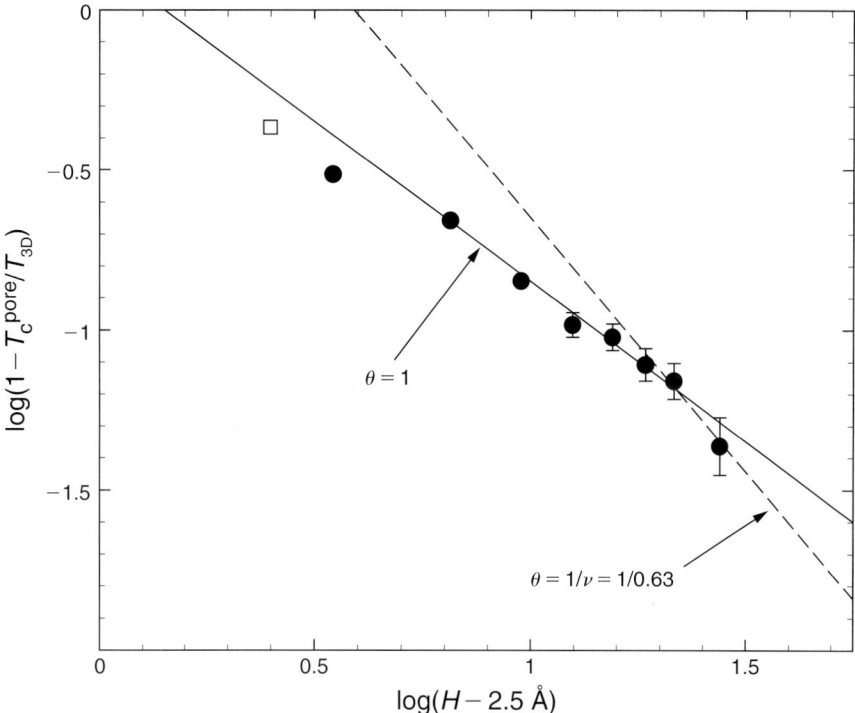

Figure 56: The dependence of the shift of the critical temperature in slit-like pores on the pore width H in double-logarithmic scale. The square corresponds to the critical temperature of the quasi-2D water, attributed to $H = 5$ Å. Reprinted, with permission, from [250].

which contain two to seven water layers, is inversely proportional to the pore size, while in the largest pore studied (about nine water layers), a crossover to a power-law dependence with $\theta = 1/\nu$ is noticeable. In Ising films such a crossover begins when their thickness achieves four to eight layers [333–336]. The critical temperatures of quasi-2D water and water in a pore, comprising single water layer ($H_p = 6$ Å), deviate from the main dependence $\Delta T_c \sim H_p^{-1}$. This deviation is in agreement with the behavior of a monolayer Ising film [333–336]. The reason of such deviations is obvious: with decreasing pore width from two to one monolayer, the system loses any features of three dimensionality and becomes essentially two dimensional.

The first-order liquid–vapor phase transition of water in cylindrical pores was found rather sharp even in pores containing just a few water

Table 2: Pseudocritical temperature T_{cp} and critical pore density ρ_{cp} of water in cylindrical pores of various radii R_p and water–wall interaction U_0 (data from [6, 28, 30, 32, 250]).

R_p (Å)	U_0 (kcal/mol)	T_{cp} (K)	ρ_{cp} (g/cm^3)
25	−0.39	555 ± 5	0.150 ± 0.007
25	−1.93	522.5 ± 5	0.37 ± 0.01
25	−3.08	502.5 ± 5	0.63 ± 0.02
25	−3.85	502 ± 5	0.70 ± 0.01
25	−4.62	502.5 ± 5	0.70 ± 0.01
20	−0.39	540.0 ± 5	0.156 ± 0.007
20	−4.62	482.5 ± 5	0.79 ± 0.01
15	−0.39	535.0 ± 10	0.172 ± 0.007
15	−4.62	355 ± 10	0.92 ± 0.07
12	−0.39	535 ± 15	0.22 ± 0.01
12	−1.93	520 ± 15	0.24 ± 0.07
12	−3.08	465 ± 10	0.42 ± 0.04
12	−3.85	307.5 ± 15	0.86 ± 0.01
12	−4.62	332.5 ± 15	0.86 ± 0.02

layers (see the previous section for more details). The coexistence curves of water in hydrophobic cylindrical pores (Fig. 54) at the first glance look like those in slit-like pores (see Fig. 43). A pseudocritical point was estimated as a temperature, where the two-phase coexistence disappears. A collection of T_{cp} values in various cylindrical pores, estimated in a similar way as in slit-like pores, are shown in Table 2. Above a pseudocritical temperature T_{cp}, only one-phase supercritical fluid exists in pore. Its density and chemical potential is determined by the average pore density in closed pore or by the state of the bulk reservoir in open pores. Below the pseudocritical temperature, the phase separation appears as alternating domains of liquid water and water vapor along the axis of the cylindrical pore (see Fig. 57).

Shift of the critical temperature in cylindrical pores of a radius R_p is comparable with ΔT_c in slit-like pore of width $H_p \approx R_p$ (see Fig. 55). However, the accuracy of data obtained in cylindrical pores does not allow valuable description of the dependence of ΔT_c on the pore radius. The largest uncertainty of T_{cp} was found in the case of narrow cylindrical pores because the decrease in the sizes of the liquid and vapor

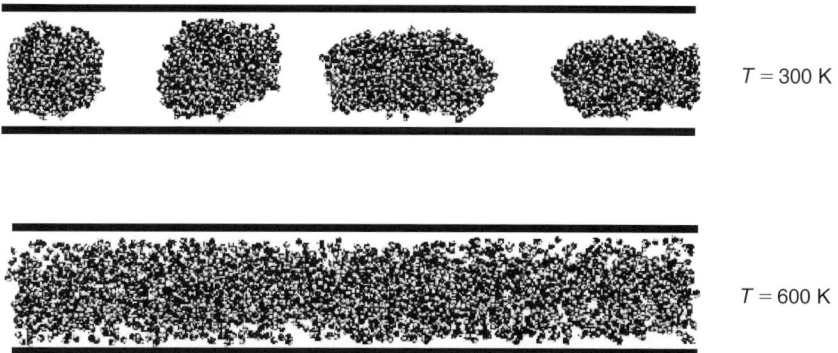

Figure 57: Arrangement of water molecules in a hydrophobic cylindrical pore with radius $R_p = 20$ Å, $U_0 = -0.39$ kcal/mol, and average density $\rho = 0.40$ g/cm^3 at subcritical ($T = 300$ K) and supercritical ($T = 600$ K) temperatures.

domains with increasing temperature prevents an accurate location of the two-phase region (more discussions on the phase transitions of water in cylindrical geometry may be found in [32]).

In hydrophilic and strongly hydrophilic pores, liquid–vapor phase transition of water occurs within the pore whose wall is already covered with two water layers (see Section 2.2). These two surface layers are completely identical in two coexisting phases and therefore may be considered "dead" layers, which create a specific "wall" for water in the pore interior. Accordingly, the effective size of pore with strongly hydrophilic walls should be smaller than real pore size and shift of the liquid–vapor critical temperature should be stronger than in the same pore with a weakly attractive walls. This effect is observed indeed when the pore critical temperatures in hydrophobic ($U_0 = -0.39$ kcal/mol) and strongly hydrophilic ($U_0 = -4.62$ kcal/mol) pores are compared (see Tables 1 and 2). Decrease in T_{cp} with decreasing pore size in strongly hydrophilic cylindrical pores is shown in Fig. 58. This dependence can be described by the equation (15), when the pore narrowing due to the dead layers is taken into account:

$$\Delta T_c = \left(T_c - T_{cp}\right) \sim R_{ef}^{-1}, \qquad (18)$$

where $R_{ef} = R_p - \Delta R$, with $\Delta R \approx 7.5$ Å (see solid line in Fig. 58). A meaningful description of T_{cp} with scaling equation (15) is not possible

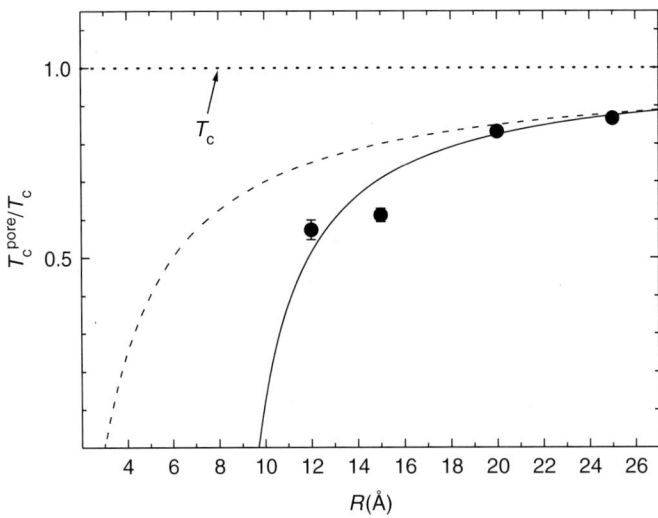

Figure 58: Pseudocritical temperature T_{cp} in strongly hydrophilic cylindrical pores normalized on the bulk critical temperature T_c as a function of pore radius R_p. Solid line represents fit to equation (18) with $R_{ef} = R_p - 7.5$ Å. Dashed line shows a fit to equation (16). The data are taken from Table 2.

(the dashed line in Fig. 58 is drawing when the presence of dead layers is neglected).

Theories predict that shift of the critical temperature should increase upon strengthening of fluid–wall interaction (see previous section). Change of T_{cp} for water confined in cylindrical pore of a radius $R_p = 25$ Å upon U_0 is shown in Fig. 59. Indeed, the predicted trend is observed for hydrophobic and moderately hydrophilic surfaces ($U_0 > -3$ kcal/mol). The shift of the critical temperature T_{cp} is about $0.956 T_c$ in pore with $U_0 = -0.39$ kcal/mol and is about $0.866 T_c$ in pores with $U_0 \leq -3$ kcal/mol. So, ΔT_c in hydrophilic pores is about factor of 3 ΔT_c in strongly hydrophobic pore. However, dead layers appear when $U_0 \approx -3$ kcal/mol; therefore, ΔT_c in pores with $U_0 \leq -3$ kcal/mol cannot be compared with ΔT_c in pores without dead layers. That is why the strongest shift ΔT_c of about 58° is observed when $U_0 \approx -2$ kcal/mol, which is about 2.3 times stronger than in strongly hydrophobic pore. This value is in a reasonable agreement with a factor 2.07 expected within framework of the scaling theory [322, 323].

Fig. 59 evidences that T_{cp} does not change upon strengthening U_0 when pore walls are covered by about two dead water layers. This indicates

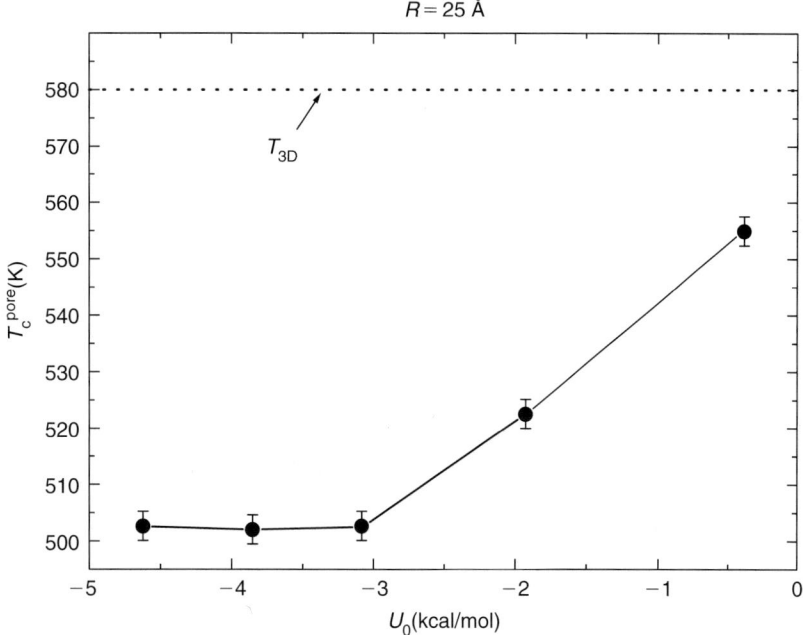

Figure 59: Change of the pseudocritical temperature T_{cp} of water in cylindrical pores of a radius $R_p = 25$ Å with strengthening of the water–wall interaction U_0. When $U_0 \leq -3$ kcal/mol, liquid–vapor phase transition occurs in a pore with walls already covered with about two water layers. The data are taken from Table 2.

that water bilayer effectively screens wall interactive potential and water molecules out of dead layers do not "feel" pore wall at all. Water in the pore interior is indeed confined in the effective pore, whose walls consist of two dead layers of water. This observation has an important consequence: liquid–vapor phase transition of water and its critical (or pseudocritical) point in hydrophilic pores should depend mainly on pore size. Properties of pore walls, their structural peculiarity, presence of charged sites and hydrophobic groups, etc. should not affect notably the liquid–vapor phase transition of water if the pore surface on average is hydrophilic enough to create two dead water layers.

The value of the critical density of confined water, ρ_{cp}, calculated as a ratio of water mass to the pore volume, depends strongly on both pore size and hydrophilicity/hydrophobicity of pore walls. In a pore with weakly attractive walls, ρ_{cp} is notably below the bulk critical density

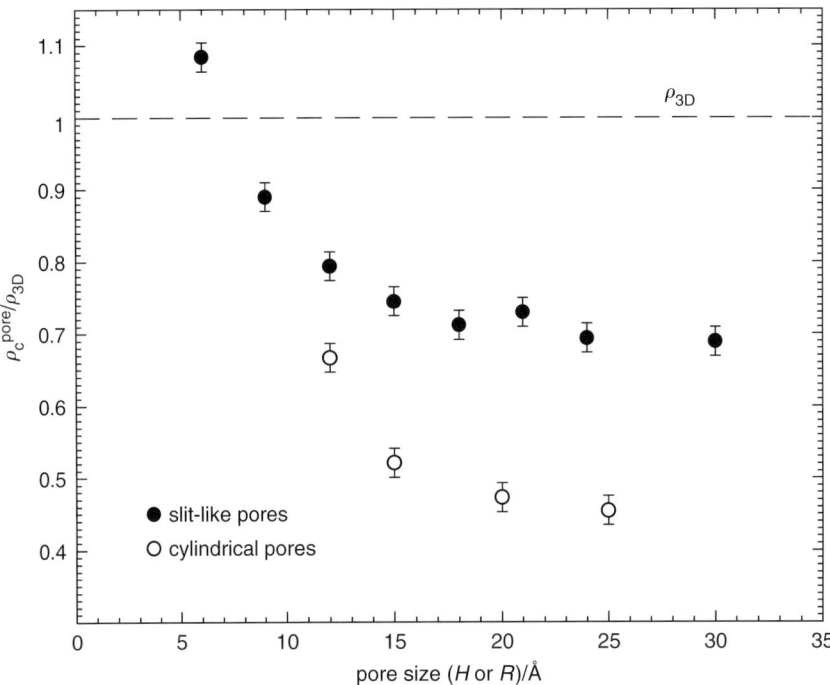

Figure 60: The dependence of the pore critical density (ρ_c^{pore}) normalized to the bulk critical density (ρ_{3D}) on the pore size. Closed symbols: slit-like pores. Open symbols: cylindrical pores. The critical density of bulk water is shown by dashed line. Reprinted, with permission, from [250].

ρ_c (≈ 0.33 g/cm^3 for discussed TIP4P water model [32]). The ratio ρ_{cp}/ρ_c decreases with increasing pore size both in slit-like and cylindrical pores (see Fig. 60). Low value of ρ_{cp} in hydrophobic pores is due to the depletion of density near pore wall caused by effect of missing neighbors (see Section 3 for more details). Note that density depletion in liquid water is pronounced in the whole temperature range of the liquid–vapor coexistence. Due to the density depletion, the average density of liquid water in pore is much lower that the density of bulk liquid water (see liquid branches at the coexistence curves shown in Figs. 43 and 54). At some level of pore hydrophobicity, water vapor also shows depletion toward a pore wall (see Fig. 52 in Section 3). As a result, the critical density in hydrophobic pore may be essentially lower than the bulk critical density.

The density depletion in both coexisting water phases in hydrophobic pores is governed by the bulk correlation length. Liquid–vapor coexistence in the extremely narrow pores is strongly shifted from the bulk in terms of chemical potential and pressure; therefore, the correlation length ξ is smaller when pore is narrower. This effect makes density depletion in small pores less pronounced than in the large pores, and ρ_{cp} is not very far from ρ_c. With increasing pore size, ξ increases, approaching the bulk value, and achieves saturation in pores, whose width $H_p \gg \xi$. Probably, this process dominates in the size dependence of ρ_{cp} in slit-like pores (solid circles in Fig. 60). In cylindrical pores, suppression of ρ_{cp} is stronger than in slit geometry (open circles in Fig. 60). It may be due to specific surface critical behavior near concave surfaces or an embryo of a drying layer may appear near cylindrical surface.

Enhancement of the pore wall hydrophilicity increases the densities of water phases in pore and accordingly ρ_{cp} approaches the bulk value. Pore with $U_0 = -1.93$ kcal/mol may serve an example of such moderate hydrophilicity, which provides $\rho_{cp} \approx 0.37$ g/cm^3 (see also Table 2). In strongly hydrophilic pores, two dead layers near the pore wall are present in vapor phase and make its average density incomparably larger than the vapor density of bulk water at the same temperature. Hence, the critical density in hydrophilic pores is much higher than ρ_c (see Table 2). Note that density of coexisting phases in the pore interior remains comparable with corresponding bulk liquid and vapor densities. This was observed in various pores with various walls [262].

A crossover from 3D to 2D critical behavior, expected theoretically for fluid phase transition in slit-like pores (see Section 4.1 for more details), may be studied using the dependence of the average order parameter $\Delta \rho^{av}$ on the reduced temperature deviation τ_{pore}. A 3D $\rightarrow$ 2D crossover appears as a change in an effective exponent β^{ef} in equation (17) from the values close to the 3D exponent $\beta \approx 0.326$ to the 2D exponent $\beta = 0.125$ upon heating. Possibility of such crossover was studied in slit pores of various sizes with hydrophobic walls [250]. In large slit-like pores of width $H_p = 21$ and 30 Å, the crossover toward two dimensionality was noticed with approaching pore critical temperature when the correlation length ξ achieves about 0.25 H_p. In a narrow pore with $H_p = 6$ Å, containing approximately one molecular layer, the 2D critical behavior holds in the whole temperature range. In the pore of the intermediate sizes ($H_p = 9, 12,$ and 15 Å, which correspond to about two to four molecular layers), behavior of the order parameter was found to be close to three dimensional in the whole temperature interval of liquid–vapor phase transition, contrary to the expectation that the region of 2D behavior should be essentially wider than in larger pores. This apparent 3D behavior seems to be a result of a competition between the trend toward 2D criticality and a progressive contribution to the average order parameter arising from the surface layers, which obey the laws of the surface critical behavior. Note that the dimensional crossover in small pores certainly needs further studies both for fluid and lattice [336] systems.

Liquid–solid phase transition of confined water was extensively studied experimentally. On freezing, water in the pore interior transforms into cubic ice [281, 338–343], and its stability with respect to the ordinary hexagonal ice increases as the pore becomes smaller [341]. There is a strong decrease in the freezing temperature of water in confinement, which is roughly inversely proportional to the pore radius (dependence is similar to equation (15)) [281, 337, 338, 344–346] and achieves 60° in the cylindrical pore of the radius $R_p = 14.4$ Å [338]. The dependence of the melting temperature of ice, confined in cylindrical mesoporous silica pores, on pore radius is shown in Fig. 61. Both sets of the experimental data (shown by closed and open circles) are consistent with equation (15), when a layer of "unfreezable" or "bound" water is assumed to be in contact with a pore wall. Accordingly, the thickness of the "bound" water layer may be estimated as 4 Å [337] or 6 Å [338]. In other experimental studies of water in porous silica glasses, Vycor glasses, silica gels, and porous silicon, the thickness of this layer was found to be of about 2.5 to 3 water monolayers [344], about 8 to 17 Å thick [347], about 10 Å [348], about 4 to 8 Å thick [349], about 4 to 6 Å [350, 351], about 5 Å [341], and about 3.5 Å [345].

For the "bound" water, the liquid–solid phase transition is smeared in the temperature interval of about 180 to 220 K, which does not depend on the pore size [281, 338, 350]. When the radius is less than approximately 10 to 12 Å, only bound water remains, and there is no sharp solid–liquid phase transition in such small pores [281, 338, 350, 352]. In simulations, freezing of water confined in narrow hydrophobic pores was obtained at very high pressures and/or by adjusting the pore size to fit some particular form of monolayer or bilayer ice [353–358]. Interestingly, studies of the liquid–vapor transition in hydrophilic pores (see Section 2.2) indicate the formation of about two water layers at the pore wall, which are in

Phase diagram of confined water 111

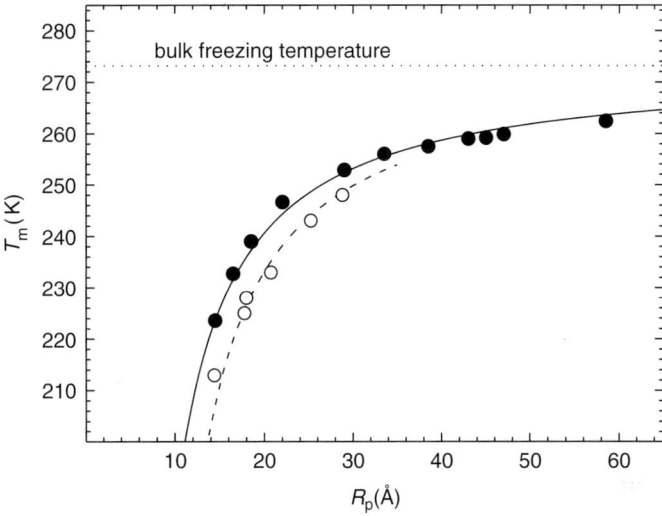

Figure 61: The dependence of melting temperature of ice in the cylindrical pores of various radii R_p. Data taken from [337] and [338] are shown by closed and open circles, respectively.

fact "dead" layers and which do not participate in the liquid–vapor phase transition in the pore interior. These two layers are more orientationally ordered than the rest of liquid water (see Section 5). In a solid state, the situation is opposite: there is a crystalline ice in the pore interior, whereas one to two water layers at the pore walls remain liquid-like and possess little short-range order, as was evidenced by X-ray diffraction experiments [341].

As the adsorbed water layers are unfreezable, they may be used for experimental test of the hypothesis concerning liquid–liquid transitions of supercooled water. Differential scanning calorimetry measurements of water freezing in incompletely filled silica pores evidence two exothermic peaks at 233 and 237 K [337, 359]. These peaks are seen at various pore fillings and should be attributed to some transitions related to adsorbed water layers. They may originate from the "delayering" transition [337], which should occur at the temperature of the triple point, where adsorbed water film coexists with bulk water and water vapor (see Figs. 22 and 25). Below this temperature, water film becomes unstable with respect to other two water phases. Such interpretation is questionable because the temperature of such transition is highly sensitive to the pore wall hydrophilicity (see the surface phase diagram of water in Section 2.4) and also depends on the pore size. Another interpretation relates these peaks to the transition of water layer itself. In the incompletely filled Vycor glass, exothermic peak upon cooling is seen at about 240 K [360, 361]. Neutron diffraction data evidence the sudden change in the density of adsorbed water between 238 and 258 K. Below and above the temperature of the transition, adsorbed water remains a liquid. Thus, the transition of water in incompletely filled pores at about 230 to 240 K may be attributed to the *liquid–liquid* transition of interfacial water. As this transition occurs at zero pressure, the critical point of this transition should be located at negative pressures (in accordance with the scenario, shown in the right panel of Fig. 7).

Calorimetric, X-ray diffraction, and neutron scattering experiments [243, 281, 338, 362] show the absence of a phase transition of water in the completely filled cylindrical pore with a radius less than about 12 Å upon cooling at zero pressure. Qualitative change in the temperature dependence of water dynamics is seen at about 225 K in cylindrical pores with radii 7 to 9 Å [243, 362–366]. In more hydrophobic pores (carbon nanotubes), such dynamic transition of water occurs at a slightly lower temperature of about 218 K [367]. Dynamic transition of supercooled interfacial water may be attributed to the distant effect of the liquid–liquid critical point of interfacial water, which is located at positive pressures. This interpretation is supported by the fact that the temperature of the observed dynamic transition of water shifts to lower temperatures with increasing pressure [243], i.e. in accordance with behavior expected for the liquid–liquid transition of supercooled water (see Section 1). Location of the liquid–liquid critical point at positive pressure is also supported by the observation that the density of liquid water confined in narrow cylindrical pore changes in a continuous way upon cooling [35]. Interestingly, on the surface of biomolecules, hydration water undergoes quite similar dynamic transition approximately at the same temperatures (at 222 and 220 K in the case of DNA [368] and lysozyme [369], respectively). This transition may be related to the dynamic transition of biomolecules, which occurs in a rather narrow temperature interval and is accompanied by the onset of biological function upon heating (see Section 6).

Liquid–liquid transitions of confined supercooled water were also studied theoretically [370] and by computer simulations [10, 357, 371,

372]. The density functional theory studies of the associating fluid show a strong effect of confinement on the liquid–liquid transition, which can be reduced only when the fluid is treated as homogeneous [370]. Simulation studies of water in a narrow hydrophobic pores indicate the disappearance of the liquid–liquid transition [371] or its shift to lower temperatures [357]. The liquid–liquid phase transition of supercooled confined water was simulated directly in slit-like pore of 24 Å width and with three levels of pore hydrophobicity/hydrophilicity [10]. In bulk ST2 water, the first (lowest density) liquid–liquid transition crosses the liquid–vapor transition at about 270 K, which results in a triple point, where two different liquid phases coexist with a water vapor. This triple point appears as a horizontal step in the liquid branch of the liquid–vapor coexistence curve (see Fig. 7 and Fig. 62). In a strongly hydrophobic pore ($U_0 = -0.39$ kcal/mol), this triple point shifts by about 7° to lower temperatures (left panel in Fig. 62). In a moderately hydrophilic pore ($U_0 = -1.92$ kcal/mol), this shift is just 2°. In a strongly hydrophilic pore ($U_0 = -3.08$ kcal/mol), the critical point of the liquid–liquid transition shifts to positive pressures and the triple point disappears. The gradual transition from normal water to tetrahedral water upon cooling

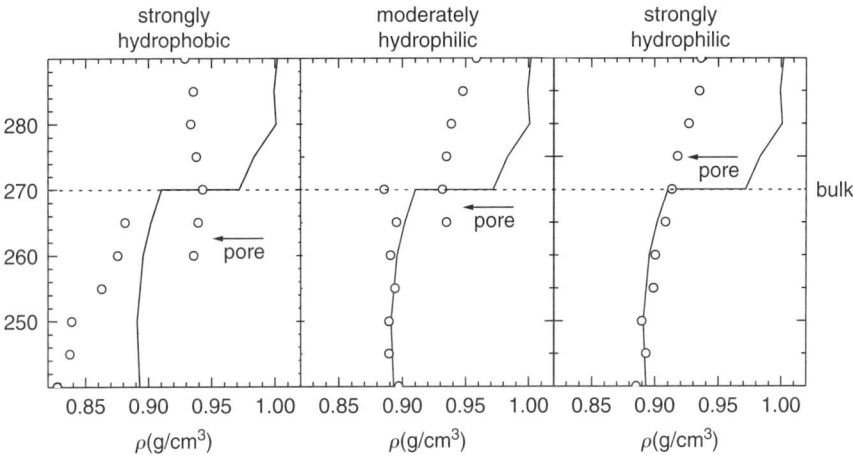

Figure 62: Shift of the liquid–liquid transition of supercooled water in pores. Solid line: liquid branch of the liquid–vapor coexistence curve of ST2 bulk water with a step at 270 K, indicating liquid–liquid transition at zero pressure (dotted line). Liquid branches of the coexistence curves of water in pores of various hydrophilicity are shown by open circles. Temperatures of the liquid–liquid transitions at ambient pressure are indicated by arrows [10].

along the liquid–vapor coexistence occurs supercritically in this case. The temperature of this transition is about 5° higher than the temperature of the liquid–liquid transition in the bulk case.

4.3 Capillary condensation and capillary evaporation

Situation when water in pore is in equilibrium with a bulk water occurs frequently in various processes in the earth's conditions. When air humidity and temperature vary, water may condense in porous materials or may evaporate from them. These processes may, for example, cause aging and destruction of building materials [373]. Evaporation of liquid water from porous media upon heating may be used for heat storage. Attraction of extended hydrophobic surfaces and repulsion of extended hydrophilic surfaces in liquid water are directly related to the situation when a confined fluid is in an *open* pore. To understand driving forces of these and related phenomena and to clarify the peculiar role of water in such processes, regularities of the capillary condensation/evaporation of water should be studied.

A liquid–vapor phase transition in an open pore may be induced by varying pressure (chemical potential) of a bulk fluid. This transition appears as a step in adsorption isotherms or isobars. Liquid–vapor phase transitions at several temperatures are shown schematically in Fig. 63. In the bulk, liquid–vapor condensation of a fluid occurs along horizontal line binding the liquid and vapor coexistence densities at a given T (left panel in Fig. 63). The same transitions appear in isotherms, as it is shown by the vertical lines at the right panel of Fig. 63. With increasing temperature, condensation occurs at higher pressures, and the difference between vapor and liquid densities becomes smaller. Chemical potential of a fluid in an open pore is equal to the chemical potential of the external (infinite) bulk fluid. So, the state of confined liquid is controlled by the thermodynamic conditions of external fluid. Being in equilibrium with the bulk reservoir, a pore may be filled with a liquid or with a vapour. A phase state of fluid in pore depends not only on the thermodynamic state of a bulk fluid, but on the pore properties (hydrophilicity/hydrophobicity, size and shape) as well. The most interesting practical situation corresponds

Phase diagram of confined water

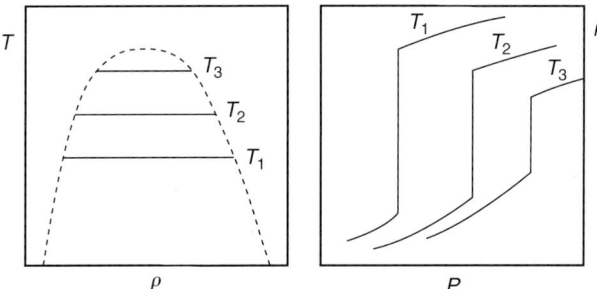

Figure 63: Left panel: liquid–vapor phase transitions of bulk fluid (horizontal lines) shown in $T - \rho$ coordinates together with the bulk coexistence curve (dashed line). Right panel: the same transitions shown as isotherms at $T_1 < T_2 < T_3$ in $\rho - P$ coordinates (solid lines).

to the equilibrium of a confined fluid with a bulk fluid at saturated vapor pressure, i.e. the bulk fluid at the liquid–vapor coexistence. In this case, the phase state of a fluid in open pore depends on the strength of the fluid–wall interaction only and does not depend on the pore size [205, 206, 374]. In particular, for weakly attractive walls, only the vapor phase is stable for any distances H_p between the confining (infinite) walls.

Neglecting the hysteresis phenomenon, adsorption isotherm in a single pore is shown schematically in Fig. 64. If the fluid–wall interaction is strong (hydrophilic pores), condensation transition in pore occurs at undersaturated bulk pressure $P < P_0$ (dashed line). This phenomenon is called capillary condensation. In the pore with hydrophobic walls, oversaturated bulk pressure $P > P_0$ is required to push a liquid inside the pore (dotted line). At a bulk saturated pressure, such pore is filled with a vapor only, and this phenomenon is called capillary evaporation. The lines of the phase transitions of confined fluids, corresponding to the capillary condensation and capillary evaporation, are shown in the right panel of Fig. 64 in $P - T$ coordinates by dashed and dotted lines, respectively. Both lines terminate at lower temperatures in comparison with the line of the bulk liquid–vapor transition. There are no grounds to expect that the lines of the phase transitions of bulk and confined fluids may cross at some temperature. With increasing pore size at fixed fluid–wall interaction, line of the phase transition approaches (but never crosses) the line, corresponding to the bulk liquid–vapor phase transition. At some particular strength of a fluid–wall interaction, condensation in pore of

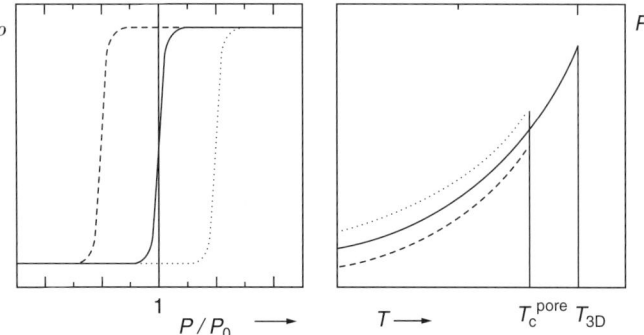

Figure 64: Liquid–vapor phase transition of bulk fluid (right panel, thin solid line) and in hydrophilic pore (dashed lines) and hydrophobic pore (dotted lines). Condensation in pore with a "neutral" wall, which happens at the same pressure as the bulk transition, is shown by solid line in left panel.

any size occurs at the same pressure as a condensation in bulk fluid (thin solid line at the left panel of Fig. 64). It is important to estimate the value of this critical interaction for various fluids, including water.

Numerous experimental and simulation studies were performed to study the effect of capillary evaporation of water. This phenomenon was observed for water in porous media such as activated carbon, silica glasses, mesoporous silica pores, zeolites, and many other porous media. Experimental studies show that at some level of the pore wall hydrophobicity, liquid water does not penetrate into the pore [346, 375, 376]. Simulation studies evidence evaporation of liquid water from hydrophobic pores [206–208, 377, 378]. Knowledge of the critical water–wall interaction, which separates regime of the capillary condensation from the regime of capillary evaporation, is important for understanding the properties of confined water and for prediction of its behavior in various confinements. In simulations, an equilibrium between confined fluid and saturated bulk fluid can be provided by specifying common values of the temperature and chemical potential for both systems. This can be achieved by simulations of the confined fluid in the grand canonical ensemble with the value of the chemical potential of a bulk fluid, which must be determined in advance. Alternatively, confined fluid can be equilibrated directly with a bulk vapor [379, 380] or bulk liquid [205, 206, 208] by simulations in the Gibbs ensemble [187]. The latter approach is the most accurate, as the available force fields for fluids were parameterized mainly for bulk liquid phase.

A critical water–wall interaction was determined by complementary studies of the condensation/evaporation transition in open pores and of the liquid–vapor coexistence in closed pores [30, 32, 205, 208]. Such studies were performed for water confined in cylindrical pores of the radius $R_p = 12$ Å. The liquid density in open pore was obtained by direct equilibration in the Gibbs ensemble of confined water and a bulk liquid water at ambient pressure, starting from completely filled pores [208]. The liquid density obtained is a monotonic function of the water–wall potential U_0, and the values obtained at $T = 300\,K$ are shown in Fig. 65 (solid squares). In parallel, the density of liquid water at equilibrium with saturated vapor in pore was calculated (Fig. 65, open circles). It also depends on U_0, however, weaker than the density in an open pore. Liquid densities in closed and open pores become equal at $U_0^c \approx -1.0\,\text{kcal/mol}$ (see crossing point in Fig. 65). If $-U_0 \leq 1.0\,\text{kcal/mol}$, only water vapor is stable in the pore. Accordingly, liquid is stable in the pore, if $-U_0 \geq 1.0\,\text{kcal/mol}$. Thus, the value of $U_0^c \approx -1.0\,\text{kcal/mol}$ separates regimes of capillary evaporation and capillary condensation [32, 208]. This value approximately coincides with the critical water–wall interaction, which provides an absence of wetting or drying transitions up to the liquid–vapor critical point (see Section 2.4).

When bulk fluid is oversaturated, so that its pressure is higher than at the liquid–vapor coexistence, and a hydrophobic pore is filled with liquid, capillary evaporation ultimately occurs when capillary width is less than some critical distance H_{crit} between the walls [377, 381, 382], which depends on the magnitude of the bulk pressure applied. A simple estimation of H_{crit} at ambient conditions was done [237], assuming that liquid water is under conditions of oversaturation due to the additional pressure of about 1 bar caused by the earth's atmospheric gases. The value H_{crit} of about 1000 to 10000 Å was predicted using the Kelvin equation. Note, however, that such simple consideration neglects natural presence of atmospheric gases in liquid water, which makes a real water to be a fluid mixture, which possesses its own phase diagram different from that of pure water. So, we cannot expect oversaturation of real liquid water at ambient conditions, and accordingly, there is no critical distance for capillary evaporation. In equilibrium with a saturated bulk, a liquid water is

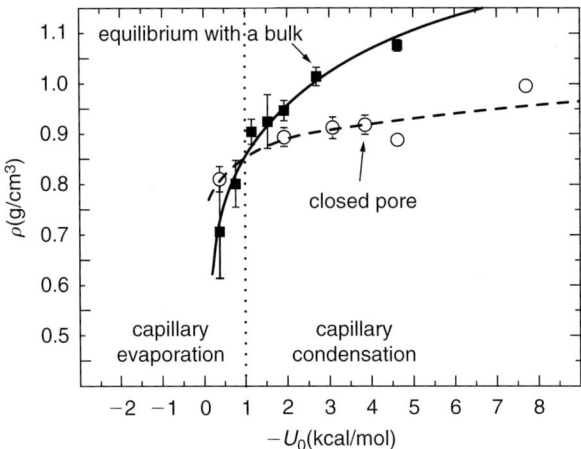

Figure 65: The average water density in cylindrical pores with $R_p = 12$ Å as a function of the water–wall interaction U_0. Closed squares: confined water in open pore is in equilibrium with saturated bulk water. Open circles: confined water in closed pore at the pore coexistence curve. Crossing point of two dependences indicates a critical water–wall interactions, which separates regimes of capillary condensation and capillary evaporation.

always metastable between the walls with $U_0 > -1$ kcal/mol. Of course, in a wide hydrophobic pores, this metastable state may be long lived. With pore narrowing, metastable liquid water approaches the stability limit, and cavitation becomes unavoidable [383]. In narrow cylindrical pores, the liquid–vapor transition is strongly rounded, and this results in an intermittent permeation of water through narrow channels [384–386]. This behavior should be attributed to the thermally induced transitions between vapor-like and liquid-like states, separated by small energetic barrier [385].

Metastability of liquid water between hydrophobic surfaces [381, 387] and/or liquid density depletion near these surfaces [388, 389] give rise to the long-range attractive forces (hydrophobic forces) between them. Both metastability of a liquid between weakly attractive surfaces and depletion of a liquid density near a weakly attractive surface are general phenomena, and they should be relevant to all fluids. The only peculiar feature of water is abundance of weakly attractive (hydrophobic) surfaces for water on the Earth. The phenomenon of hydrophobic attraction was extensively studied experimentally [212]. Less is known about the phenomenon of hydrophilic repulsion [390], which was mainly observed for interactions between soft amphiphilic surfaces in liquid water [391]. Similar to hydrophobic attraction, hydrophilic repulsion is not related to peculiar water properties but is a general phenomenon for all fluids between strongly attractive surfaces [389, 392]. As we can see from Fig. 65, due to the equilibrium with a saturated bulk, a liquid water density in hydrophilic pore should increase by 10 to 20%. This may cause a drastic repulsion between hydrophilic surfaces in liquid water. In particular, hydrophilic repulsion may be responsible for the destruction of building materials, including marbles [373].

It is natural to attribute the attraction between large hydrophobic objects in liquid water to the phenomena described above. When we consider *two* objects with extended hydrophobic surfaces in liquid water, then the use of the analogy with a liquid in a pore geometry may be fruitful. However, in some other cases, use of such analogy leads to misleading conclusions. When dealing with a macroscopic number of hydrophobic particles in liquid water, their aggregation (clustering) is determined by the location of the considered state point to the two-phase region [393] and not by mythical "hydrophobic forces." In one-phase region, this clustering continuously increases when the system approaches the two-phase region due to the variation of temperature, pressure, or concentration, and this effect is universal for all binary mixtures. Of course, clustering of hydrophobic particles in water has no relation to the attraction between extended hydrophobic surfaces in liquid water.

5 Water layers at hydrophilic surfaces

Near hydrophobic surfaces, density of liquid water is depleted and its structure becomes less ordered (Sections 2.3 and 3.2). Quite similar behavior is seen for other fluids (for example, LJ fluid) near weakly attractive surfaces (Section 3.1). More peculiar behavior of water may be expected near hydrophilic surfaces, as strong localization of molecules due to the attraction to the surface causes specific rearrangement of water–water H-bonds. In this section, we characterize the arrangement of water molecules in various phase states of water: vapor, monolayer, bilayer, and liquid water. In particular, percolation transition, that is a continuous transition between a low density vapor and a complete monolayer, is considered in Section 5.1. A specific orientational ordering of water near the surface and its intrusion into a bulk liquid water is analyzed in Section 5.2.

5.1 Percolation transition of hydration water

The existence of an infinite (spanning) network of H-bonded water molecules strongly affects the properties of aqueous systems and plays an important role in various technological and biological processes. Percolation transition is directly related to the respective phase transition, whose critical point is a percolation threshold of physical clusters [23]. Percolation of hydration water, i.e. formation of an infinite H-bonded network of water molecules adsorbed on the surface, is related to the layering transition (quasi-2D condensation) of water at the same surface. Therefore, percolation transition should be observed above the critical point of the layering transition. Besides, we may expect this transition when the layering transition is smeared out due to the surface heterogeneity. Percolation transition of hydration water is quasi-2D since even at a smooth surface, the adsorbed water molecules are not restricted to a single plane parallel to the adsorbate surface. Therefore, some deviations of the percolation transition of hydration water from conventional percolation in strict 2D systems can be expected. In this section, we show how the percolation transition of water can be studied by computer simulations. First, we consider the case of a smooth hydrophilic plane and describe the methods that allow location of the percolation threshold. Then, these methods are applied to characterize percolation on the surface of a finite object (sphere). The methods developed are applied for biological systems in Sections 7 and 8.

The critical temperature of the layering transition of water at the smooth planar surface, with water–surface interaction strength $U_0 = -4.62$ kcal/mol is about 400 K (Section 2.2). Therefore, we can study the percolation transition at $T = 425$ K. This temperature is notably lower than the bulk critical temperature of 3D water; therefore, the H-bonded water cluster is a rather good approximation for the physical cluster of hydration water. Water molecules were considered to belong to the same cluster if they are connected by a continuous path of H-bonds [26, 100, 204, 395]. H-bond between two water molecules may be defined in different ways. A double distance-energy criterion assumes the H-bond to exist when the distance between the oxygen atoms <3.5 Å and the water–water interaction energy <-2.4 kcal/mol. The distance ~ 3.5 Å corresponds to the first minimum of the oxygen–oxygen distribution function, and this value is not sensitive to the water model, and it is commonly used for the analysis of hydrogen bonds in computer simulations of water. The energy -2.4 kcal/mol corresponds to the minimum of the distribution of the water–water pair interaction energies at $T = 425$ K, and it varies slightly with temperature.

Percolation transition of hydration water is intrinsically a site-bond percolation problem. At some temperature, percolation transition occurs upon increase in the surface coverage C, which is analogue of the occupancy variable p. At low coverages, only finite clusters are present in the system, whereas there is an infinite cluster above the percolation threshold. In Fig. 66, typical arrangement of water molecules, adsorbed at hydrophilic plane, is shown for three surface coverages. Visual inspection does not allow determination of the percolation threshold. This can be done by the analysis of various cluster properties for a system of a given dimensionality [396]. As hydration water is not a strict 2D system, the reliable estimation of a percolation threshold assumes an independent use of several criteria.

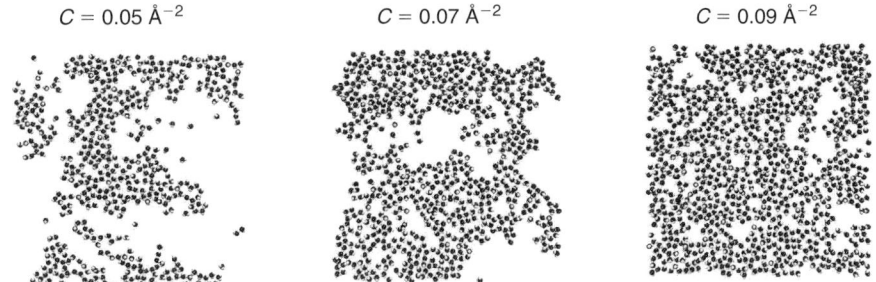

Figure 66: Arrangement of water molecules adsorbed at hydrophilic surface with $U_0 = -4.62$ kcal/mol at $T = 425$ K and various surface coverages C.

a) Cluster size distribution

The cluster size distribution n_S is an occurrence frequency of water clusters of sizes S. Right at the percolation threshold, the cluster size distribution obeys the universal power law:

$$n_S \sim S^{-\tau}, \qquad (19)$$

with exponents $\tau = 187/91 \approx 2.05$ [396] and $\tau \approx 2.2$ [397] in the case of random 2D and 3D percolation, respectively. In an infinite system, this universal behavior should be valid for all cluster sizes S. In any finite system, n_S follows equation (19) in a broad range of S, up to large clusters, whose linear extension becomes comparable with the system size. Due to the fact that clusters with the linear extension larger than the size of the system simulated cannot be observed, they effectively contribute to the probabilities of smaller clusters, whose population therefore is overrepresented and hence a hump appears on the n_S distribution. This hump strongly affects n_S and makes its use inconvenient to locate a percolation threshold in small systems [25]. With increasing system size, the hump at the n_S distribution shifts to larger S, which enables observation of a power a behavior equation (19) in wide range of S (see Fig. 67).

When approaching the percolation threshold via increase of the surface coverage, the cluster size distribution undergoes qualitative changes. At low surface coverage, most of the water molecules belong to small clusters and n_S shows a rapid exponential decay with increasing S. Upon increasing the hydration level, a hump appears in n_S at large $S(C = 0.047$ Å^{-2} in Fig. 68). At the percolation threshold, the cluster size distribution n_S

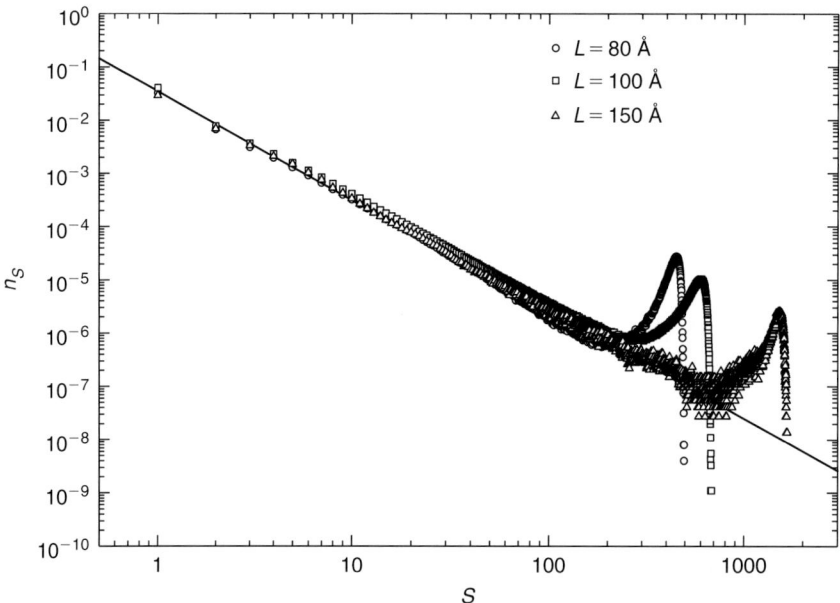

Figure 67: Probability distribution n_S of clusters with S water molecules at planar surfaces of various sizes at surface coverages close to the percolation thresholds: $C = 0.078$ Å^{-2} (circles), 0.070 Å^{-2} (squares), and 0.078 Å^{-2} (triangles). The critical power law $n_S \sim S^{-2.05}$ is shown by a solid line. Reprinted, with permission, from [394].

follows the power-law behavior $\sim S^{-\tau}$ in the widest range of cluster sizes with $\tau = 2.05$ for 2D percolation (see $C = 0.074$ and 0.078 Å^{-2}). When crossing the percolation threshold, deviations of n_S from the power law at large S before the hump change the sign from positive to negative (compare $C = 0.074$ and 0.078 Å^{-2} in Fig. 68). The negative deviations of n_S increase rapidly with increasing hydration above the percolation threshold ($C = 0.082$ Å^{-2}). Thus, evolution n_S shown in Fig. 68 evidences that the percolation threshold of the adsorbed water C_p at the plane with $L = 80$ Å occurs close to the surface coverage $C = 0.078$ Å^{-2} or slightly below. This estimation is valid also for larger surfaces, taking into account rather coarse variation of the surface coverage (Fig. 67). So, the left and middle pictures in Fig. 66 show arrangement of water molecules below the percolation threshold, whereas a spanning water network is present in the right picture.

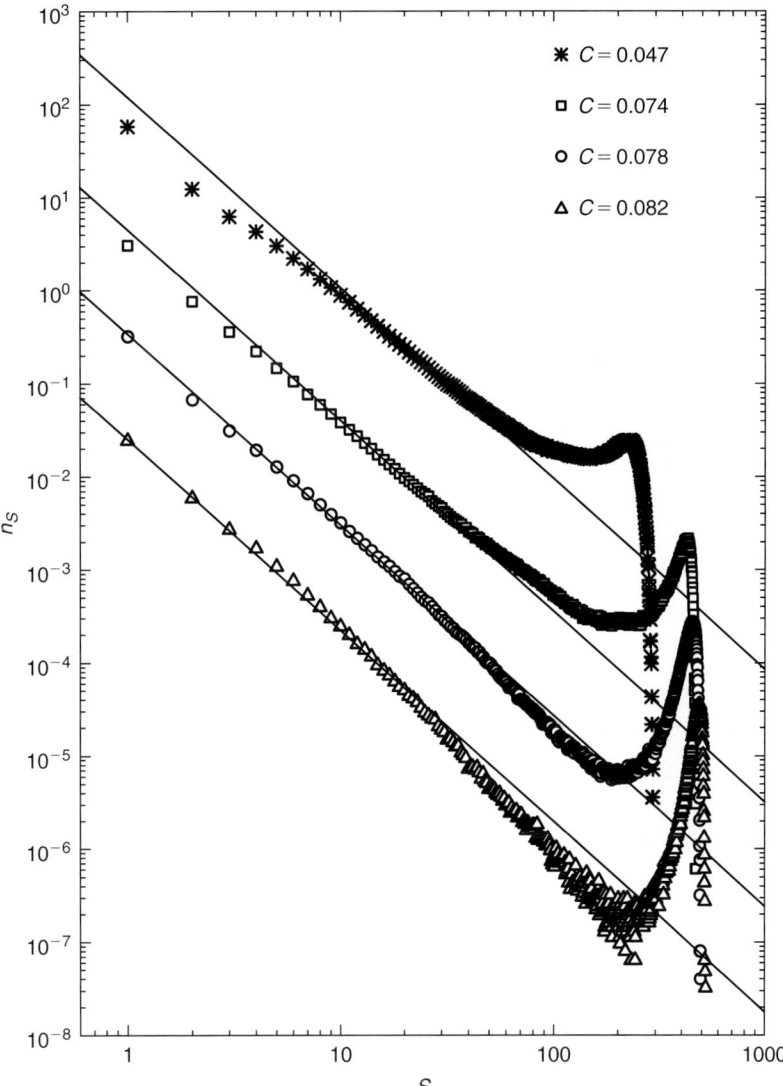

Figure 68: Probability distribution n_S of clusters with S water molecules for several surface coverages C (in Å^{-2}) below and above the percolation threshold ($C_p \approx 0.078$ Å^{-2}) at the plane with $L = 80$ Å. The critical power law $n_S \sim S^{-2.05}$ is shown by the solid lines. The distributions are shifted vertically by one order of magnitude consecutively. Reprinted, with permission, from [394].

b) Mean cluster size
Mean size of the water clusters:

$$S_{mean} = \frac{\sum S^2 n_S}{\sum S n_S}, \qquad (20)$$

where the largest cluster is excluded from the sum, diverges at the percolation threshold in an infinite system. In finite system, S_{mean} passes through a maximum at some hydration level below the percolation threshold [396]. Such a maximum of S_{mean} with increasing surface coverage is indeed observed for hydration water near a planar hydrophilic surface. In Fig. 69, we compare the normalized mean cluster sizes $S^*_{mean} = S_{mean} * L^2/(80\ \text{Å})^2$, indicating that $S^*_{mean} = S_{mean}$ for the smallest planar surface with $L = 80$ Å. The maximum of S_{mean} becomes narrower and approach the percolation threshold with increasing system size.

c) Spanning probability
Spanning probability R is a probability that system percolates, i.e. contains an "infinite" cluster [396]. In an infinite system, $R = 1$ above and $R = 0$ below the percolation threshold. In a finite system of linear dimension L, the probability of a spanning cluster to be present in the system

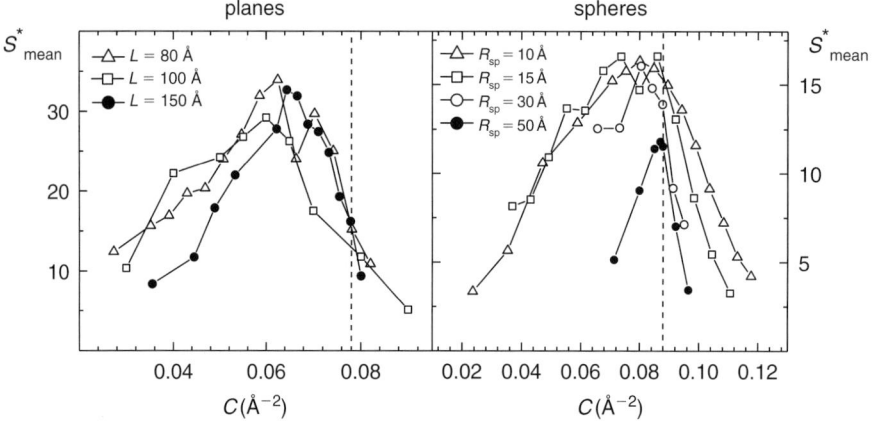

Figure 69: Mean size S^*_{mean} of water clusters on the surface of planes (left) and spheres (right) as a function of hydration level. S^*_{mean} is normalized by the ratio of the surface of a plane/sphere to the surface of the smallest plane/sphere. The percolation threshold C_p is indicated by vertical dotted lines. Data are taken from [394, 398].

is described by the function $R(C, L)$. Near the percolation threshold C_p and for large L, function $R(C, L)$ exhibits the universal behavior as a function of the scaling variable $(C - C_p)L^{1/\nu_p}$ (ν_p is a critical exponent) when neglecting irrelevant variables [396, 399]. The scaling function is universal for the systems of given spacial dimension and boundary conditions [399], but it depends on a spanning rule, which is applied to the definition of an infinite (or spanning) cluster. The most widely used spanning rules for fluid systems are based on the spatial extension of the cluster: the cluster is *crossing* if the maximal distance between some pairs of its particles is greater than L or it connects the opposite borders of the systems either in vertical or horizontal direction. We call further such a cluster a spanning cluster and probability to observe it in the particular system a spanning probability.

The spanning probability R calculated at the planar surfaces of various size and fits of the data to sigmoid function are shown in Fig. 70. Below the percolation threshold, the probability to observe a spanning cluster of hydration water is higher for smaller system. Right at the percolation threshold, R exceeds 95% [394], indicating almost permanent existence of a spanning water cluster even in very large systems.

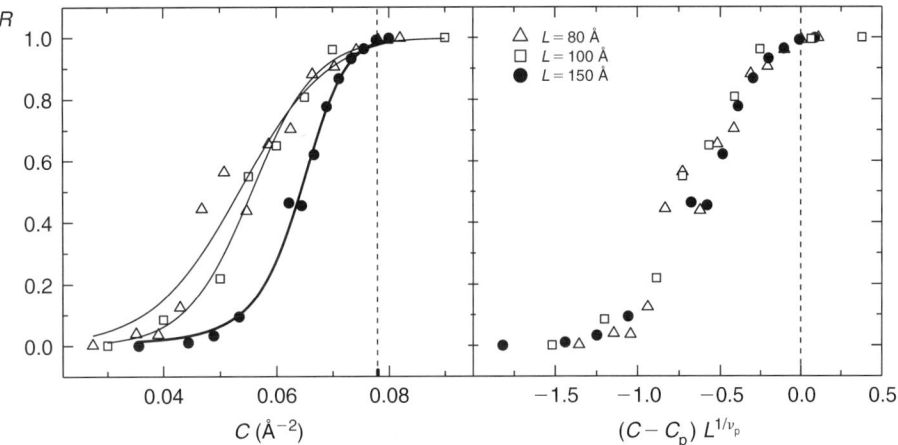

Figure 70: Spanning probability R for water adsorbed at the planar surfaces of size L as a function of the surface coverage C (left panel) and scaling variable $(C - C_p)L^{1/\nu_p}$ (right panel). The percolation threshold $C_p = 0.078$ Å^{-2} is indicated by vertical dashed lines. Data are taken from [394].

A scaling function of spanning probability of hydration water may be constructed when scaling variable $(C - C_p)L^{1/\nu_p}$ is calculated. Assuming that the percolation transition of hydration water obey the universal percolation laws for 2D system, we impose $\nu_p = 4/3$ [396]. The simulated data for small planar surfaces show rather arbitrary scattering around R for the largest surface of $L = 150$ Å (Fig. 70, right panel), so that the dependence $R((C - C_p)L^{1/\nu_p})$ in the latter case may serve a rough estimation of the scaling function for spanning probability of hydration water.

d) Fractal dimension of the largest cluster

At the percolation threshold, the largest cluster has a specific structure, which may be characterized by the fractal dimension that is universal for all systems of a given dimensionality. The universal value of the fractal dimension of the infinite cluster in 2D systems $d_f^{2D} = \frac{91}{48} \approx 1.896$ [396]. No clusters with the fractal dimension lower than d_f^{2D} can be infinite in the 2D space. The fractal dimension d_f of a cluster can be evaluated by fitting the cumulative radial distribution of its molecules

$$m(r) \sim r^{d_f}, \tag{21}$$

where $m(r)$ is the number of molecules that belong to the cluster and are located closer than the distance r from a given molecule of this cluster. Any finite cluster cannot be a strict fractal object, which should be essentially infinite. However, the mass distribution $m(r)$ within the cluster may be described by the power law (21) in some range of r. Alternatively, the clusters of various sizes may be characterized by the effective value of d_f obtained from the fits of $m(r)$ in the same range, such as $L/2$. The fractal dimension d_f of the largest water cluster calculated by latter method is shown in Fig. 71 (left panel). As the largest cluster evolves to the true fractal object, only at the percolation threshold, d_f values obtained in systems of different sizes should coincide at $C = C_p$. Such behavior is indeed observed for water near planar surfaces. It is interesting that the fractal dimension of the largest cluster of hydration water at the percolation threshold is indistinguishable from d_f^{2D} predicted by the percolation theory for 2D lattices [396].

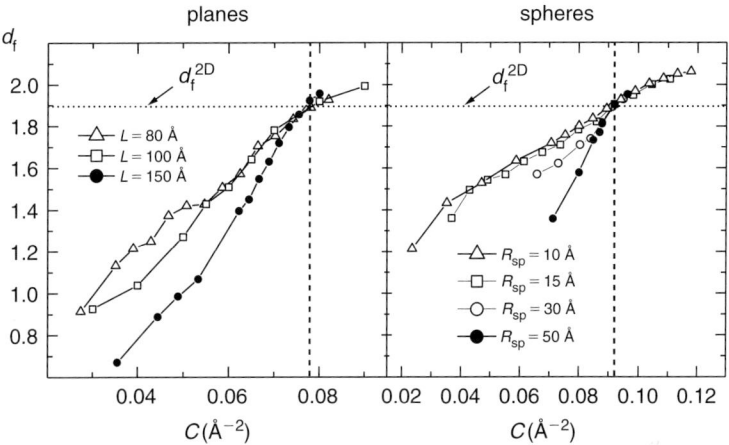

Figure 71: Fractal dimension d_f of the largest water cluster at the planar (left) and spherical (right) surfaces of various sizes. The percolation threshold $C_p = 0.078$ Å^{-2} is indicated by vertical dashed lines. Data are taken from [394, 398].

e) Size distribution of the largest cluster

The probability distribution $P(S_{max})$ of the size S_{max} of the largest water cluster near a planar surface shows a specific behavior with increasing surface coverage [394]. Far below the percolation threshold, $P(S_{max})$ shows a maximum at low S_{max}, whereas a sharp maximum of $P(S_{max})$ at large S_{max} is seen well above the percolation threshold. The widest distribution $P(S_{max})$ is observed below the percolation threshold when the spanning probability $R \approx 50\%$. At small planar surfaces, $P(S_{max})$ shows a characteristic two-peak structure: the peak at small S_{max} corresponds to the finite (nonspanning) largest clusters, whereas the peak at large S_{max} represents the spanning clusters (see Fig. 72). This two-peak structure of $P(S_{max})$ smears out and disappears with increasing system size. The ratio of the average sizes of spanning and nonspanning largest clusters close to the percolation thershold is about 1.8 for all planar surfaces studied so far [394]. Note that two-peak structure of $P(S_{max})$ may be enhanced moving from periodic to open boundary conditions [400] and on the surface of a finite object (see below).

f) Percolation transition at the spherical surface

An infinite cluster cannot appear at the finite surface of a sphere. A percolation threshold at the spherical surface in the limit of an infinitely

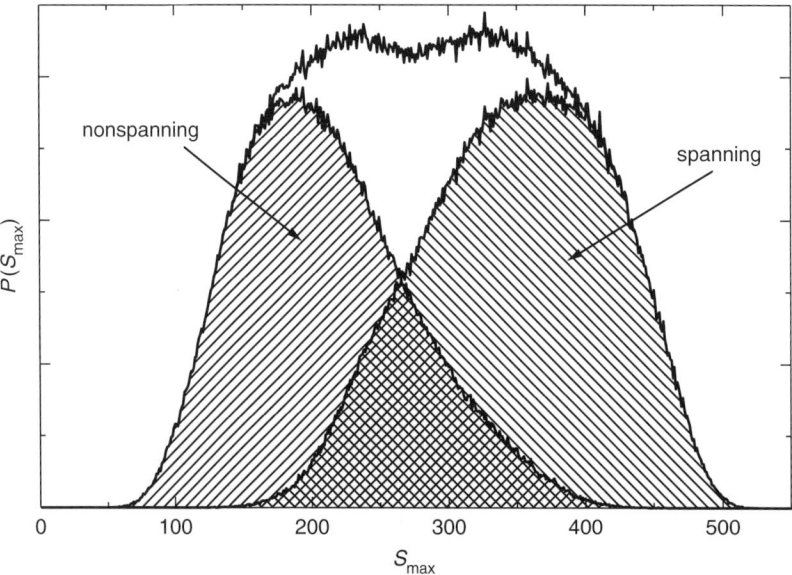

Figure 72: Probability distribution $P(S_{max})$ of the size S_{max} of the largest water cluster on the planar surface with $L = 100$ Å at the surface covereage $C = 0.055$ Å^{-2}, where spanning and nonspanning largest clusters exist with comparable probabilities. Size distribution of spanning and nonspanning largest clusters, normalized on their respective probabilities, are shown by two dashed areas. Reprinted, with permission, from [394].

large radius should coincide with that at the planar infinite surface. When considering structure and clustering of the hydration water, no essential difference is expected between the planar surface and spherical surface of large radius (weak curvature). Various cluster properties, such as cluster size distribution, mean cluster size, fractal dimension of the largest cluster may be studied at the spherical surface similarly to the planar surface. For example, mean cluster size shows a maximum with increasing surface coverage (Fig. 69, right panel). The effective fractal dimension d_f of the largest water cluster increases with hydration and achieves the universal d_f^{2D} value at $C = 0.092$ Å^{-2} (Fig. 71, right panel). Staring the percolation threshold, d_f values coincide for all spheres, indicating the percolation threshold at $C_p \approx 0.092$ Å^{-2}, in agreement with the behavior of S_{mean}.

Some cluster properties at the spherical surface look rather different from those at the planar surface. A pronounced two-peak structure of

Water layers at hydrophilic surfaces

the size distribution $P(S_{max})$ of the largest cluster is the most impressive peculiarity of the spherical surface [401]. When considering the surfaces of comparable surface area at the same hydration, two peaks of $P(S_{max})$ at the spherical surface are well separated by a pronounced minimum, whereas it is much less pronounced in the case of a planar surface (Fig. 73). Similarity between $P(S_{max})$ distributions at planar and spherical surfaces allows assignment of the left and right peaks to the nonspanning and spanning largest water clusters, respectively. It is interesting that the two-peak structure of $P(S_{max})$ seems to be not sensitive to the size of a spherical surface and remains almost the same on the surfaces of radius $R_{sp} = 10, 15, 30, 50$ Å [402]. Note that the surface of the largest sphere is huge (about 35000 Å^2). Obviously, two-peak structure reflects not the finite size effect, as at the planar surface, but the specific

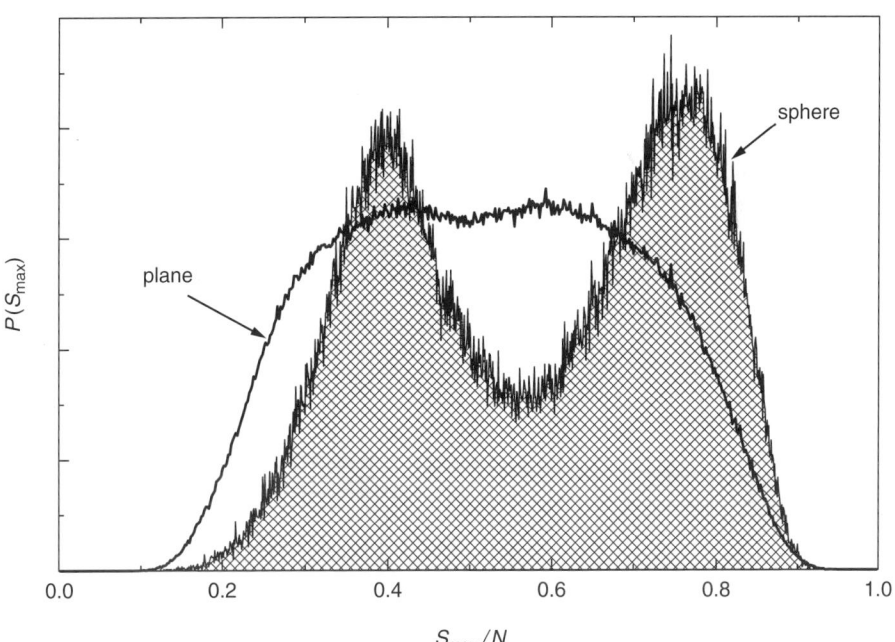

Figure 73: Probability distribution $P(S_{max})$ of the size S_{max} of the largest water cluster at the planar surface with $L = 100$ Å and on the surface of a sphere of radius $R_{sp} = 30$ Å at $C_p = 0.088$ Å^{-2}. The size S_{max} is normalized on the total number of water molecules at the respective surface. Reprinted, with permission, from [394].

closed surface topology, or, other words, a specific boundary conditions at the spherical surfaces.

The ratio of the average sizes of spanning and nonspanning water clusters reflects the distance between two peaks of $P(S_{max})$, and it was found about 2 at the spherical surfaces. This means that a spanning water cluster contains about twice more water molecules than a nonspanning largest cluster. A deep minimum between two peaks of $P(S_{max})$ indicates that the largest water cluster, which contains at about 50–60% of all water molecules at the spherical surface, is rare. In other words, the largest water clusters of intermediate sizes are unstable on the surface of a sphere. Two most probable configurations of the largest water cluster are shown in Fig. 74.

Two well-separated peaks of $P(S_{max})$ allow estimation of a probability for the largest cluster to be a spanning cluster at the spherical surface. One may assume that all largest clusters of size $S_{max} > S_{max}^t$ are spanning, where S_{max}^t is an estimation for the minimum of $P(S_{max})$ or for the midpoint between two peaks. This approach fails if the spanning or nonspanning largest cluster has a very low probability, but it seems to be reasonable near the midpoint of the percolation transition, when both exist with comparable probabilities. The spanning probability R, calculated as an integral of $P(S_{max})$ for $S_{max} > S_{max}^t$, is shown in the left panel of Fig. 75 for various spheres as a function of a surface coverage. These

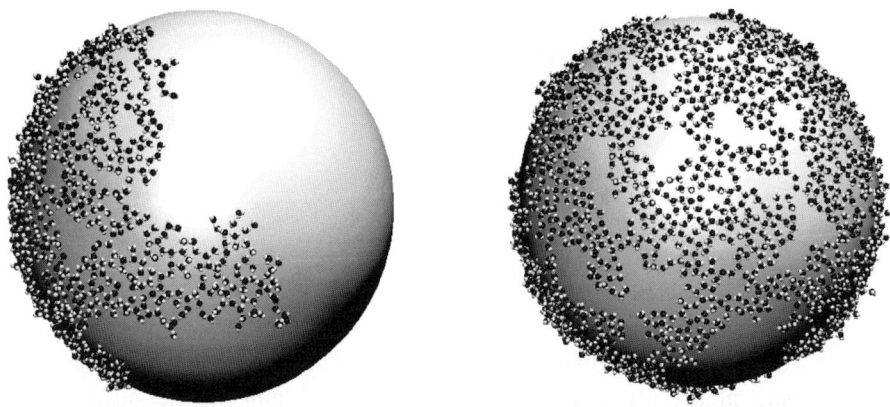

Figure 74: Arrangement of water molecules on the surface of a hydrophilic sphere of radius $R_{sp} = 50$ Å at the hydration level, where the probability to find a spanning water cluster is about 50%. An example of a nonspanning and spanning largest cluster is shown in the left and right panel, respectively.

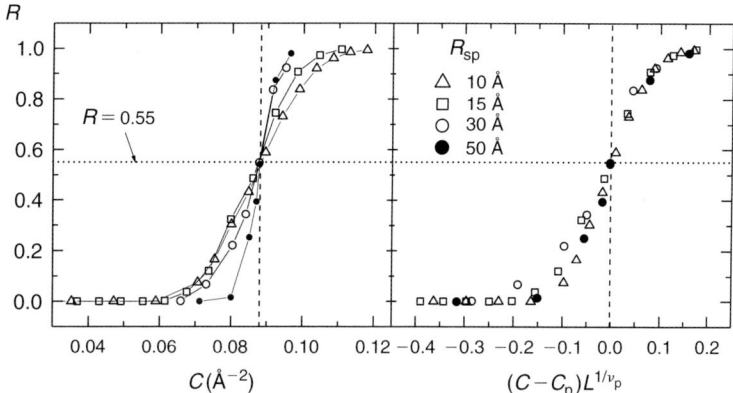

Figure 75: Spanning probability R at the spherical surfaces as a function of a surface coverage C (left panel) and scaling variable $(C - C_p)L^{1/\nu_p}$ (right panel). The crossing point of spanning probabilities at $C = 0.088$ Å^{-2} is indicated by a vertical dashed lines (data from [398]).

dependences cross at one point at about $C \approx 0.088$ Å^{-2} and $R \approx 0.55$. A scaling plot for the spanning probability R, obtained when the percolation threshold is identified with the crossing point of R, is shown in Fig. 75 (right panel). It does not look like a single master curve, probably due to some arbitrariness in the definition of R. We cannot exclude that $R(C)$ for two largest spheres studied cross at the point where $R \approx 90\%$ or higher (see squares and solid circles in Fig. 75), i.e. close to the values, obtained for planar surfaces.

The properties of a spanning cluster at the spherical surfaces are not yet studied within the percolation theory. The value $R \approx 55\%$ obtained at the crossing point for hydration water at the spherical surface is close to the $R^{(1)} = 50\%$ for crossing probability in one chosen direction in 2D lattices with open boundary conditions [399] and $R^{(1)} \approx 52\%$ for wrapping probability in 2D lattices with periodic boundary conditions [403]. Although the mapping of R for spherical surfaces onto the conventional crossing or wrapping probability for 2D lattices is questionable, the two-peak structure of $P(S_{\max})$ allows unambiguous determination of the hydration level where spanning and nonspanning largest clusters are equally populated ($R \approx 50\%$). Besides, even approximate separation of the spanning and nonspanning largest clusters enables a comparative study of their various properties, which will be discussed below.

g) Specific properties of a spanning water cluster at the spherical surface

Two-peak structure of the size distribution $P(S_{max})$ of the largest cluster allows distinguishing between the spanning and nonspanning largest clusters at the spherical surface based on the cluster size S_{max}. This allows finding of the structural properties of the largest cluster, which are peculiar for the spanning clusters only. For example, the spanning character of the largest cluster can be characterized by the distance H_{max} between the center of mass of the largest water cluster and the center of a sphere. The center of mass of the largest cluster is located at $\vec{r}_0 = \sum m_i \vec{r}_i / \sum m_i$, where $\vec{r}_i$ is a vector that defines the position of the ith water molecule and where m_i is the mass of water molecule. If the center of a sphere is in the origin of the coordinate system,

$$H_{max} = \sqrt{\vec{r}_0}. \qquad (22)$$

The probability distribution $P(H_{max})$ of the largest water cluster at the spherical surface shows a two-peak structure in a wide range of a surface coverage [402]. A joint probability distribution $P(H_{max}, S_{max})$ of the distance H_{max} and size S_{max} of the largest cluster indicates that two-peak structure of $P(H_{max})$ is intrinsically related to two-peak structure of $P(S_{max})$ (Fig. 76). Thus, one peak $P(H_{max})$ represents a spanning, whereas another one a nonspanning largest water clusters. Water clusters, covering the spherical surface completely or homogeneously, are represented by a sharp peak of $P(H_{max})$ close to zero values of H_{max} (left peaks in Fig. 77). In this case, the center of mass of the largest water cluster should be close to the sphere center (see right picture in Fig. 74). The second peak, situated at higher values of H_{max}, obviously represents the nonspanning largest clusters (see left picture in Fig. 74). With decreasing water coverage, the right peak approaches $H_{max}/(R_{sp} + 3 \text{ Å}) = 1$, which corresponds to the largest cluster containing a single water molecule, whose center of mass is located at the distance $(R_{sp} + 3 \text{ Å})$ from the center of a sphere. (Centers of oxygen atoms of water molecules adsorbed at the smooth surface are located at about 3 Å from the surface.) A clear minimum of $P(H_{max})$ is observed near $H_{max} \approx 0.3(R_{sp} + 3 \text{ Å})$. It was argued in [402] that this minimum evidences low probability of the largest clusters to cover homogeneously the area of about semisphere. This fact could be treated as instability of the largest clusters of such

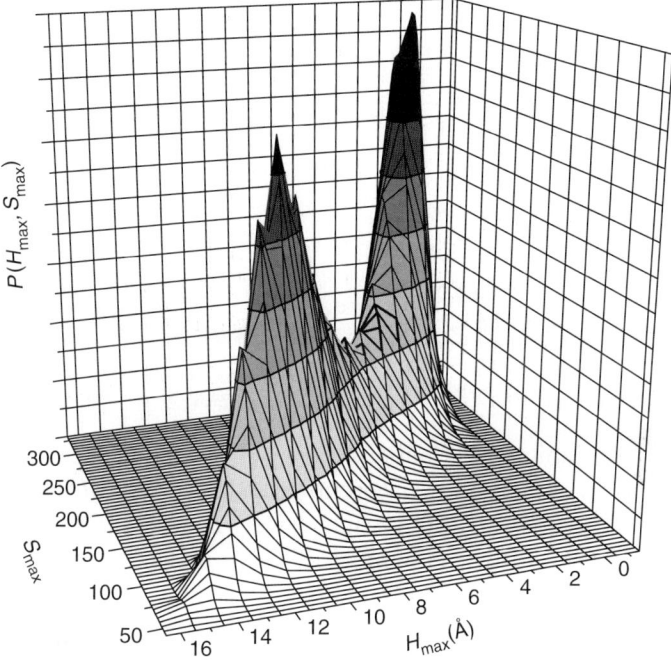

Figure 76: Joint probability distribution $P(H_{max}, S_{max})$ of the distance H_{max} and size S_{max} of the largest water cluster at the spherical surface of radius $R_{sp} = 15$ Å at $T = 425$ K and surface coverage $C = 0.086$ Å^{-2} ($N = 350$). Reprinted, with permission, from [398].

sizes. Presence of a small peak of $P(H_{max})$ at $H_{max} \leq 0.1(R_{sp} + 3$ Å$)$ already on the surface coverage noticeably below the percolation threshold evidences that any spanning largest cluster spans essentially more than half of the spherical surface [402]. Obviously, small largest cluster should be strongly rarefied to span such large area.

The compactness of the largest clusters can be measured by the radius of gyration R_g

$$R_g = \sqrt{\sum_{i=1}^{N} m_i (\vec{r}_i - \vec{r}_0)^2 / \sum_{i=1}^{N} m_i}, \qquad (23)$$

where $\vec{r}_i - \vec{r}_0$ is a distance of ith water molecule to the center of mass of the cluster. The joint probability distribution $P(R_g, S_{max})$ shown in

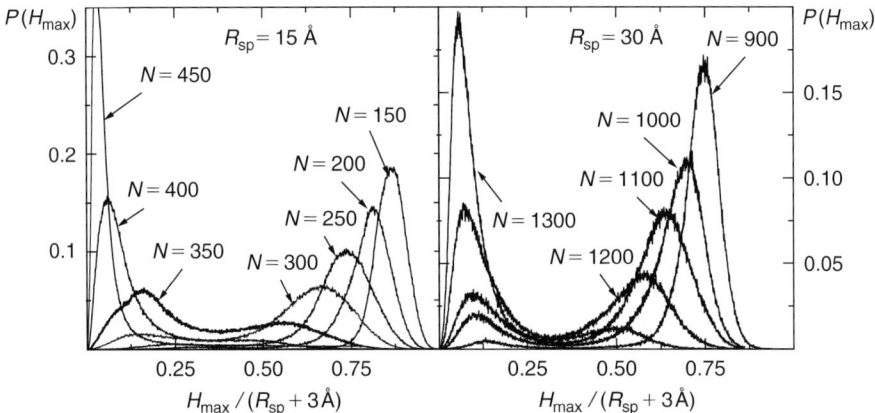

Figure 77: Probability distributions $P(H_{max})$ of the distance H_{max} between the center of mass of the largest water cluster and the center of spheres of radius $R_{sp} = 15$ Å (left panel) and 30 Å (right panel) at $T = 425$ K. The hydration levels are characterized by the number of water molecules on the surface N shown in the figure (data from [402]).

Fig. 78 evidences a pronounced two-peak structure of the probability distribution $P(R_g)$ of the radius of gyration R_g. Indeed, it is observed in a wide range of hydrations at the spherical surfaces of various size [402]. This indicates that spanning and nonspanning water clusters are characterized by very different values of R_g. The sharp peak of $P(R_g, S_{max})$ at large S_{max} evidences that the radius of gyration of a spanning largest cluster is close to the effective radius of a sphere ($R_{sp} + 3$ Å), and it is not sensitive to the hydration level. The low and wide peaks situated at smaller S_{max}, correspond to nonspanning clusters. Note that R_g of nonspanning largest cluster continuously increases with increasing surface coverage [402].

Probability distribution $P(L_{max})$ of the maximum linear extension L_{max} of the largest water cluster on the hydrophilic spheres also shows a two-peak structure [402]. However, L_{max} is comparable with the diameter of a sphere even for nonspanning largest clusters dominated at the surface coverage $C = 0.06$ Å^{-2}. At the surface coverage, where the probability R for observing a spanning cluster is about 50%, for the vast majority of the largest clusters, L_{max} exceeds $2(R_{sp} + 3$ Å), indicating that even small largest clusters extend through the essential part of the spherical surface. Such behavior of L_{max} is obviously caused by dominating chain-like

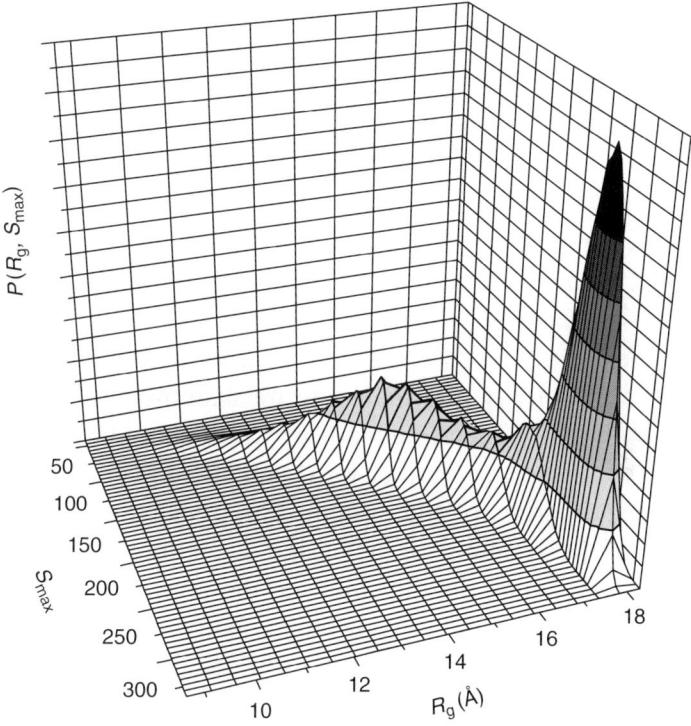

Figure 78: Joint probability distribution $P(R_g, S_{max})$ of the radius of gyration R_g and size S_{max} of the largest water cluster on the surface of a sphere with radius $R_{sp} = 15$ Å at $T = 425$ K and surface coverage $C = 0.086$ Å^{-2} ($N = 350$). Reprinted, with permission, from [398].

structure of the surface water even when the surface is covered by a complete monolayer of water (Section 5.2). The spanning water cluster with a fractal dimension close to $d_f \approx d_f^{2D}$ is indeed a strongly rarified object, which mainly consists of H-bonded water chains [394, 398, 402]. Note, that L_{max} may noticeably exceed $2(R_{sp} + 3$ Å$)$ because the largest cluster includes water molecules, which do not belong to the first surface layer. Two peaks of $P(L_{max})$ merge together with decreasing sphere radius and become indistinguishable for $R_{sp} < 15$ Å.

Above, we have considered the percolation transition of hydration water at idealized smooth surfaces. Real surfaces are structured at nanoscopic level and may be strongly heterogeneous. This can suppress the critical temperature of the layering transition and/or causes smearing out of this transition. However, we can expect percolation transition of water on real hydrophilic surfaces, at which a layer of adsorbed water can be formed. This is not possible for the hydrophobic surfaces, where adsorbed water forms droplets instead of a film. The surface phase diagram of water, described in Section 2.4, helps determine the level of hydrophilicity, which is required for the percolation transition of hydration water. At ambient temperatures, this transition is expected on the surfaces with the average water–wall potential of about -3 kcal/mol or stronger.

In experiments, the percolation transition of hydration water can be detected by conductivity and dielectric measurements. Formation of a condensed hydrogen-bonded network of water should provide a media for the charge transport (proton or ions) and should change qualitatively the dielectric properties of the system. Sharp stepwise increase of the conductivity of the system with increasing water content at some threshold hydration level may directly indicate the appearance of an infinite hydrogen-bonded water network via a percolation transition. The dielectric response is also expected to increase drastically at the percolation threshold. Note, however, that the strongly attractive sites on the surface, which immobilize water molecules, may complicate interpretation of the results. As these effects occur on the surfaces, their experimental observation is possible first in the system with high surface/volume ratio (in various porous media).

Importance of the water networks for the conductivity of substances such as cellulose, silk, and wood was recognized long time ago. Increase of the conductivity with hydration is due to the appearance of the "complete chains of water molecules connecting neighboring ion-generating sites" [404–406], and this mechanism provides power-law dependence of the conductivity on the hydration level h. Dielectric constant of silica gel is not affected by adsorbed water up to $h \approx 0.02$ g/g [407]. Differential heat of adsorption is very high at low hydrations, indicating adsorption of water molecules on the strongly attractive sites. Within hydration range of 0.02 to 0.09 g/g, dielectric constant increases linearly with h, which should be attributed to the appearance of water molecules, which are only weakly bound to the surface. Starting from $h \approx 0.09$ g/g, dielectric constant starts to increase faster with h, and this may indicate the appearance of a condensed water layer. Dielectric constant of the powdered magnesium hydroxide starts to increase stepwise at $h \approx 0.012$ g/g [408], which is close to about monolayer water coverage, estimated from the adsorption isotherm. Water adsorption on porous alumina provides monolayer water coverage at about $h = 0.05$ g/g [409]. Dielectric constant increases several times within rather narrow hydration range around this value. The dielectric properties of the low-porosity polycrystalline marbles change qualitatively at $h \approx 0.0004$ g/g [410], which was attributed to the formation of the adsorbed water layer.

Electrical conductivity of various sulfonated proton-conducting membranes, such as Nafion membranes, is strongly sensitive to the water content and develops stepwise, clearly indicating percolation transition of water [411–415]. In accordance with the percolation theory, above the percolation threshold h_c, conductivity σ varies as:

$$\sigma(h) \sim (h - h_c)^t, \tag{24}$$

where t is 1.3 for 2D system and about 2.0 for 3D system [416]. In experiments with Nafion and other sulfonated membranes [411, 412, 415], the value of the exponent t varied from 1.3 to 1.5. This evidences quasi-2D character of the percolation transition of water in the considered systems.

5.2 Structure of water layers at hydrophilic surfaces

Near strongly attractive surfaces, liquid structure differs noticeably from the bulk one. This is caused by the packing effect due to the localization of molecules in a plane(s) parallel to the wall and by specific fluid–wall interactions, such as H-bonds. Density oscillations of liquids near solid substrates were observed in experiments [143, 144, 417–419] and in numerous computer simulations of confined fluids. Besides, fluids with strongly anisotropic interactions (such as water) unavoidably undergo orientational ordering near the wall. It is important to know the character of this ordering and its intrusion into the bulk liquid. In the present section, we consider structural properties of adsorbed water layers in the liquid, bilayer, and monolayer phases.

Numerous simulation studies of liquid water near various smooth and structured surfaces were reported (see Refs. [28, 32, 206] for more details). Near the surface, water molecules show orientational ordering with a tendency to keep the orientation of the dipole moment parallel to the surface in order to maximize the number of water–water hydrogen bonds [420–425]. Strong layering of liquid water with a few distinct layers was observed in the case of strongly hydrophilic metallic and polar substrates [424, 426–428]. Obviously, the degree of water layering and degree of its orientational ordering depend on the strength of water–wall interaction and on the details of the wall structure. Besides, water structure strongly depends on the thermodynamical conditions. For example, with increasing pressure, packing effects become more important, and strong density oscillations may be obtained near any surface. It is of practical importance to consider confined fluid in equilibrium with a bulk reservoir and confined fluid in the incompletely filled pore, i.e. usually at the conditions of the two-phase coexistence. Hence, correct equilibration of confined and bulk fluids in simulations [205–207] and knowledge of the phase diagrams of simulated confined fluids [29, 30, 32, 249, 250] are of crucial importance for meaningful studies.

Density profiles of liquid water, which coexists with a vapor at $T = 300$ K near surfaces of various hydrophobicity/hydrophilicity, are shown in Fig. 79. In the case of a strongly hydrophobic surface

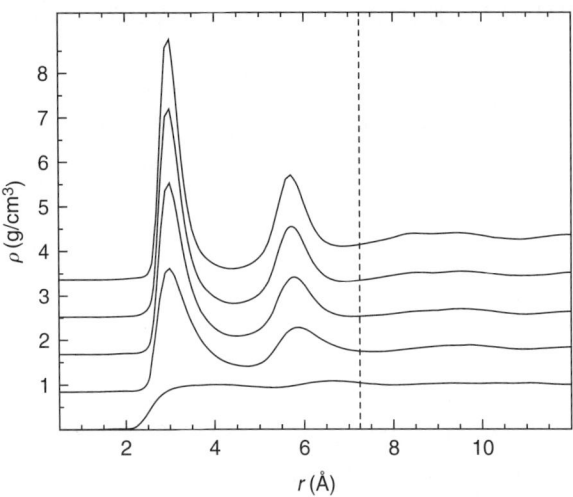

Figure 79: Density profiles of liquid water near the surfaces of the cylindrical pores with $R = 25$ Å at $T = 300$ K. The well depth U_0 of the water–wall interaction (from bottom to top): -0.39; -1.93; -3.08; -3.85, and -4.62 kcal/mol.

($U_0 = -0.39$ kcal/mol), density oscillations near the wall, are almost absent. For more hydrophilic surfaces, there are two pronounced density oscillations, whereas further from the wall, the density of water is already close to the bulk value. Thus, for a wide range of a water–wall interactions, liquid water near the wall may be approximately divided into two subsystems, separated by the vertical dashed line in Fig. 79: a) two surface layers, where water molecules are localized in two planes, parallel to the wall ($r < 7.5$ Å); b) liquid water with bulk properties ($r > 7.5$ Å).

Orientational ordering of water molecules in two surface layers can be seen from the distribution of the hydrogen atoms. Two strong maxima in these profiles coincide with the maxima in the liquid density profiles, indicating preferential formation of H-bonds within each of the layer. Besides, there are two pronounced peaks between two surface water layers (see Fig. 80). These peaks reflect orientational ordering of water molecules due to the formation of H-bonds between molecules in different layers. Similar profiles were observed in computer simulations of water at metallic surface [424]. As can be seen from Fig. 80, the degree of the orientational ordering is more sensitive to the strength of the water–wall interaction, than the degree of the localization in layers

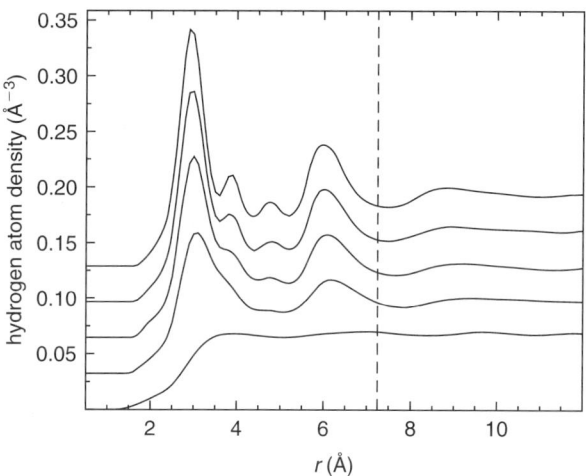

Figure 80: Hydrogen atom density profiles of liquid water near the surfaces of the cylindrical pores with $R = 25$ Å at $T = 300$ K. The well depth U_0 of the water–wall interaction (from bottom to top): -0.39; -1.93; -3.08; -3.85; and -4.62 kcal/mol.

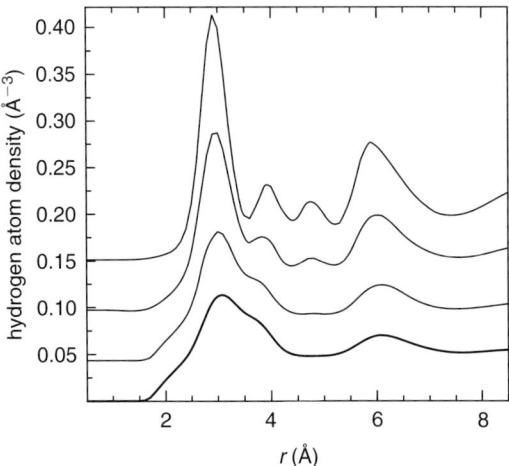

Figure 81: Hydrogen atom density profiles of liquid water near the surface of the cylindrical pore with $R = 25$ Å and $U_0 = -3.08$ kcal/mol. Temperature (from top to bottom): $T = 200, 300, 400$, and 475 K.

(see Fig. 79). Temperature affects the orientational ordering of water molecules in a drastic way: it strongly enhances in supercooled region and almost disappears above 400 K (see Fig. 81). Specific orientational ordering of water molecules in the two surface layers could be also seen from the probability distributions of the angle α between the water dipole moment and the pore radius and of the angle β between the water OH vector and the pore radius (Fig. 82). In the first surface layer, most of the OH vectors of water molecules are oriented parallel to the surface or toward the second layer (as there are no orientational water–wall interactions in the considered system). Accordingly, in the second layer, these vectors are oriented toward the first layer. Starting from the third layer the preferential orientational ordering of water molecules practically disappears.

Ordering of water within the surface layer can be characterized by the in-plane pair correlation functions $g_{O-O}(r)$ between the water oxygens in this layer only. In a bulk liquid, there are three main maxima of $g_{O-O}(r)$: maximum at about 2.75 Å corresponds to the H-bonded neighbors in the first coordination shell; maxima at about 4.5 Å and 6.8 Å reflect tetrahedral arrangement of neighbors in the second and third coordination shells (see lower curve in Fig. 83). Additional maximum of $g_{O-O}(r)$ appears in

Water layers at hydrophilic surfaces

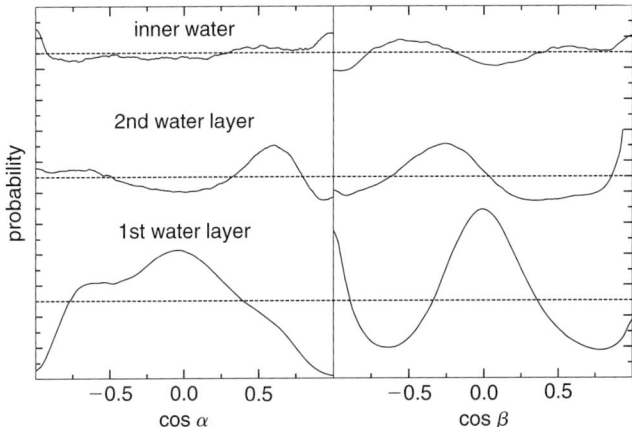

Figure 82: Angular distributions for water molecules in the liquid phase in the spherical pore with $R = 12$ Å and $U_0 = -4.62$ kcal/mol at $T = 300$ K: α is the angle between the dipole moment and the pore radius; β is the angle between the OH vectors and the pore radius. Horizontal lines indicate probability equal to 0.5.

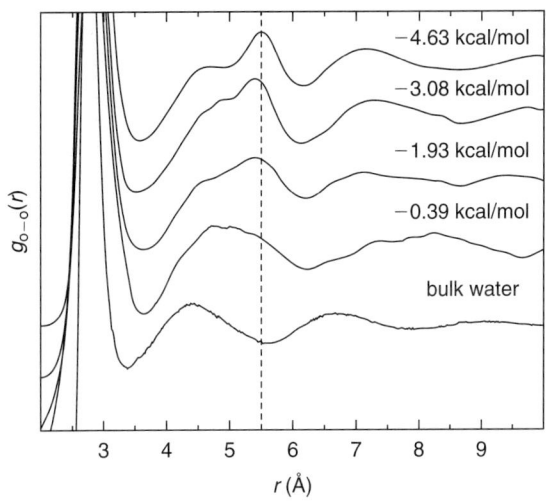

Figure 83: In-plane pair correlation functions $g_{O-O}(r)$ of water in the first surface layer in cylindrical pores with $R = 12$ Å and various strengths U_0 of water–wall interaction, indicated in the figure. Vertical dashed lines correspond to $r = 5.5$ Å, i.e. to the doubling of the first maximum of $g_{O-O}(r)$.

the first surface layer near a strongly hydrophilic surface at about 5.5 Å (upper curve in Fig. 83). This maximum corresponds to the doubling of the first maximum $g_{O-O}(r)$ and reflects linear arrangement of three H-bonded water molecules. Such arrangement evidences the chain-like structure of surface water. Tetrahedral ordering of water molecules diminishes near any boundary, and to preserve maximal number of H-bonds, water molecules form H-bonded chains. With the weakening of the water–wall interaction, the maximum of $g_{O-O}(r)$ at 5.5 Å becomes wider. So, the chain-like water structure directly correlates with the degree of the localization of water molecules in the surface layer. The specific maximum of $g_{O-O}(r)$ at 5.5 Å was also observed in computer simulations of water at various structured surfaces [425, 426, 429–431].

Near strongly hydrophilic surfaces, water structures in the two surface layers are not very sensitive to the presence of bulk liquid water. In-plane pair correlation functions $g_{O-O}(r)$ between the water oxygens in the first surface layer are shown in Fig. 84. When the first adsorbed water layer is covered by the second and subsequent water layers, its structure practically remains the same. Second water layer also shows maximum of $g_{O-O}(r)$ at about 5.5 Å when only two water layers are adsorbed at the

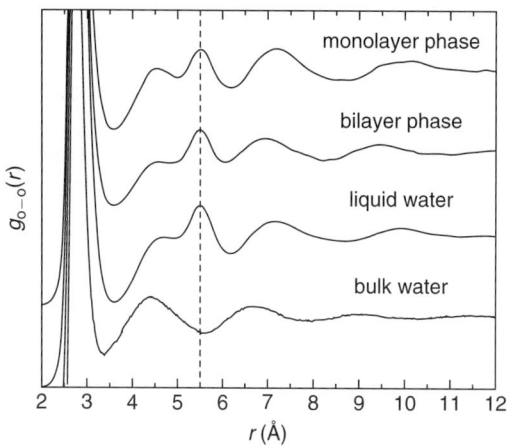

Figure 84: In-plane pair correlation functions $g_{O-O}(r)$ of water in the first surface layer in various water phases in cylindrical pores with $R = 12$ Å and $U_0 = -4.62$ kcal/mol. Vertical dashed lines correspond to $r = 5.5$ Å, i.e. to the doubling of the first maximum of $g_{O-O}(r)$.

Water layers at hydrophilic surfaces

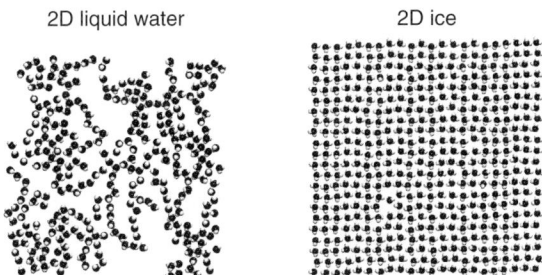

Figure 85: Arrangement of water molecules in the solid and liquid states of 2D water. Left panel: 2D liquid, $T = 320$ K. Right panel: 2D ice, $T = 255$ K.

wall. However, contrary to the first layer, adding of the subsequent layers makes the structure of the second layer much closer to the bulk one.

We consider the effect of temperature on the tetrahedral and chain-like ordering of water molecules in the surface layer, using the monolayer water phase. The model 2D water undergoes a solid–liquid transition at about 285 K (Section 3.2). Arrangement of 2D water in solid and liquid phases is shown in Fig. 85. 2D ice represents a square-like lattice, where several kinds of the ordered structures with different symmetries may be seen (Fig. 85, right panel). Similar structures were observed in experimental studies of adsorbed water monolayer (see, for, example, Refs. [180, 432] and in the simulations of water monolayers at various structured adsorbing substrates (see, for example, Refs. [433–438]) and in confined water layer [355]. In the liquid 2D water, the chain-like structures of the water–water H-bonds dominate (Fig. 85, left panel). Arrangement of water molecules in the surface layer in hydrophilic slit-like pore with $H = 24$ Å and $U_0 = -4.62$ kcal/mol is shown for three temperatures in Fig. 86. These states correspond to the first layering transition of water at the hydrophilic surface (see Section 2.2). Formation of a solid ice phase with long-range order is not seen at $T = 200$ K (it is not observed even after extensive equilibration in the Gibbs ensemble at $T = 100$ K). The structure of water monolayer at low temperatures is characterized by the combination of the chain-like and tetrahedral structures: both H-bonded chains and non-short-circuited H-bonded polygons (hexagons, pentagons, et.) are present (see left panel in Fig. 86). Upon heating, tetrahedral elements of water structure disappear, and the chain-like structures dominate (see middle and left panels in Fig. 86).

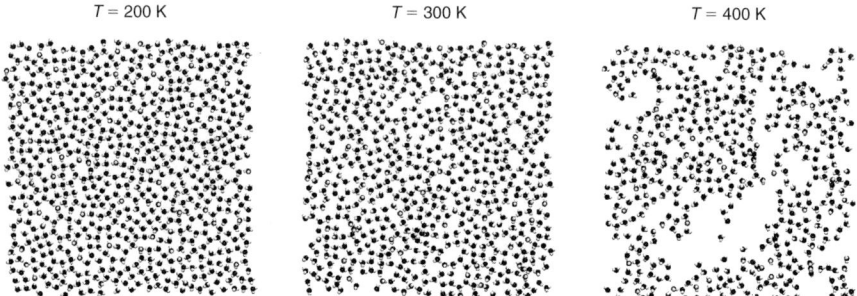

Figure 86: Arrangement of water molecules in the surface layer at hydrophilic surface with $U_0 = -4.62$ kcal/mol.

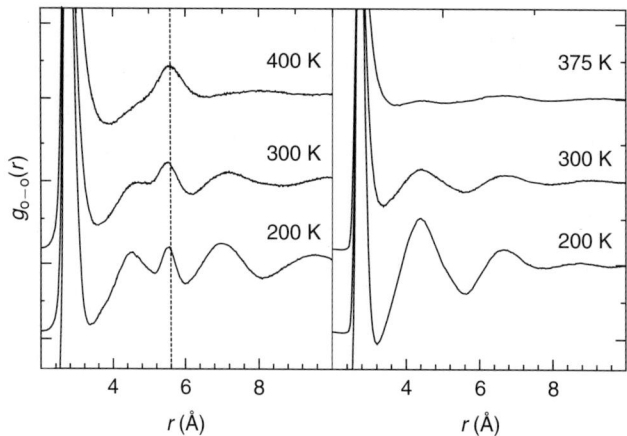

Figure 87: In-plane pair correlation functions $g_{O-O}(r)$ of water in the first surface layer in cylindrical pores with $R = 12$ Å and $U_0 = -4.62$ kcal/mol (left panel) and radial distribution of bulk liquid water (right panel). Vertical dashed lines correspond to $r = 5.5$ Å, i.e. to the doubling of the first maximum of $g_{O-O}(r)$.

Ordering of molecules in a monolayer phase can be characterized by the in-plane pair correlation functions $g_{O-O}(r)$ (Fig. 87). Upon heating, the tetrahedral arrangement of water molecules, represented by the maximum of $g_{O-O}(r)$ at about 4.5 Å, strongly diminishes both in the bulk liquid water (right panel in Fig. 87) and in the surface water layer (left panel in Fig. 87). However, this is not the case for the maximum of $g_{O-O}(r)$ at about 5.5 Å, which reflects chain-like ordering of water molecules. With increasing temperature, this kind of ordering even enhances and is clearly seen at $T = 400$ K. As the water molecules are localized in a plane, parallel to the wall, even at high temperatures, chain-like water structures should be preserved even in supercritical region. Note that these structures dominate in the structure of a bulk supercritical water [24, 25]. Probability distributions of the angles between the water dipole moment and OH vectors and the pore radius for monolayer water phase at various temperatures are shown in Fig. 88. These distributions are similar to those for water in the first surface layer of a liquid water phase in the same pore (Fig. 82). Upon heating, oreintational ordering of water molecules diminishes but remain clearly seen even at $T = 400$ K.

Structure of surface water could also be characterized by the distributions of the non-short-circuited hydrogen-bonded polygons [207, 439, 440], and first by the distributions of the most populated hexagons and pentagons. Due to the strong density oscillations of water near hydrophilic surface, polygons are preferentially oriented parallel to the surface. In Fig. 89, the hexagon density profile is shown with the respect to water density profile. Hexagon density shows four maxima in between the first and second density maxima of liquid water. Geometrical analysis reveals that the first maximum (closest to the surface) corresponds to rings of molecules, which are all located in the first water layer. The next maximum represents hexagons with five oxygens in the first layer

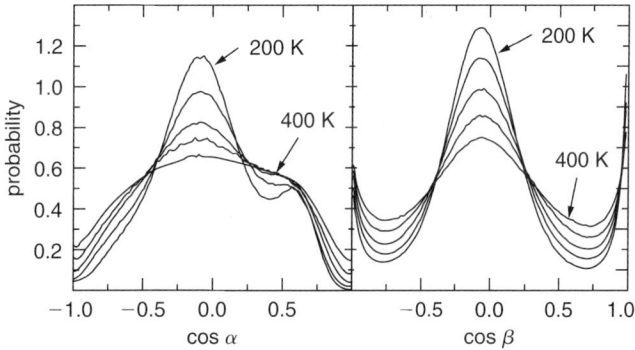

Figure 88: Angular distributions for water molecules in the monolayer phase on the surface of the cylindrical pore with $R = 15$ Å and $U_0 = -4.62$ kcal/mol: α is the angle between the dipole moment and the pore radius; β is the angle between the OH vectors and the pore radius.

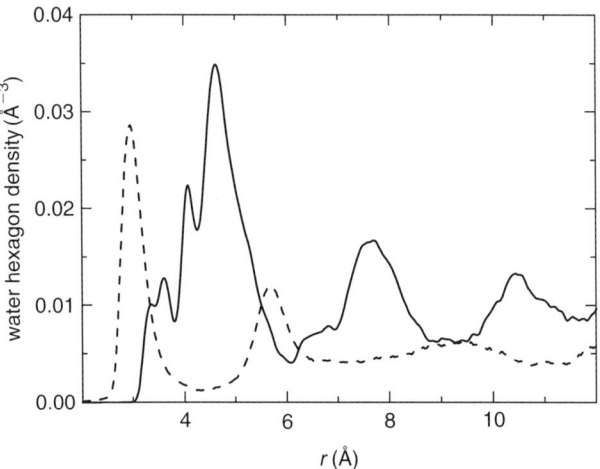

Figure 89: Water hexagon density profiles near the surface of the cylindrical pore with $R = 20$ Å and $U_0 = -4.62$ kcal/mol at $T = 300$ K. Water density profile is shown by the dashed line in arbitrary scale.

and one in the second layer and so on. The similar behavior is observed for pentagons. The increase in the pore size allows the centers of mass of the hexagons and pentagons to be located closer to the surface, whereas the total density of the polygons in the surface layer decreases. Most of the water hexagons near the surface have three water molecules in the first layer and three molecules in the second layer. This structure is a known chair-form hexagon, which is a basic elementary structure of a hexagonal ice. The most populated pentagons contain three water molecules in the first layer and two molecules in the second layer. Note that both pentagons and hexagons are the basic elementary structures of amorphous ices.

The results presented above evidence that distortion of a bulk liquid water structure near hydrophilic surfaces does not extend on more than two surface water layers. There are three main structural features of these layers: localization of molecules parallel to the wall, chain-like structures of water molecules within the layers and orientational ordering of molecules. These features strongly appear in the first layer and noticeably weaker in the second layer. Except for these two layers, the rest of liquid water is similar to bulk liquid. This observation corroborates various manifestations of the specific "bound" water at hydrophilic surfaces.

"Unfreezable" water in the solid–liquid transition of confined water is about two layers thick (Section 4.2). There are two "dead" water layers in the liquid–vapor transition of water near strongly hydrophilic surfaces (Section 2.2). At mineral surfaces, one to two water layers are highly ordered [441]. Numerous experimental and simulation studies indicate quite different dynamic properties (diffusion coefficient, residence time, reorientational time, etc.) of liquid water within the first 1-2 layers.

6 Role of interfacial water in biological function

Importance of water in biology is well known: life on the earth cannot exist without water. There is a large amount of water in living organisms (about 60% by weight in human body), both inside and outside the biological cells. Water is involved in various biochemical reactions and acts as a solvent for biomolecules. Despite the relatively high water content in living organisms, pure liquid water is practically absent in biosystems. Both intracellular and extracellular liquids consist mainly of water, but the concentration of organic compounds, including large biomolecules, is very high (about 20 to 30%). The central role of water in biological function is recognized [442, 443], but the numerous questions concerning the physical mechanisms behind the importance of water for life remain unanswered. There are several important physical phenomena, which should be taken into account when considering water properties in biosystems and the role of water in biological function.

First phenomenon is related to the bulk phase transitions in aqueous mixtures. In biosystems, water is a component of a multicomponent fluid mixture with various biomacromolecules, small organic molecules, ions, etc. This complex mixture unavoidably possesses a rich phase diagram with numerous phase transitions and respective critical points, which may occur close to the thermodynamic conditions typical of living organisms on the earth. The general features of these phase transitions are similar to the ones of the liquid–liquid transitions of binary mixtures of small organic molecules with water. However, there are several factors that make the phase transitions in biological liquids much more complex. Multiplicity of the transitions in a multicomponent mixture assumes multiplicity not only of the stable but also of the metastable states, which may exist during a long period of time. Phases enriched with macromolecules are usually not liquids but solid-like structures with some level of ordering at the mesoscopic or macroscopic scales (micelles, fibrils, etc.). Biomolecules have variety of conformational states, which are strongly coupled with the phase state of a system. Strictly speaking, conformational transition of a single biomolecule and the phase transition, which involves an ensemble of such molecules, cannot be considered separately. Finally, situation is complicated by the possible chemical reactions in complex biosystems.

The phase state of the aqueous mixture, in particular its location with respect to the phase transitions, governs the clustering of both water and organic molecules. For example, being inside the two-phase region, two phases may appear as two macroscopic clusters of like molecules. In the system being in the one-phase region, the clustering of like molecules (water or biomolecules) is determined by the proximity to the phase transition. When the phase transition is approached, clustering of the minor component enhances. This approaching may be achieved by varying temperature, pressure, pH and by adding some cosolvents, ions, etc. Majority of aqueous solutions of organic molecules show a closed-loop phase diagram, which terminates by the lower critical solution temperature (LCST) and upper critical solution temperature from low and high temperature sides, respectively. For example, the system in a one-phase region below LCST separates into two phases upon heating. Accordingly, the trend of the biomolecules to form clusters intensifies when the system approaches solution temperature upon heating. In chemical literature, clustering of solute molecules in water is often described as a manifestation of "hydrophobic interactions." Note that the phase transition and related clustering of biomolecules inside the relatively small biological cells may be affected by the finite size effect [332], which should suppress aggregation of biomolecules [444].

Second phenomenon is related to the surface phase transitions. It is natural to expect preferential adsorption of water or another component of the biological liquids on the cell wall or other biosurfaces. Obviously, this adsorption strongly affects the properties of biological liquids near the walls. In particular, adsorption of biomolecules may facilitate formation of their ordered aggregates. If the effective attraction of biomolecules to a surface is strong enough, we may expect a surface phase transition, which results in the formation of a specific surface phase. Description of the biological fluids based on the statistical theory of the bulk and surface phase transitions should be very useful for understanding their properties. Due to the extremely complex character of these systems, full application of such approach seems to be possible in the long-term perspective only. However, the phase behavior and properties of water in biosystems may be studied by the experimental and simulation methods available.

Biological liquids contain small solvent molecules (water) and high concentration of large solute molecules (biomolecules). Due to the strong difference in the sizes of typical biomolecules and water molecules, a high fraction of water molecules belongs to the hydration shells of biomolecules, as just one to three water layers separate biomolecules in living cells. Accordingly, water in biosystems exists mainly as interfacial (hydration) water, which is located in a close vicinity of the surfaces of biomolecules, cell walls, etc. This emphasizes the role of *interfacial water in biological function*. To describe the properties of interfacial water in a systematic way, we have to characterize its possible states, taking into account the effect of the phase transitions. For example, layering transition of hydration water (Section 2.2) is closely related to the formation of the hydrogen-bonded water network, which covers some surface homogeneously (Section 5.1). This network breaks upon heating or upon dehydration, indicating qualitative changes of the state of hydration water. Liquid–liquid transition(s) of hydration water (Sections 1 and 4.2) may affect its properties upon cooling and pressurization. Analysis of the possible states of hydration water should help clarify how the presence of water makes the biological function possible. In this section, we consider how biological function depends on hydration level, temperature, and pressure. Formation of the spanning water network upon hydration and its effect on the properties of biosytsems are analyzed in Section 7. Properties of hydration shell in fully hydrated biosystems are considered in Section 8.

To clarify the role of water in biofunction, it is reasonable first to consider the relation between the *hydration level* and various manifestations of biological activity. Experimental studies of some biosystems show that their physiological activity appears rapidly at some critical hydration level. At the cellular and multicellular levels, biological function of living organisms appears as metabolism, which includes a set of chemical reactions and transport of metabolites. The possibility to study these processes upon dehydration/hydration of living organisms is limited by the fact that most of them die when the water loss exceeds some critical level. For most organisms, this level is 50% of body water (about 14% for humans). However, some unicellular organisms, plants, and invertebrates (seeds of plants, fungal spores, lichens, cysts of embryos, nematodes, rotifers, tardigrades, etc.) remain viable after almost complete dehydration (95 to 99%) [445–451]. After dehydration, metabolism is completely shutdown and organisms can stay in such state of a temporary death for many years, but they cannot function untill some hydration level is restored. The first observation of this phenomenon was described by the pioneering microscopist Antony van Leeuwenhoek in 1702 [452]. The ability of organisms to survive in anhydrobiotic state may be explained by the water-replacement hypothesis [453]. This hypothesis assumes that under dehydration, some polyhydroxyl compounds, such as glycerol, cucrose, and theralose, substitute intracellular water, preserving macromolecular integrity and preventing cells from destruction. Experimental studies of the dehydration/hydration processes of anhydrobiotic organisms give unique possibility to follow decline/restoration of metabolism in living organisms with hydration level. Understanding of the microscopic mechanisms of these "hydration-dependent metabolic transitions" should clarify the role of water in biofunctions [453].

There is a clear correlation between the water content and metabolism in living organisms. For example, the metabolism of tardigrades drastically declines with decreasing humidity, and when humidity is below 48%, oxygen consumption is below 0.035% of its value for hydrated animals [456]. The most detailed experimental studies of the interrelationship between hydration and metabolism in a living organism were performed for *Artemia salina* cysts [453–455, 457–462]. Biological activity of these cysts develops upon hydration in a stepwise fashion. There are no emergence of larvas below the hydration level h (gram of water per gram of organics) of about 0.46 g/g, whereas at $h = 0.72$ g/g, already 22% of cysts produce swimming larvas [454] (see Fig. 90). The onset of various important biochemical processes is seen in the vicinity of this interval of hydrations. At the critical hydtation level $h \approx 0.60$ g/g, conventional cellular metabolism develops in a stepwise fashion. In particular, mass of the cysts starts to decrease, indicating oxidation of their endogeneous reserves of carbohydrate [457]; cellular respiration appears [455] (Fig. 90); amount of adenosine triphosphate starts to increase and the total content and composition of free amino acids start to change [461]; and incorporation of CO_2 into proteins and RNA begins [460]. Another critical hydration level $h \approx 0.30$ g/g indicates initiation

Role of interfacial water in biological function

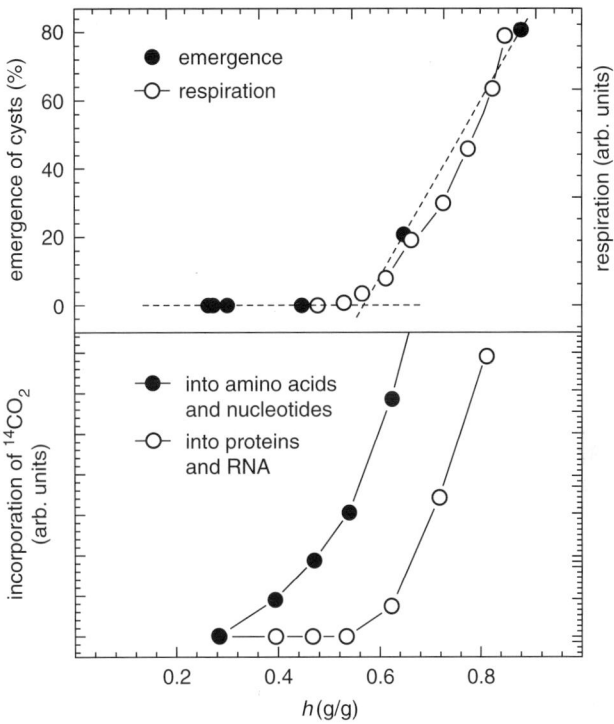

Figure 90: Hydration-induced metabolic transition of *Artemia* cysts. Upper panel: emergence of cysts [454] and respiration [455]. Lower panel: incorporation of radioactivity into amino acids, nucleotides, proteins, and RNA [453].

of intermediary metabolism, which involves some particular amino acids [461] and causes incorporation of CO_2 into amino acids and nucleotides [459, 460] (Fig. 90).

Respiration rate of the yeast cells linearly decreases with water content upon dehydration and apparently stops at hydration level $h \approx 0.20$ g/g [463] (Fig. 91). For lichens, two "switching points" in the hydration-induced metabolism were found [464]. Limited metabolism appears when water content is below 10% of the fully hydrated samples, and at hydrations above 20%, another class of enzymes becomes active. Seeds of plants may stay for years in dehydrated state but germinate promptly upon hydration. This makes the analysis of the evolution of physiological activities of seeds with increasing hydration possible. The rate of O_2 consumption and the rate of CO_2 evolution by dry seeds are very low, indicating an absence of mitochondrial metabolism. It

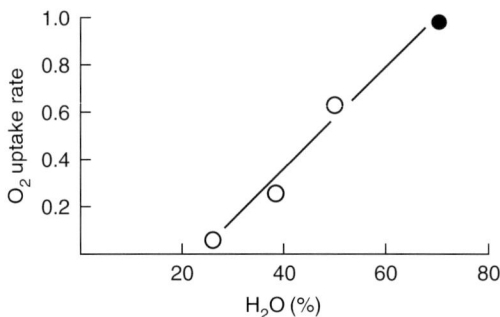

Figure 91: Respiration rate of partially dried yeast cells. Oxygen uptake rates at 30°C are plotted in relative units. The closed circle represents the internal respiration rate of the native cells. Reprinted, with permission, from [463].

increases dramatically in a stepwise manner at some critical hydration level [465–468]. This level is $h \approx 0.14$ g/g for apple, 0.20 g/g for corn, 0.24 g/g for soybean, and 0.26 g/g for pea. Additionally, some other physiological activities (photosynthetic electron transport, transfer of light-excited states) start at lower hydration levels (about two times lower than those given above).

At molecular level, the manifestations of the biological activity appear in specific biochemical reactions, conformational behavior, and dynamical properties of biomolecules. Experimental studies of various partially hydrated enzymatic proteins show that their activity accelerates rapidly at some critical hydration levels. Onset of the enzymatic activity of urease occurs at $h \approx 0.15$ g/g [469]. In the presence of chymotrypsin, the acylation reaction is undetectable at hydrations $h < 0.12$ g/g, but its rate grows sharply above this critical hydration level [470]. The rate of enzymatic activity of glucose-6-phosphate dehydrogenase, hexokinase, and fumarase becomes detectable and start to increase sharply at $h \approx 0.20$ g/g, whereas this critical hydration is about 0.15 g/g for phosphoglucose isomerase [471]. Enzymatic activity of lysozyme can be detected only when hydration level achieves $h \approx 0.20$ g/g [472, 473] (see Fig. 92).

Existence of the critical hydration level h_c for enzymatic activity may reflect the fact that hydration water can serve as a transport media for the substrates and/or for the products of the reactions only above h_c [471]. This possibility was explored by the experiments with gas-phase

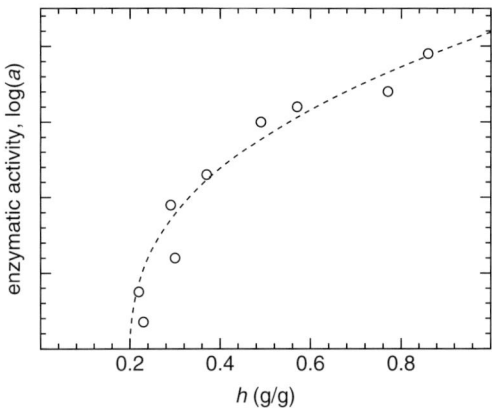

Figure 92: The rate a of the enzymatic activity of lysozyme at various hydration levels [473].

substrates [474–476] and by the experiments with enzymes in nonaqueous fluid environment [477, 478]. Activity of alcohol dehydrogenase from bakers yeast with respect to substrate vapor appears when hydration level reaches 0.16 g/g [474]. In other studies, nonzero enzymatic activity of lipase and esterase was detected for gas-phase substrates at extremely low hydrations [475, 476]. However, in these cases, a noticeable increase of enzymatic activity is also seen in the hydration range 0.10 to 0.20 g/g. Activity of laccase [478] and subtilisin [477] in organic solvents appears only at some critical hydration level of added water, which depends on solvent. Obviously, in experiments with enzymes in organic solvents, the critical water level is determined by the miscibility of water and solvent and by the difference in the water–protein and solvent–protein interactions. Clearly, less water amount is necessary to provide the same coverage of protein molecules in hydrophobic solvents. When the enzymatic activity is analyzed as a function of water bound to enzyme, the critical water level does not depend noticeably on the solvent and is close to about 0.10 g/g for yeast alcohol oxidase in various solvents [479].

Bacteriorhodopsin is an intramembrane protein, which uses adsorbed light energy to transfer a proton through the membrane. The microscopic mechanism of the proton pumping is based on the set of isomerization processes initiated by the light adsorption. Upon dehydration, photoisomerization of bacteriorhodopsin reduces [480–484] and proton pumping stops below 60% relative humidity [483–486]. The above examples show direct correlation between the hydration level and the biological activity of biomolecules.

In most cases, it is not easy to get explicit dependence of some form of biological activity on the hydration level even at molecular level. However, we may consider effect of hydration on the properties of biomolecules, which are known to be necessary for their functionality. Biomolecules in biologically active state are characterized by the specific conformation and by some level of internal conformational dynamics. Conformational stability of DNA double helix strongly depends on hydration water. DNA exists in biologically relevant B-form until the hydration Γ, measured as a number of water molecules per nucleotide, exceeds $\Gamma \approx 20$ [487, 488]. In the B-form, DNA is a right-handed double helix, which makes a turn every 34 Å, and the distance between two neighboring base pairs is 3.4 Å. At lower hydrations, DNA undergoes different conformational transitions depending on its sequence, bound metal ions, and other environmental conditions. The most studied is the transition from B- to A-form [489], with the midpoint at about $\Gamma = 15$ [487, 490, 491]. In the A-form, DNA helix remains right handed but becomes shorter and broader (Fig. 93). Dehydration of B-DNA may be achieved not only in the vapor phase by decreasing the relative humidity but also in a liquid phase by adding some organic solvent. For instance,

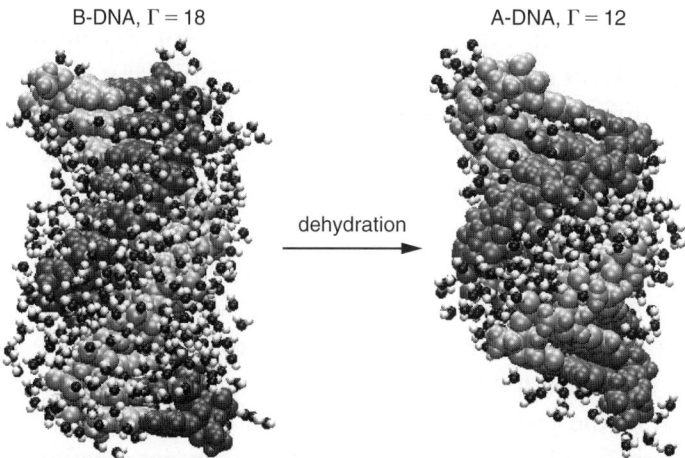

Figure 93: DNA exists in a biologically relevant B-form at high hydrations and undergoes conformational transition into A-form upon dehydration.

B to A transition was also observed in concentrated solutions of some nonelectrolytes miscible in water [492, 493].

Proteins and polypeptides also undergo conformational changes upon dehydration [494]. For example, a Raman spectrum of a dry lysozyme powder differs from a spectrum of a solution. The parameters of the main structure-sensitive spectral bands achieve their values in solution at hydration $h \approx 0.20$ g/g [495, 496], which coincides with the onset of the enzymatic activity of lysozyme [472, 473]. Experimental studies of NMR spectra of a lysozyme powder also evidence conformational changes within the hydration range from 0.1 to 0.3 g/g [497]. The hydration-induced conformational changes of lysozyme are fully reversible, whereas in some other proteins, these changes are stronger and only partially reversible [498]. Lyophilized subtilisin undergoes conformational transition in organic solvent, when water content increases from 0.15 to 0.35 g/g [499]. Conversion of hemichrome to methemoglobin with increasing water content shows sigmoid dependence on hydration level, with an inflection point at about 0.25 g/g [500, 501]. It is well known that conformation of a biomolecule may be strongly affected when it is adsorbed on the surface (for example, on the surface of a membrane). Apart from various factors that affect conformation of a biomolecules in this case, "dehydration" due to the direct contact with a surface should also play a role. Similar effect may result from the crowding of biomolecules in a cell.

A biologically relevant lamellar phase of biomembranes exists only when hydration level exceeds some critical value, typically about $h \approx$ 0.20 to 0.30 g/g [502–504]. For example, this hydration level is required to suppress the leakage from seeds and pollen [502, 505]. Neutron scattering studies evidence "hydration-induced flexibility" of biomembranes [484, 506, 507]. Slower motions are more strongly influenced by the hydration level, and for the purple membrane samples, they increase when hydration increases from about 0.3 to 0.4 g/g.

Internal dynamics of biomolecules is practically frozen without water. Upon increasing hydration level, it develops in a stepwise fashion [508]. At $h \approx 0.15$ g/g, internal protein motion, monitored by hydrogen exchange, achieves its solution rate [509]. Full internal dynamics of lysozyme is restored at $h \approx 0.38$ g/g [510]. Mossbauer spectroscopy studies evidence restoration of the internal dynamics of lysozyme molecules when hydration level achieves 0.1 to 0.2 g/g [511]. Neutron and light scattering experiments indicate the appearance of a slow relaxation process in lysozyme powder at about 0.20 g/g [512, 513]. Experiments with lysozyme in glycerol show the onset of its dynamics at about 0.1 g/g of water with saturation at $\approx$ 0.4 g/g [514–516]. Elastic properties of elastin strongly depends on its hydration. At room temperature, its elongation under constant load increases drastically at $h \approx 0.25$ g/g [517]. Upon hydration to 0.2 g/g, only backbone motion of elastin slightly increases, whereas above 0.3 g/g hydration there are large-amplitude motions of both the backbone and the side-chains [518].

Importance of hydration water in the dynamics and functions of biomolecules is also seen from the studies of hydrated biomolecules at low temperatures. In the temperature interval from about 180 to 230 K, dynamics of biomolecules show rapid increase. Experimental studies show dynamic transition of crystalline ribonuclease A at about 220 K [519], and this temperature corresponds to the onset of its enzymatic activity upon heating [520]. Approximately at the same temperature, enzymatic activity of elactase [521] and myoglobin [522] starts to develop upon heating. The dynamic transition of chromatophore membrane occurs at about 180 K, and at the same temperature, the efficiency of the photoinduced electron transfer starts to increase upon heating [523]. The dynamic transition of biomolecules was detected by various experimental methods: Mossbauer scattering [523, 524], neutron scattering [525–528], X-ray crystallography [519, 529], infrared spectroscopy [530], etc. Besides, this transition is clearly seen in computer simulations of hydrated biomolecules [531–537]. The temperature of the dynamic transition is not very sensitive to the biomolecular structure and for various biomolecules (ribonuclease [519, 520], DNA [526–528, 535], bacteriorhodopsin [484, 486, 507, 538], myoglobin [524, 525, 530, 531, 533, 534, 536, 537], lysozyme [514, 515, 528, 539], carbohydrates [540], etc.) varies within relatively narrow temperature interval.

Dynamic transition does not occur when biomolecules are dry as it requires some minimal amount of water [514, 527, 528, 530, 538, 540–542] and may be strongly affected by the presence of cosolvents [514, 515, 539]. The apparent temperature of the dynamic transition increases with the lowering of hydration level or with adding of cosolvents [514, 539, 540, 542–545]. The most drastic increase in this temperature occurs when hydration level increases from 0.1 to 0.3 g/g [514, 543–545]. These facts indicate that the temperature-induced dynamic transition of biomolecules is governed by hydration water.

There is some upper temperature limit for life. Some microorganisms remain viable at 121°C [547], but in most cases, this temperature is below 100°C. This upper limit is closely related to the loss of the ordered structures of biomolecules upon heating. Activity of biomolecules depends on their flexibility, and a less flexible biomolecule should be more stable against heating [548]. Dehydration of biomolecules or removing of water by adding some organic solvents increases their thermal stability [549–552]. Irreversible thermal inactivation of trypsin and ribonuclease is strongly suppressed by drying [549]. Thermal stability of some enzymes is enhanced when they are suspended in anhydrous organic solvents [550, 551]. The denaturation temperature of bacteriorhodopsin increases by more then 50°C upon dehydration [552]. For lysozyme, an increase in the denaturation temperature exceeds 90°C [546, 553, 554] and becomes noticeable when the hydration level h is below 0.4 g/g [515, 546] (see Fig. 94). The temperature width of the denaturation peaks

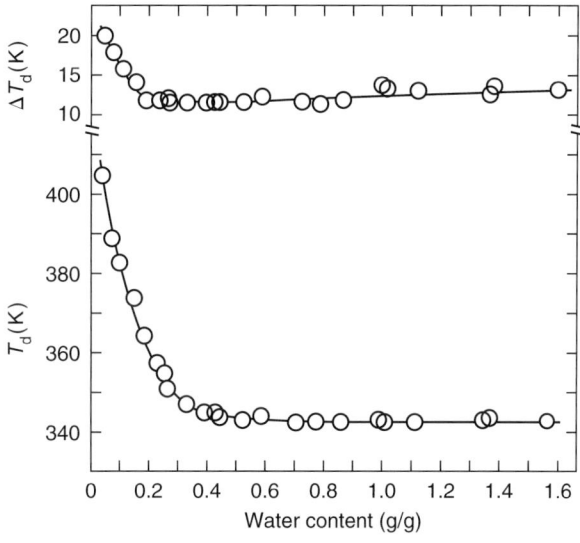

Figure 94: The temperature of denaturation, T_d, and temperature width of denaturation peak, ΔT_d, of lysozyme as function of water content. Reproduced, with permission, from [546].

in the different scanning calorimetry thermogram of lysozyme starts to increase when h is below 0.2 g/g [546, 553]. Similar to lysozyme, the denaturation temperature of ovalbumin starts to increase at $h < 0.4$ g/g [555]. The denaturation temperature of elastin and collagen increases upon dehydration by more than 150°C, and this effect is noticeable when water weight concentration is below 50% [556]. The chain melting temperature of biological membranes increases by more than 30°C upon dehydration, and this increase starts when the hydration level is below about 15 water molecules per lipid molecule [557].

Temperature-induced unfolding of fully hydrated biomolecules is usually accompanied by their aggregation. Upon heating, aqueous solutions of some polypeptides (for example, large ELP [558]) separate into water-rich and organic-rich phases, thus possessing a LCST. The temperature of this phase transition depends on the peptide composition, its concentration, addition of cosolvents, pH, etc. Similar to other macromolecules whose aqueous solutions show an LCST [559–561], polypeptides drastically change their conformational distribution when crossing the phase separation temperature. The origin of the LCST in aqueous solutions is often considered in relation to the ordered character of the hydration shell, surrounding the solute molecule [64, 562–565]. At low temperatures, a solute molecule is covered by ordered hydration shell, which promotes its solubility in water. This shell becomes less ordered upon heating, that causes demixing at some temperature [566]. So, even in the case of a fully hydrated biomolecule, the state of the hydration shell can noticeably affect its properties.

Biosystems and their functions can be strongly affected also by pressure [567–570]. Activity of some microorganisms may increase upon applying pressure and reach a maximum in some pressure range [569]. However, above some pressure, activity of all living organisms decays. For example, a noticeable decay of cellular activity of yeasts starts at $P \approx$ 1 kbar, and yeast cells are killed at $P \approx 2$ kbar [571]. Upon pressurization to about 6 kbar, DNA molecules undergo the conformational transition from the native B-form to left-handed double-helical Z-form [572]. The chain melting temperature of biological membranes increases by about 22°C, when pressure increases by 1 kbar [570]. The melting temperature of DNA is also sensitive to pressure and may increase or decrease upon pressurization [573]. There were extensive experimental studies of the pressure-induced protein unfolding (see [568–570] and references therein). In the temperature–pressure plane, there is a closed-loop region of the protein stability. Inside this region, proteins mainly preserve their native conformations and are miscible with water, whereas outside this region, they undergo unfolding/denaturation, accompanied by the protein aggregation. For example, staphylococcal nuclease (S Nase) undergoes unfolding at about 50°C at ambient pressure. At $T = 25$°C, the qualitatively similar unfolding transition occurs at about 2 kbar [574, 575]. Biomolecules, which are insoluble in water and form aggregates at ambient pressure, may dissolve with increasing pressure. For example, pressure of about 300 bar is sufficient to prevent aggregation of insulin [576].

Closed-loop temperature–pressure stability diagram of some protein in water should be directly related to the phase diagram of the protein/water mixture. It is well known that the phase diagrams of aqueous solutions are highly sensitive to pressure [63, 577–580], which may either promote miscibility or induce demixing. For some aqueous solutions, which show immiscibility gap at zero pressure, increasing pressure causes extension of this gap in concentration and temperature range (some pyridines [577]). Similar trend can be seen for some solutes, which are completely misscible with water at zero pressure: upon pressurization, aggregation of solute molecules enhances, indicating approaching immiscibility (methanol [581]). For other solutes, the effect of pressure is opposite (tetrahydrofuran [579], alkanes, noble gases). In some solutions, changes of the solubility with pressure are even nonmonotonous [579, 580]. Therefore, various evolutions of the phase diagrams with pressure can be expected for aqueous solutions of various biomolecules. When considering the effect of pressure on hydrated biomolecules, we have to take into account possible changes of the phase state and thermodynamic properties of a bulk liquid water upon pressurization (see Section 2), which should also affect hydration water at biosurfaces.

We have considered various manifestations of the importance of water in biological function. In most cases, there are clear indications on the crucial role of *interfacial* water in life. Two main aspects of the phase behavior of interfacial water can be distinguished: a) condensation of a layer of hydration water at biosurfaces and b) effect of temperature and pressure on the state and properties of this hydration layer. These two aspects are considered in the Sections 8 and 9, respectively.

7 Water in low-hydrated biosystems

A step-like growth of various forms of biological activity occurs, when the coverage of a biosurface by water approaches about one layer. Formation of a condensed water monolayer on the surface of various biosystem may be expected at the hydration below about 0.4 g/g. It is natural to relate appearance of a biological function to some qualitative change in the state of the interfacial water, which causes essential changes in water properties. Formation of a condensed water monolayer indicates transition of a hydration water from the gas-like state, where only small hydrogen-bonded water clusters are present, to the condensed state. In the case of idealized smooth surfaces, this transition may occur via a first-order layering transition (Section 2.2) or continuously via a percolation transition (Section 5.1). On strongly hydrophilic and heterogeneous biosurfaces, the critical temperature of the layering transition may be below the ambient temperatures. Besides, this transition may be smeared out if the surface heterogeneity is strong enough. In both cases, at ambient temperatures, we may expect the formation of a condensed water layer at biosurfaces via a percolation transition. In the Section 7.1, we consider percolation transition of water in low-hydrated biosystems, and its effect on the properties of the system is analyzed in Section 7.2.

7.1 Percolation transition of water in low-hydrated biosystems

Formation of the hydrogen-bonded water networks may affect conductivity of a system in a drastic way, as these networks provide the paths for the conduction of protons, ions, or other charges in the system. So, the qualitative changes in the conductivity may be expected at hydrations, close to the percolation transition of water. Surface conductivity of quartz increases relatively slowly with increasing hydration level until the completion of the adsorbed water monolayer, but much faster at higher hydrations [582]. The hydration dependence of the dielectric losses of hydrated collagen

[583] may be described by a power law (equation 24) with $h_c = 0$ and exponent $t \approx 6$. Dependence of such kind was observed also for cellulose ($t \approx 9.3$), silk ($t \approx 16.0$), and wood ($t \approx 16.4$) [404–406], where it was attributed to the developing network of conducting water chains. The qualitative changes in the conductivity of a biosystem at some threshold hydration level were first reported for albumin [584, 585] (Fig. 95). Steady-state conductivity σ strongly depends on hydration level, and this dependence changes at $h \approx 0.07$ g/g. Change in the conductivity–hydration relationship at this particular hydration was attributed to the change in the "configuration of the water molecules bound to the protein surface" [585]. Note that the use of the power-law hydration dependence (equation 24) with $h_c = 0$ to describe the data, shown in Fig. 95, gives high values (6 to 9) of the exponent t. Sharp increase in the conductivity of melanin upon hydration is observed at $h_c \approx 0.11$ g/g [586]. This increase may be described by a power law with $t \approx 11$. The conductivity exponents, obtained in the considered cases, exceed noticeably the values 1.3 and 2.0 expected for 2D and 3D percolation [416]. This may be related to the underestimation of the threshold hydration level or to the setting it to zero in the studies described above. DC conductivity of hydrated triticale seed samples and their counterparts increases in a stepwise manner

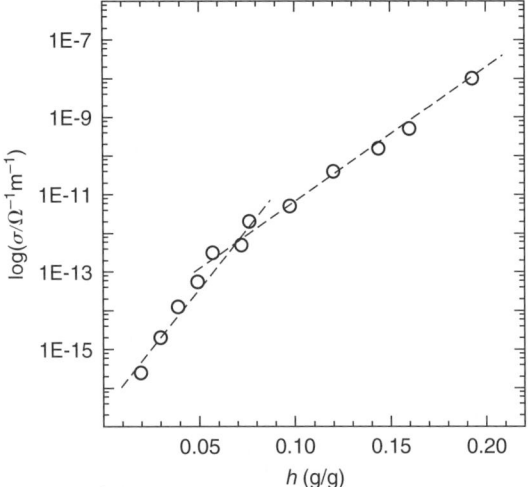

Figure 95: Variation of the steady-state conductivity σ with hydration of albumin [585].

by about 4 orders of the magnitude within hydration range $h = 0.15$ to 0.30 g/g [587, 588].

Capacitance measurements of hydrated lecithin, cytochrome-c, and hemoglobin at low frequencies (from 10^3 to 10^5 Hz) show its strong increase at $h \approx 0.04$, 0.06 and 0.12 g/g, respectively. These values are close to the hydration levels, corresponding to monolayer coverage, which were estimated from adsorption isotherms [589]. In the case of a lysozyme powder, increase in the dielectric losses with increasing hydration was attributed to the protonic conductivity, which is "restricted to the surface of individual macromolecules and involves shifting of protons between ionizable side chain groups of the protein" [590]. So, this conductivity appears due to the formation of mesoscopic water networks, which spread over the distances comparable with the size of one macromolecule. Experimental studies of the capacitance C of a lysozyme powder at various hydration levels h allowed approximate estimations of the threshold hydration, corresponding to the percolation transition of water [591]. The threshold hydration h_c was estimated by the extrapolation of the derivative dC/dh to zero. The average value of h_c, obtained from the measurements in the frequency range from 10 kHz to 4 MHz and for pH from 3.11 to 7.0, was 0.152 ± 0.016. The critical hydration practically does not depend on pH = 3 to 8 but increases to 0.24 at pH = 9.9.

Experimental measurements of the capacitance of the hydrated lysozyme powder were used to calculate the hydration dependence of the dc protonic conductivity [592]. The obtained dependence is shown in Fig. 96 in the linear (left panel) and in the double-logarithmic (right panel) scales. Analysis of the conductivity in the vicinity of the percolation threshold allows its localization and also gives information about the dimensionality of the percolating cluster [416] (see equation (24)). In accordance with the results, shown in the right panel of Fig. 96, just above the percolation threshold, water molecules form 2D networks, which are characterized by the conductivity exponent t of about 1.3. These networks are mesoscopic and provide possibility of the long-range protonic displacements along the surfaces of lysozyme molecules. When h exceeds $h_c + 0.03$, 2D water networks transform into 3D networks and conductivity exponent crosses over to the value of about 2.0.

For hydrated sample of purple membrane, the percolation threshold of water was reported at $h = 0.0456$ g/g [593] and at $h = 0.06$ g/g [594].

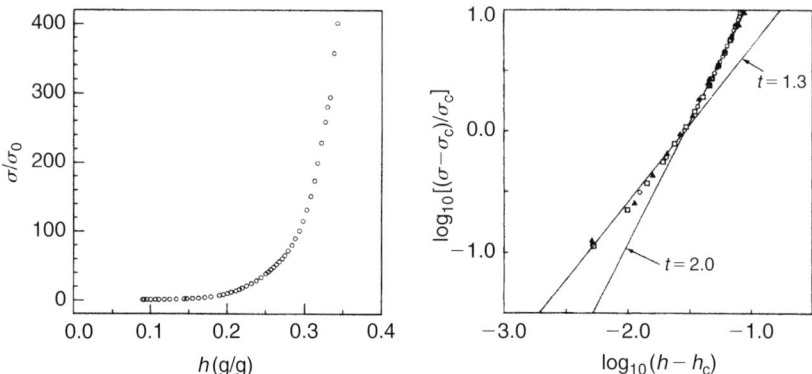

Figure 96: DC conductivity, calculated from capacitance data measurements, vs hydration level h of the lysozyme powder. Left panel: dc conductivity σ normalized by the dc conductivity σ_0 of the dry sample. Right panel: dc conductivity vs hydration level in a double-logarithmic scale. Solid lines show the slopes, corresponding to the critical exponents 1.3 and 2.0 for 2D and 3D percolation, respectively. Reprinted, with permission, from [592].

The conductivity exponent of about 1.23 indicates 2D character of the percolation transition. Similar values of the conductivity exponent were obtained for the hydration dependence of the conductivity of embryo and endosperm of maize seeds [595, 596], where the percolation threshold is $h = 0.082$ and 0.127 g/g, respectively. In hydrated bakers yeast, protonic conductivity evidences 2D percolation transition of water at $h = 0.163$ g/g, and the value of the conductivity exponent is about 1.08 [597]. In this system, increase in conductivity due to 3D water percolation is observed at essentially higher hydration level $h = 1.47$ g/g, where conductivity exponent is about 1.94, i.e., close to the 3D value $t = 2.0$. Conductivity measurements of *Artemia* cysts at various hydrations show strong increase in conductivity starting from the threshold hydration $h = 0.35$ g/g [598] (see Fig. 97). The conductivity exponent in this system is 1.635, which is in between the values expected for 2D and 3D systems. DC conductivity of lichens, evaluated from the dielectric studies at frequencies between 100 Hz and 1 MHz [599], shows strong enhancement at some hydration level. Fit of the conductance–hydration dependence to equation (24) gave the following parameters: $h_c = 0.0990$ g/g, $t = 1.46$ for *Himantormia lugubris* and $h_c = 0.0926$ g/g, $t = 1.18$ for *Cladonia mitis*. So, percolation transition of water in lichens has 2D character and reflects the formation of a spanning network of hydration water.

Low-frequency dielectric measurements (0.1 Hz–1 MHz) of hydrated lysozyme, ovalbumin, and pepsin were used to estimate the fractal dimension for the random walk of protons through the hydrogen-bonded network of water molecules by fitting the shape of the dielectric loss peak [600, 601]. In all three systems, a crossover from 2D to 3D water network occurs within the interval of hydration from 0.05 to 0.10 g/g. For lysozyme, these values are noticeably below the value of about 0.17 g/g, reported in [592]. This difference may be attributed to the presence of about 0.07 g/g of strongly bound water, which presumably was not taken into account in [601]. When hydration exceeds the threshold value, dielectric losses increase almost linearly with temperature. With approaching $T \approx 310$ to 330 K, this increase slows down and dielectric losses turn to decrease with temperature. This behavior may reflect thermal break of the spanning hydrogen-bonded water network, which will be considered in Section 8.1.

At low temperatures, radiation-induced conductivity critically depends on the water content and appears only above the critical hydration levels 0.41 and 0.79 g/g for collagen and DNA, respectively [602, 603]. The critical hydration level for DNA corresponds to about 15 water molecules per phosphate group ($\Gamma = 15$). The effects of various additives on the conductivity evidence charge migration in the hydration shell of DNA [604]. At much lower hydrations (0.12 to 0.22 g/g), conductivity of hydrated DNA shows exponential dependence on h [605], which may be attributed to the intrinsic semiconductivity of the DNA backbone. More detailed experimental studies of DNA hydration [606] show that radiation-induced conductivity starts not strictly at $\Gamma = 15$ [602, 603] but via a sigmoid-like increase within hydration range from $\Gamma = 11$ to $\Gamma = 16$ with subsequent stepwise increase at $\Gamma \approx 24$.

The above experimental studies of the conductivity of hydrated biosystems directly evidence the formation of a spanning network of hydration water via percolation transition. The charge transfer itself may play a crucial role in biofunction [607]. In most of the cases, described above, the percolation transition of water occurs at the hydration level, where various forms of biological activity develop in a stepwise manner (see Section 6). In particular, the following biological processes starts close

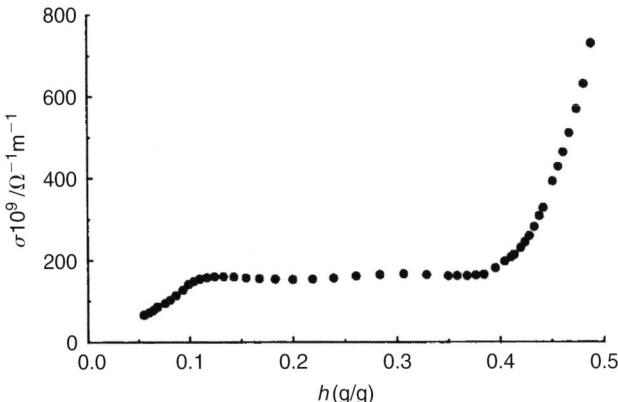

Figure 97: The dc conductivity of *Artemia* cysts as a function of hydration level h. Reprinted, with permission, from [598].

to the percolation transition of water: premetabolism of *Artemia* cysts, enzymatic activity of lysozyme, respiration of yeasts, photoelectric response of purple membrane, germination of seeds, conformational transition of DNA to biologically relevant B-form. Besides, various conformational and dynamical properties of biomolecules necessary for biological function also develop in the hydration range, where percolation transition is seen or may be expected in biosystems.

Below we show how the appearance of spanning water networks may be detected in computer simulations. In particular, a percolation transition of water upon hydration was studied by simulations in model lysozyme powders and on the surface of a single lysozyme molecule. In protein crystals, increase in hydration of a biomolecular surface may be achieved by applying pressure. In some hydration range, pressurization leads to the formation of spanning water networks enveloping the surface of each biomolecule. Finally, the formation of the spanning water network is shown for the DNA molecule at various conformations and for different forms of DNA.

a) Percolation transition of water in lysozyme powders
The structure of amorphous lysozyme powder, used for experimental studies, is not available. In order to do simulations, which may be at least qualitatively related to the experiment, the density of the lysozyme powder should be as close to the experiment as possible.

In low-humidity tetragonal crystal with the partial density of lysozyme of about 0.80 g/cm^3, approximately 120 water molecules are in the first hydration shell of lysozyme molecule. In order to explore a wide range of hydration level up to monolayer coverage (about 300 water molecules), partial density of lysozyme in powder should be $< 0.80 \text{ g/cm}^3$. In Ref. [401], two models for protein powder were studied: densely packed powder with the density of dry protein 0.66 g/cm^3 and loosely packed powder with a density 0.44 g/cm^3. In loosely packed powder, the percolation transition of water was noticeably (by a factor of two) shifted to higher hydration levels compared with experiment. The fractal dimension of the water network at the percolation threshold as well as other properties evidenced that the percolation transition of water in this model was not two dimensional. The spanning water network consists of the 2D sheets at the protein surface as well as of the 3D water domains, formed due to the capillary condensation of water in hydrophilic cavities. The latter effect causes essential distortion of various distribution functions of water clusters in loosely packed powder. Therefore, below we present an overview of the results obtained for the densely packed model powder.

Spanning probability R, defined as a probability to observe a water cluster that crosses the model system at least in one dimension, shows sigmoid dependence on the mass fraction C of water (Fig. 98, upper panel). At ambient temperature ($T = 300$ K), the inflection point of this dependence corresponding to $R = 50\%$ is located at about $C = 0.122$. This hydration level is close to that where the mean cluster size S_{mean} passes through a maximum (Fig. 98, middle panel). Fractal dimension of the largest water cluster achieves the value d_f^{2D} at $C \approx 0.155$ (Fig. 98, lower panel). Summarizing, the percolation transition of water may be attributed to the hydration level $C \approx 0.155$. The cluster size distribution n_S supports this conclusion [401].

The percolation threshold of water found in simulations should be compared with experiments performed in the lysozyme powder [591]. At ambient temperature, it was observed at the hydration level $h = 0.152 \pm 0.016$ g of water per gram of dry lysozyme, which corresponds to the water mass fraction $C \approx 0.132$ [591]. The percolation threshold seen in simulations thus occurs at $h \approx 0.183$, i.e., at slightly higher hydration level than in experiment. This should be considered as rather good

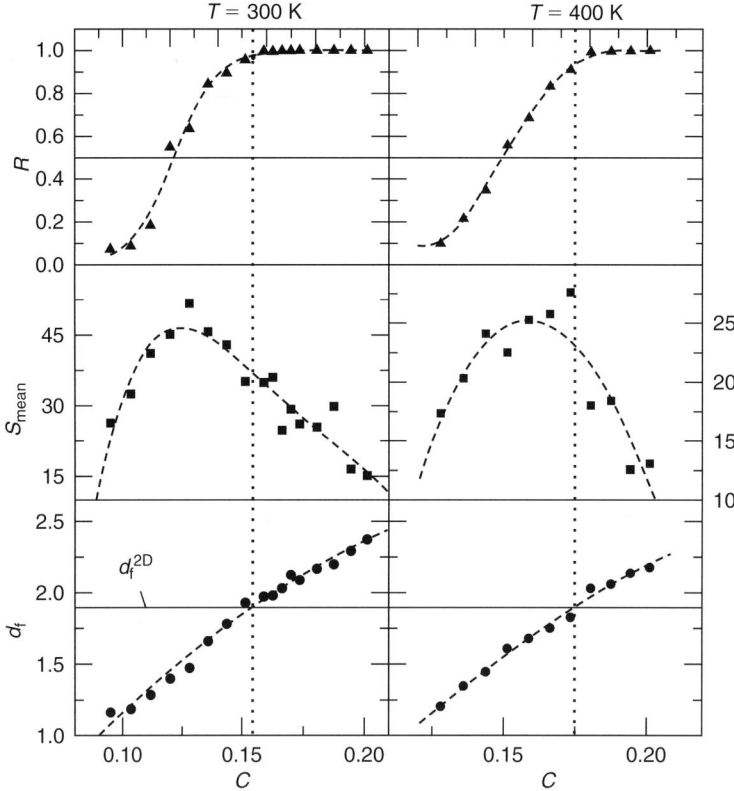

Figure 98: 2D percolation transition of water in the hydrated densely packed powder of lysozyme at two temperatures. Spanning probability R (upper panel), mean cluster size S_{mean} (middle panel), and fractal dimension of the largest cluster d_f (lower panel) are shown as functions of a mass fraction of water C. The dashed lines are guides for eyes only. Vertical lines indicate the 2D percolation threshold. Reprinted, with permission, from [401].

agreement between experiment and simulations in view of rather crude model of lysozyme powder, which was imposed to have a rather arbitrary density. Probably, the more densely packed model powder may show better agreement with experiment. Besides, in experiment, lysozyme molecules expose to water more and more their surfaces upon increasing hydration [508]. That is not the case for the simulated powder, which was frozen so that the rearrangement of lysozyme molecules upon hydration was not allowed.

Simulations enable to explore the arrangement of water molecules near hydrated proteins and their evolution during the percolation transition. Experimental studies provide the average number of water molecules per one protein ($N_w/N_p \sim 120$ [591]) at the percolation threshold. However, this number could not be equal to the average number of water molecules N_w^1 in the first hydration shell of each lysozyme in powder. In the model powder, $N_w/N_p \sim 146$ and $N_w^1 \sim 149$ at the percolation threshold at 300 K. Close values of N_w/N_p and N_w^1 could mean that most water molecules belong to the hydration shell of one protein molecule only. These numbers are significantly smaller than numbers $N_w = 450$ and $N_w^1 \approx 336$ in the case of the percolation threshold at the surface of a single lysozyme molecule at the same temperature (see below). Such strong difference could be attributed to the significant decrease in the accessible surface area of proteins in powder due to close contacts. So, the 2D percolation transition of water in protein powder appears as a formation of water network, which spans the extended "collective" surface, created by aggregated protein molecules, covering each protein molecule only partially.

Temperature affects strongly the percolation threshold of hydration water. Changes of the various cluster properties with hydration are compared for $T = 300$ and 400 K in Fig. 98. Increasing temperature notably shifts the percolation threshold to higher hydration level. It was found at $C = 0.175$ ($h = 0.212$ g/g) at $T = 400$ K. The hydration dependence of the spanning probability R becomes more rounded with increasing temperature. Cluster size distributions n_S of water in lysozyme powder at $T = 400$ K are shown in Fig. 99 at some hydration levels. A hump of n_S at large S appears at hydration where $R \approx 50\%$ and reflects a cut in the water clusters spanning the simulated system (see Section 5.1). Right at the percolation threshold the cluster size distribution n_S should follow the power law (19) in a widest range of cluster sizes. At $T = 400$ K and $C = 0.173$, n_S follows a power law in the range of cluster sizes up to 200 molecules (see squares in Fig. 99). Wave-like deviations from the power-law behavior could not be eliminated by improving statistics and should be attributed to the peculiar arrangement of lysozyme molecules in the model powder. Note that deviations of n_S from the power-law behavior become larger when the temperature decreases to the ambient. This makes the use of n_S distribution for location of a water percolation

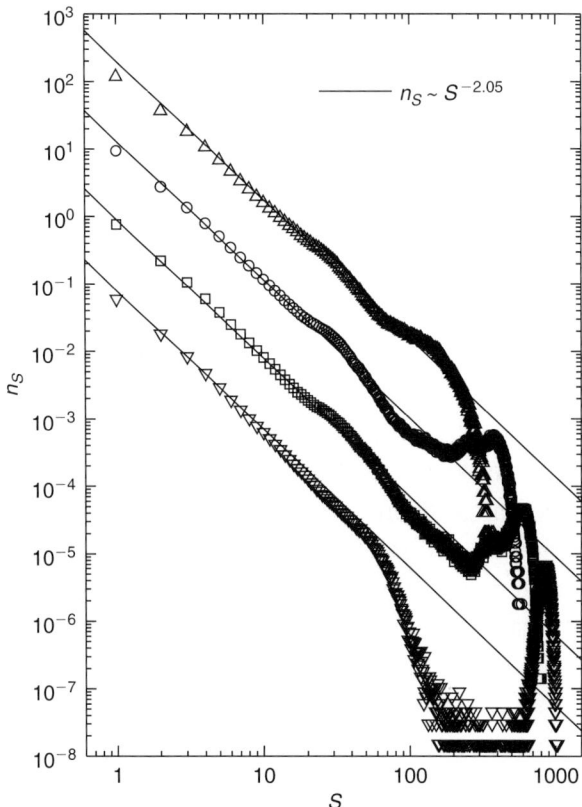

Figure 99: Distributions n_S of clusters with S water molecules in densely packed lysozyme powder at $T = 400$ K. Mass fraction of water increases from $C = 0.128$ (top) to 0.201 (bottom). Circles represent n_S at $C = 0.151$, when the spanning cluster exists with probability of about 50%, while squares correspond to $C = 0.173$, when the fractal dimension of the largest cluster is close to the 2D percolation threshold value. The distributions are shifted consecutively by one order of magnitude each, starting from the bottom. Reprinted, with permission, from [401].

threshold in powders to be not very fruitful at low temperatures. For large S left to the hump, the distribution n_S deviates from the power law upward below the percolation threshold and downwards above the percolation threshold. So, the cluster size distributions indicate the percolation threshold at $T = 400$ K at water mass fraction $C_p \approx 0.17$, that is quite close to the threshold value $C_p \approx 0.175$ estimated from the behavior of the fractal dimension d_f of the largest cluster (Fig. 98, lower panel).

b) Percolation transition of water at the surface of a single lysozyme
As we discussed above, the spanning water network in the lysozyme powder exists when each lysozyme molecule has about $N_w^1 \sim 149$ water molecules in its first hydration shell. The surface area of native lysozyme may be estimated as a surface-accessible area derived from a crystal structure (~ 6000 Å^2) or obtained in simulations (~ 6900 Å^2) [401]. A complete monolayer coverage of a smooth surface corresponds to about 10 Å^2 per water molecule (Section 2.2). This means that less than 25% of the surface of each lysozyme molecule is covered by hydration water at the percolation threshold. Formation of water network, which envelopes each lysozyme molecule, is crucial for protein dynamics [508]; therefore, this process was also studied by computer simulations.

Clustering of water molecules at the surface of a single lysozyme is rather similar to the water clustering at the smooth hydrophilic surface, described in Section 5.1. At $T = 300$ K, the fractal dimension d_f of the largest cluster achieves the critical value d_f^{2D} of 2D percolation threshold, when about 450 water molecules surround the lysozyme molecule and ~ 336 of them belong to the first hydration shell (Fig. 100, lower panel). Approximately at the same surface coverage, the cluster size distribution n_S obeys power law in the largest range of the cluster sizes [401]. This gives the minimum water coverage ($C = 0.065$ Å^{-2}), which provides permanent spanning water network at the surface of lysozyme molecule at $T = 300$ K.

The distribution $P(S_{max})$ of the size S_{max} of the largest water cluster at the surface of single lysozyme behaves with hydration similar to $P(S_{max})$ at the spherical surface. The distribution $P(S_{max})$ is the widest and has two peaks of comparable heights at hydration $C \approx 0.057$ Å^{-2}, which corresponds to $N = 375$ water molecules (Fig. 101). To be in line with approach used for hydrophilic spherical surface, one may assume that all largest water clusters which size $S_{max} > 250$ are spanning (see a vertical dashed line in Fig. 101, left panel). This assumption allows the calculation of the spanning probability R as an integral of $P(S_{max})$ for $S_{max} > 250$. Dependence R on hydration shown in Fig. 101 (right panel) evidences a midpoint of the percolation transition at $C \approx 0.057$ Å^{-2}, where $R \approx 50\%$. The mean cluster size S_{mean} shows a maximum at almost the same hydration level $C \sim 0.056$ Å^{-2} (Fig. 100, upper panel).

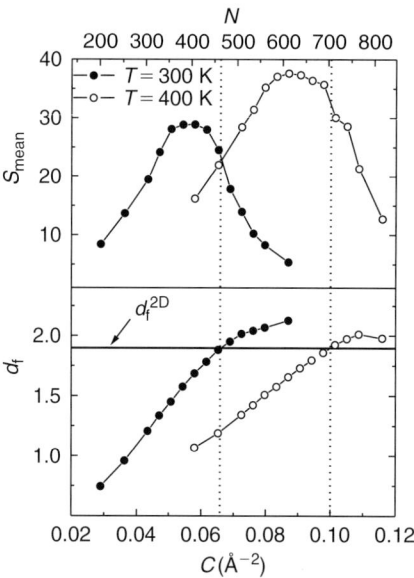

Figure 100: Mean cluster size S_{mean} (upper panel) and fractal dimension d_f of the largest water cluster (lower panel) at the surface of a single lysozyme molecule as functions of surface coverage C (lower axis) or number of water molecules N (upper horizontal axis). The percolation thresholds for two temperatures are shown by dotted lines. Reprinted, with permission, from [401].

Increase in temperature up to $T = 400$ K shifts the percolation threshold to higher hydration: $N = 690$ ($C = 0.10$ Å^{-2}). The change in the fractal dimension d_f of the largest water cluster shown in Fig. 100 (lower panel) serves for illustration of this result. A similar temperature shift is observed for the midpoint of the percolation transition, which is close to the maximum of S_{mean} (see Fig. 100, upper panel). This shift originates from the decrease in the number of water–water hydrogen bonds with increasing temperature (more details will be given below in this section).

What results obtained for the rigid-model lysozyme molecule will be valid in real world? The answer may be found by comparison of water clustering on the rigid-model lysozyme with that on the flexible-model lysozyme molecules. Flexibility of lysozyme molecule affects clustering of hydration water in several ways. First, the accessible area of lysozyme surface slightly changes (decreases) upon hydration. This effect shifts the water percolation threshold to slightly lower hydrations. In particular, the midpoint of the percolation transition where spanning probability

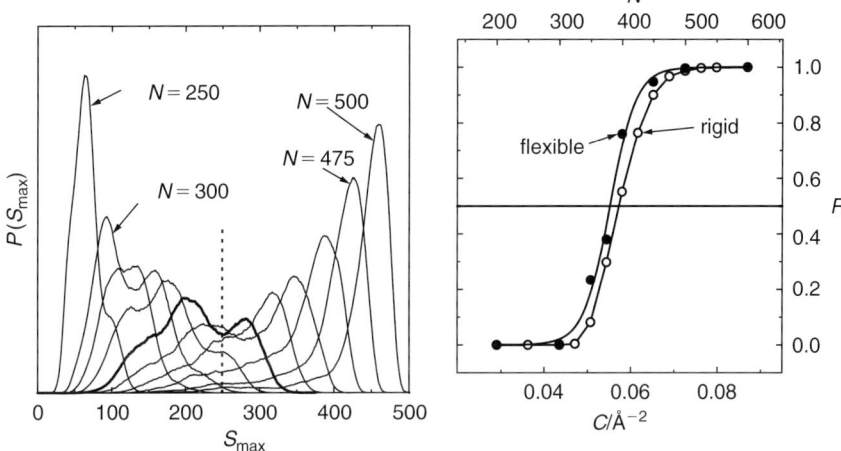

Figure 101: Left panel: the distribution $P(S_{max})$ of the size S_{max} of the largest water cluster at the surface of single rigid lysozyme molecule at various hydration levels. The number of water molecules is shown in figure for some distributions. The distribution closest to the midpoint of the percolation transition ($N = 375$) is shown by a thick line. The vertical dashed line separates roughly spanning and nonspanning largest clusters. Right panel: spanning probability R at the surface of the rigid and flexible lysozyme molecules as a function of the surface coverage C (lower axis) and N (upper axis) (data from [398]).

is close to 50% is shifted to the lower hydration (see Fig. 101, right panel) [398]. This shift, however, is rather small (~ 0.003 Å^{-2}, or about 20 water molecules). Analysis of the cluster size distribution and various properties of the largest water cluster has shown that motions of a lysozyme molecule causes some injuries to the largest water cluster only, whereas their effect on small clusters is negligible.

It is possible to estimate which part of lysozyme surface is covered by water at the percolation threshold by simulation studies. Water monolayer with bulk-like structure (corresponding to numerical density of about 0.033 Å^{-3} at ambient conditions) gives the surface coverage of about 0.1 Å^{-2}. This coverage does not change noticeably due to the packing effect near planar smooth surfaces [32] or near model protein surfaces [608, 609]; therefore, 10 Å^{-2} could be used as an average area occupied by one water molecule at the surface. At the percolation threshold on the surface of single lysozyme at $T = 300$ K, water covers about 50% of total lysozyme surface, or about 66% of the hydrophilic part of the lysozyme surface, assuming that about 74% of lysozyme surface, which consist of polar and charged residues, is hydrophilic [401]. At ambient temperature, about 75% of water molecules belong to the first hydration shell of a single lysozyme molecule at the percolation threshold, and this fraction noticeably decreases with further increase of hydration level. For comparison, this value is about 90% for hydrophilic sphere at $T = 425$ K. This evidences that lysozyme surface is noticeably less hydrophilic than smooth hydrophilic surface described above. The layering transition of water could not be expected at the lysozyme surface.

The lifetime of the spanning water network at the midpoint of the percolation transition, corresponding to 50% probability to observe a spannning cluster, is comparable with the average lifetime of a single water–water H-bond [402]. This indicates that each water molecule of the largest cluster breaks and creates on the average one H-bond before this cluster changes its character from spanning to nonspanning or vice versa. This time is significantly smaller than the residence time of water molecules at the lysozyme surface even in the case of full hydration [610]. So, break and creation of a spanning water network is related not to the translational movement of water molecules between the hydration shells, but rather with their rotational motions. Lifetime of the spanning water network decays to zero upon dehydration and drastically increases with increasing hydration level above the percolation threshold [402].

c) Percolation transition of water upon pressurization
Hydration of biomolecules may be increased by applying pressure. If the average hydration level of a system is not very high, there is no spanning water network that envelopes single biomolecule at low pressures. With increasing pressure, the number of water molecules in the hydration shell increases and spanning water network may appear. Such scenario was realized during pressurization of the hydrated crystalline Snase [611, 612]. At ambient pressure, more than 95% of water molecules belong to the hydration shells of protein molecules. However, only 40% of them belong to the hydration shells of two and more proteins simulteneously. As a result, each Snase molecule has about 670 water molecules in its hydration shell, which is not sufficient to form a water network spanning over the whole Snase molecule. So, at ambient pressure, the largest water cluster in the hydration shell of protein is nonspanning. Increasing pressure makes more water molecules to be shared by the hydration shells of

different proteins, and the largest water cluster in the hydration shell of a protein molecule becomes spanning.

Percolation transition at the surface of protein molecule may be detected using the distance H_{max} of the center of mass of the largest water cluster from the center of mass of protein. Probability distribution of H_{max} changes with pressure, in a way similar to that with increasing hydration (see Fig. 102, upper panel). The first peak at about $H_{max} = 2$ Å corresponds to the largest clusters, which span the surface of protein

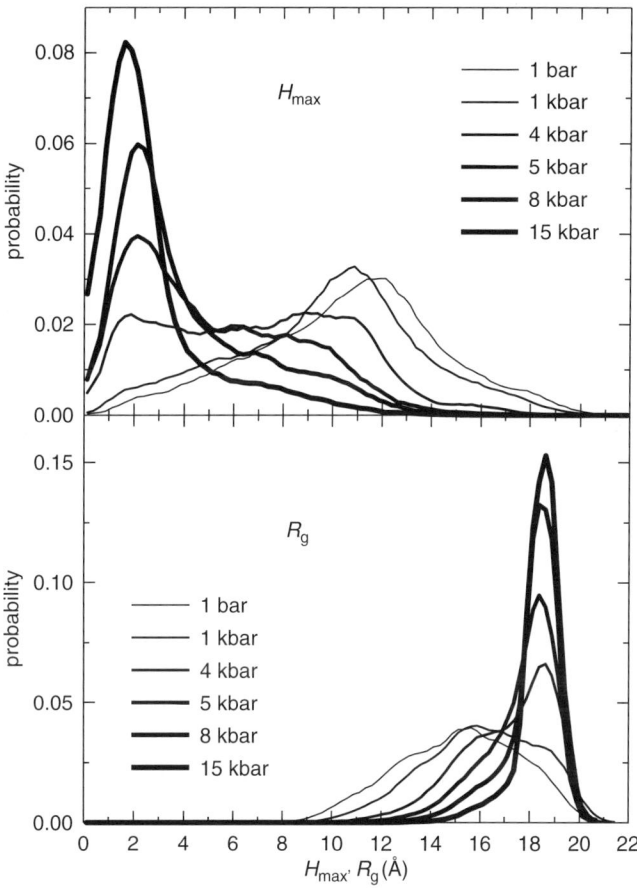

Figure 102: Distributions of the distance H_{max} of the center of mass of the largest cluster of hydration water to the center of mass of the Snase molecule (upper panel) and of the radius of gyration R_g of the largest water clusters (lower panel) at various pressures P.

molecule homogeneously. The second peak at about 11 Å corresponds to the nonspanning largest water clusters, which cover only part of the protein surface. The widest probability distribution of H_{max} is observed at $P \approx 4$ kbar, where the clusters with H_{max} between two peaks are equally populated. Behavior of H_{max} indicates close population of spanning and nonspanning clusters at about 4 kbar.

Distributions of the radius of gyration R_g of the largest water cluster at various pressure are shown in Fig. 102 (lower panel). At ambient pressure, this distribution is broad with a maximum at about 16 Å. Upon pressurization, it shifts to higher R_g values, and sharp peak at about $R_g^0 \approx 18.6$ Å is developed. Further increase in pressure makes this peak narrower and higher but does not influence its location. The value R_g^0 equals to the radius of gyration of a complete monolayer of hydration water, and it simply relates to the radius of gyration of protein R_g^p via $R_g^0 = R_g^p + 3$ Å [402]. Behavior of R_g evidences that the largest cluster of hydration water covers the protein surface rather homogeneously at $P \geq 5$ kbar when the majority of the largest clusters contribute to the peak R_g^0.

Probability distribution of the size S_{max} of the largest water cluster at the surface of a single Snase molecule shows behavior characteristic for the percolation transition of hydration water observed with increasing hydration level. In particular, a maximum of distribution moves upon pressurization from smaller to larger S_{max} in such a way that its width changes nonmonotonously. The widest distribution is observed when the existence probability R of spanning water network is about 50% [612]. The value $S_{max}^c = 350$ of water molecules in the largest cluster can be used to separate approximetely spanning and nonspanning water clusters at the surface of Snase molecule. The probability R to observe a spanning water cluster around a single protein molecule shows a sigmoid dependence on pressure P with inflection point between 4 and 5 kbar (Fig. 103). At this pressure, the number N_1 of water molecules in the hydration shell is about 723, and about 350 to 600 of them are members of the spanning water cluster. Using more natural occupancy variable N_1, the midpoint of percolation transition, where $R \approx 50\%$, was estimated at $N_1 \approx 700$. Note that Snase molecules in crystal are not equally hydrated. Therefore, midpoint of percolation transition varies on 20–30 water molecules depending on the particular position in crystalline structure. Accordingly, the percolation threshold is rounded in the

Water in low-hydrated biosystems

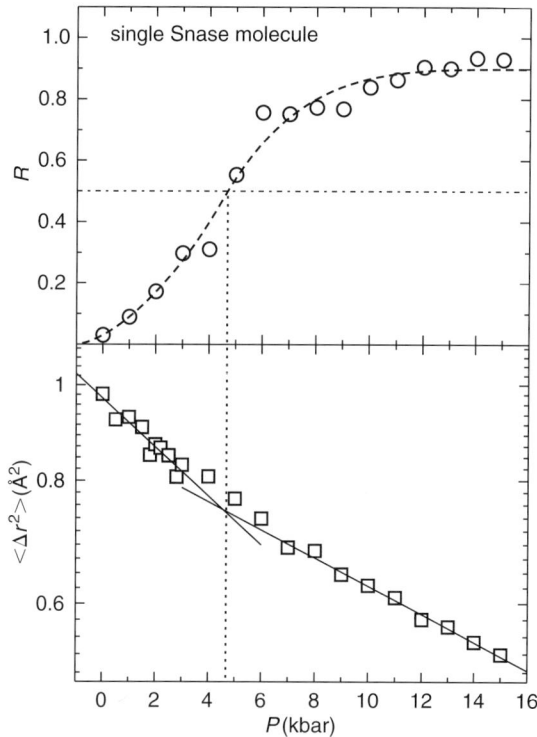

Figure 103: The average probability R to observe a spanning water cluster in the hydration shell of each Snase molecule (open circles, upper panel) and average mean-square displacement (MSD) $\langle \Delta r^2 \rangle$ of hydrogen atoms of proteins in crystal (lower panel) as a function of pressure P. Fit of R to Boltzmann function is shown by the dashed line. The pressure when $R \approx 50\%$ is denoted by dotted line. The fits of $\langle \Delta r^2 \rangle$ in the ranges $P \leq 3$ and $P \geq 7$ kbar to the Arrhenius equation are shown by solid lines (data from [612]).

crystalline system within the range ±1 kbar. Note that percolation transition of water around a single molecule was found to be the origin of the dynamic transition of crystalline Snase [611, 612].

d) Percolating water networks on the DNA surfaces
The double-helical DNA may adopt a number of different structural forms [487]. The B-form is the dominant biological conformation, while other forms probably exist *in vivo* as high energy states adopted temporarily as shown for A- and Z-DNA [613–615]. The A-form has the same handeness, topology, and hydrogen bonding as the B-form.

Its conformation is compatible with any base pair sequence, and it is separated from the B-form by only a modest energy barrier [616]. Due to all these features, reversible local B ↔ A transitions represent one of the modes for governing protein–DNA interactions [617]. The B ↔ A transitions can be also induced *in vitro* by changing the DNA environment [488, 618–620]. In condensed preparations, that is, in crystalline and amorphous fibers as well as in films, DNA adopts the B-form under high relative humidity, but it can be reversibly driven to the A-form by placing the samples under relative humidity below 80% [488, 618, 620]. DNA molecules exhibit reversible B ↔ A transition in aqueous solutions upon addition of some organic solvents [613, 619]. In all cases, the transition occurs at about the same water activity, suggesting that the B ↔ A conformational switch is driven by the hydration state of the double helix [492].

Hydration of nucleic acids has a number of distinctions due to their polyionic character and uneven nonspherical shapes [487]. In physiological conditions, the double-helical DNA directly interacts with solvent ions in several water layers from its surface; therefore, the functional DNA hydration shell is very thick. Under limited hydration, there is a strict relationship between the state of DNA and hydration number Γ measured as the number of water molecules per nucleotide (or phosphate). When Γ is reduced below 30, the common B-form of DNA is already perturbed, but it is maintained until $\Gamma \approx 20$ [487, 488]. Below this hydration, DNA undergoes different conformational transitions, among which the transition from B- to A-form [489] with a midpoint at about $\Gamma = 15$ is the most studied (see Section 6).

Formation of a spanning network of hydration water at the DNA surface upon hydration was studied by computer simulations [200, 621] using the water drop methods [622, 623]. Simulations were carried out for a rigid dodecamer fragment of double-helical DNA. The structures of the canonical B-DNA and A-DNA [624] were fixed in space. The system involved 24 bases and 22 phosphate groups in two DNA strands surrounded by a mobile hydration shell of 22 Na^+ ions and 24Γ water molecules. Evolution of the cluster size distribution n_S on the surface of B-DNA upon increasing hydration is shown in Fig. 104. At low hydrations ($\Gamma = 12$, 13, and 14), n_S shows deviations upward from the power law (19) at the intermediate cluster sizes S. At high hydrations ($\Gamma = 17, 18, 19$, and

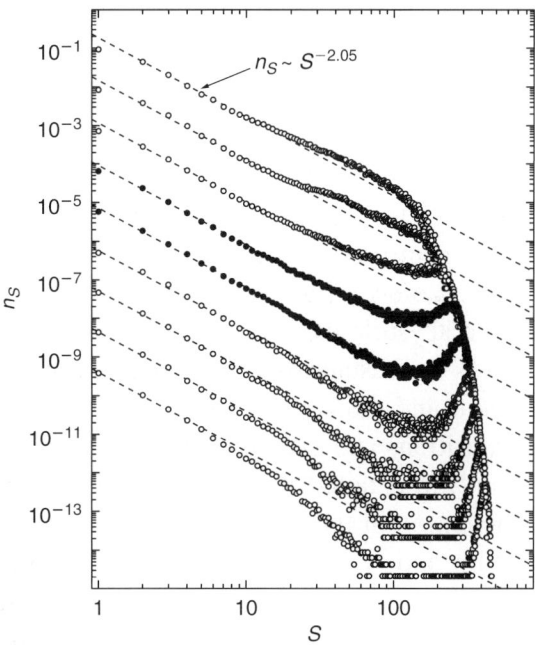

Figure 104: The size distributions n_S of water clusters at various hydrations Γ from 12 (top) to 20 (bottom). The distributions are shifted consecutively, each by one order of magnitude starting from the top. The hydration levels $\Gamma = 15$ and 16, closest to the percolation threshold, are shown by closed symbols. Reprinted, with permission, from [200].

20), a drop of n_S is clearly seen before the hump at large S. The size distribution n_S follows the universal law (19) in the widest range of S when $\Gamma = 15$ and 16 (closed symbols in Fig. 104). Note that this conclusion does not depend on the assumed dimensionality of the system being studied, that is water adsorbed on the DNA surface. Due to the groove shape of the DNA double helix, the 2D character of its hydration water is not obvious. The mean cluster size shows a skewed maximum at $\Gamma = 14$ and suggests that the percolation threshold is located above this hydration level [200]. Probability distribution of the size S_{max} of the largest water cluster allows calculation of the spanning probability R, which achieves 50% at $\Gamma = 14.3$. So, analysis of the various cluster properties evidences the percolation transition of hydration water at the surface of B-DNA when $\Gamma \approx 15.5$ and midpoint of the percolation transition at $\Gamma \approx 14$.

The primary water shell around B-DNA is usually estimated as about 20 water molecules per nucleotide [490]. Therefore, the percolation threshold of hydration water on the surface of rigid B-DNA corresponds to about 80% of one full hydration layer. Approximately 65% and 50% of a monolayer coverage is necessary to form a spanning hydration network on smooth hydrophilic surfaces [394] and the surface of the lysozyme molecule [401, 508], respectively. It is reasonable to attribute a relatively high percolation threshold for B-DNA to the presence of Na^+ ions in a hydration shell. The key role of free metal ions in low-hydration polymorphism of DNA is well established by experimental studies [625]. By changing the amount and the type of ions, one can shift the midpoints of polymorphic transitions and even their pathways [626]. Almost nothing is known about the detailed mechanisms involved in such effects. A step toward elucidation of these problems is study of water clustering and percolation with and without free ions. As small hydration shells around charged DNA fragments are inherently unstable [622], DNA molecules should be neutralized artificially. Neutralization of DNA has been used in simulations since long ago [627], and usually this is done by reducing phosphate charges. For electrostatically neutral B-DNA obtained by reducing charges of phosphate oxygens, water does not show a percolation transition in the course of gradual hydration. The probability distribution $P(S_{max})$ of the size of the largest cluster behaves as if the system consists of small water droplets that merge into one large water patch with increased hydration [621]. This scenario is also suggested by the absence of the sigmoid behavior for spanning probability R, the absence of maximum of S_{mean}, a monotonous change of ΔS_{max} etc. Formation of a large continuous water patch was found to be typical for water near hydrophobic surfaces or in mixtures with hydrophobic solutes [204], which is surprising because, even with phosphates neutralized, the DNA surface remains highly polar.

It turned out, however, that the behavior of hydration water near neutral DNA depends on how its surface was neutralized. Properties of hydration water were found to be similar in the cases when the neutralizing charge was uniformly distributed over the whole system including DNA and water and between all DNA atoms only. In both cases, water undergoes a normal percolation transition with increasing Γ. With ions removed, the percolation threshold of hydration water is shifted by $\Delta \Gamma \approx 4$ toward lower hydration. Therefore, hydration of each ion requires about four additional water molecules, which is close to the hydration number of Na$^+$. The water clustering on the surface of A-DNA molecule was studied in the presence of 22 Na$^+$ ions in a hydration shell. Spanning probabilities R for A- and B-DNA molecules are compared in Fig. 105 (left panel). Fit of the hydration dependence of R to sigmoid function suggests that the midpoint of water percolation transition at the surface of A-DNA is close to $\Gamma = 12.9$. This means that spanning water cluster on the surface of A-DNA appears at the hydration level about 1.4 lower than at the surface of B-DNA molecule. Accordingly, the fraction S_{max}^{av}/N_w of water molecules in the largest cluster, shown in the right panel of Fig. 105, drastically increases when Γ grows from about 13 to 18, indicating the midpoint of the percolation transition in A-DNA molecule at a slightly lower hydration than in B-DNA. However, the mean cluster size, which characterizes the properties of all clusters, except the largest one, passes through a maximum at about $\Gamma = 14$ for both DNA molecules (Fig. 106, left panel). This may indicate that conformation of DNA molecule affects slightly the largest water cluster only. Accordingly, evolution of n_S distributions with hydration turned out to be very similar on the surfaces of both DNA molecules, and the estimated percolation threshold was identical, i.e. $\Gamma = 15.5 \pm 0.5$ [621]. The spanning

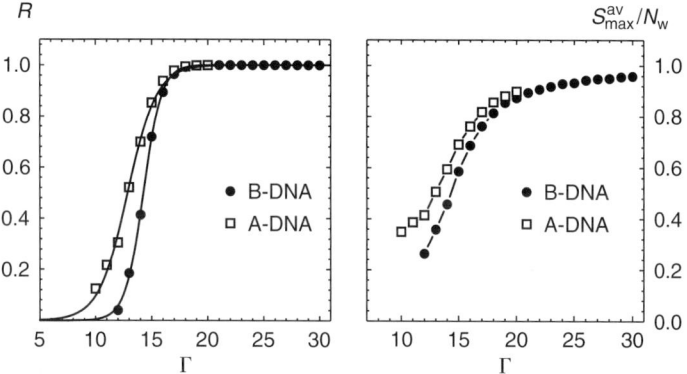

Figure 105: The probability R of observing a spanning water cluster (left panel) and the fraction S_{max}^{av}/N_w of water molecules in the largest cluster (right panel) as functions of hydration number Γ for B- and A-DNA surfaces. Sigmoid fits are shown by solid lines in the left panel (data from [621]).

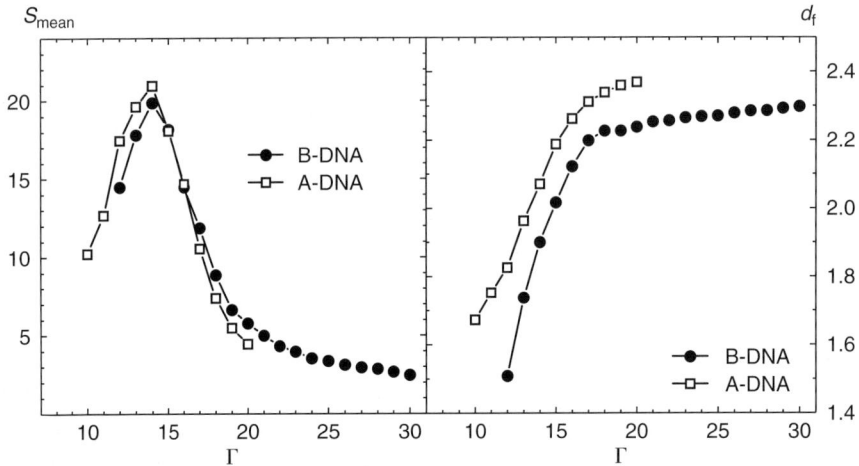

Figure 106: Mean cluster size (S_{mean}) (left panel) and fractal dimension (d_f) of the largest water cluster (right panel) at the surface of B- and A-DNA under various hydrations (data from [621]).

probability R at the percolation threshold is about 90% on the surfaces of both DNA molecules, which is close the values observed at the true percolation transition of water on the surface of a lysozyme molecule (about 90%) and on the surfaces of smooth spheres (from 70% to 90%, depending on a sphere size).

The dimensionality of the largest water cluster is characterized by the effective fractal dimension d_f shown in Fig. 106 (right panel). In ideal 2D and 3D systems, the percolation threshold is characterized by $d_f \approx 1.89$ and 2.53, respectively [396]. Fig. 106 indicates that hydration water at the B-DNA surface represents a quasi-2D system. Deviations from a 2D behavior are larger for A-DNA, indicating a more heterogeneous distribution of hydration water. At $\Gamma \approx 17$, the slopes of the $d_f(\Gamma)$ plots drastically fall for both A- and B-DNA. Apparently, a qualitative change of the internal structure of the largest water cluster takes place just above the percolation threshold.

The surface of the double-helical DNA is usually considered as involving at least two distinct nonoverlapping parts with qualitatively different properties, namely, the major and minor grooves. In B-DNA, the minor groove is narrow and deep, whereas the major groove is very wide and its surface is easily accessible from solution. In contrast, in the A-form of

DNA, the minor groove represents almost a flat exposed surface, while the major groove becomes very deep and narrow. In A-DNA, the opening of the major groove is probably blocked by free metal ions sandwiched between the two opposed phosphate arrays [628, 629]. During the B to A transition, the major DNA groove collapses around these ions, while the minor groove turns inside out, completely losing its initial properties. All these events are certainly related to changes in the water structure. To get an insight into their mechanisms, hydration and water clustering in the two DNA grooves should be studied separately. The hydration shells of A- and B-DNA may be divided in two parts, one of which contains the closed compartments where hydration conditions do not change with Γ. It was anticipated that the minor groove of B-DNA and the major groove of A-DNA probably represent the natural such compartments. Accordingly, the second part of hydration water involves all water in B-DNA major groove and, *vice versa*, in the A-DNA minor groove.

Fig. 107 (left panel) shows variation in the number of water molecules in the grooves of B- and A-DNA with Γ. As expected, the weight of the hydration shells in the minor groove of B-DNA and the major of A-DNA remains very stable. Even with Γ reduced below the percolation threshold, the number of water molecules in these compartments change insignificantly. Variations of Γ mainly affect the remaining part of water. The clustering behavior is also radically different. The probability R to observe a spanning water cluster in the major groove of B-DNA and the minor groove of A-DNA exhibits a sigmoid behavior typical of percolation transitions, with inflection points ($R \approx 50\%$) at $\Gamma \approx 15.8$ in both cases (Fig. 107 (right panels)). The spanning probabilities R in the minor groove of B-DNA and in the major groove of A-DNA show only a weak positive trend with hydration number. When the percolation transition occurs in the whole hydration shell of DNA, the probability to observe a spanning cluster in the minor groove of B-DNA and the major groove of A-DNA is about 20 and 50%, respectively. This indicates that although water in the relatively isolated B-DNA minor and the A-DNA major grooves contributes to the largest water cluster in the whole hydration shell, the spanning water cluster appears permanently due to the percolation transition in the opposite exposed grooves. This complex picture is supported by the behavior of other cluster properties [621].

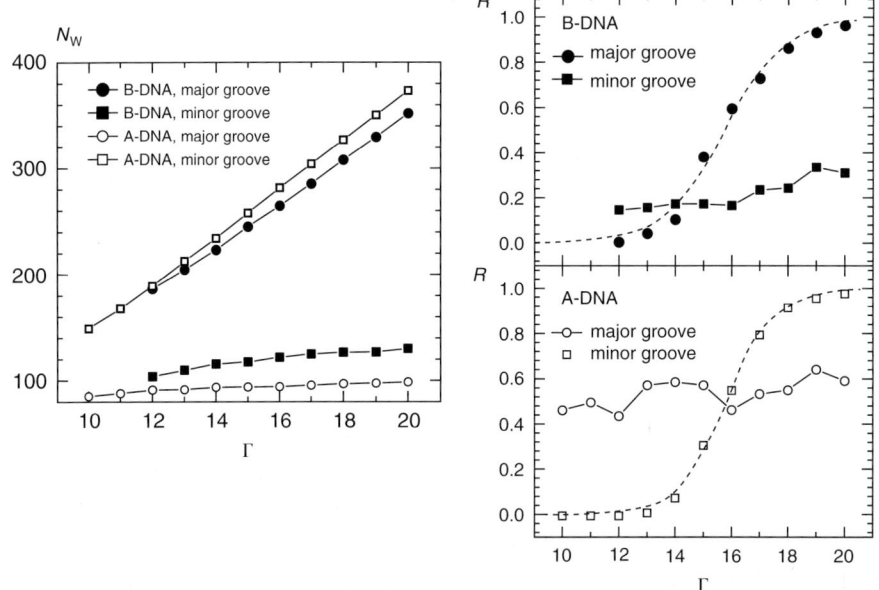

Figure 107: Left panel: number of water molecules (N_w) in the major and minor grooves of B- and A-DNA as functions of hydration number (Γ). Right panels: probability (R) of observing a spanning water cluster in the major and minor grooves of B- and A-DNA molecules as functions of hydration number (Γ). Reprinted, with permission, from [621].

e) Universality of water percolation in low-hydrated systems

As we show above, the percolation transition of hydration water follows the universal laws predicted by the percolation theory for lattices [396]. Behavior of various properties of water clusters upon increasing hydration corresponds to the site percolation problem in lattices, more correctly to the correlated site percolation problem. The laws of percolation transition are universal for all systems of a given dimensionality, but the values of critical exponents depend on the Euclidean dimension of system. Water in low-hydrated biosystems is a quasi-2D system that is not strictly two dimensional. The deviation from the strict two dimensionality is determined by degree of localization of molecules near a surface and depends on the surface structure. The hydration water on smooth planar surface can be regarded as a 2D system even at $T = 425$ K and already in systems of ~ 80 Å size. Deviation from 2D character is noticeable at the small spherical surfaces of a radius $R_{sp} = 15$ Å at

$T = 425$ K. It appears in lower dimensionality of the spanning water cluster at the percolation threshold [394, 631]. When considering a single lysozyme molecule, the fractal dimension of a spanning cluster of hydration water at the percolation threshold is indistiguishably close to $d_f^{2D} \approx 1.896$, expected at the percolation threshold in 2D lattices [401]. Increase in temperature to 400 K results in decreasing localization of water near a lysozyme surface but has a little effect on a fractal structure of the largest water cluster at the percolation threshold. This indicates that lysozyme molecule is not so small and its surface is not so rough to cause a notable deviation of water percolation transition from that in a strict 2D systems.

Percolation transition of hydration water in 3D systems like protein powders is also close to the 2D percolation in spite of the spanning water cluster extending to infinity in three spacial dimensions. Infinite H-boned water network in powder spans the extended "collective" 2D surface created by densely packed protein molecules. The fractal dimension of the largest water cluster at the percolation threshold is close to d_f^{2D}. Further increase in hydration makes larger surface area of protein to be accessible to water molecules up to the fully hydration state, when each protein possesses its own separate hydration shell and $h \approx 0.42$ g/g [401, 508]. Percolation transition of water at the DNA surface was also found close to the 2D percolation, although the spanning water network at the percolation threshold has notably higher fractal dimension than d_f^{2D}. As such trend is seen on both the DNA with Na^+ ions and the uniformly neutralized DNA molecules, it should be attributed to a specific double-groove structure of the DNA hydration shell.

A comparison of the location of the percolation thresholds in terms of hydration levels is not trivial in different systems as proteins, DNA, hydrated powders, and crystalline proteins. Biomolecules differ strongly in the structure of their surface, level of hydrophylicity, and presence of charged groups and ions. Packing of molecules is also important for the threshold hydration level. Besides, adsorption of water molecules on biosurface is not uniform so that spanning network of hydration water includes also the water molecules from the second hydration shell, which are not directly adsorbed on the surface. Note that percolation threshold in lattices is essentially system-dependent parameter which is determined by lattice structure. It may be expressed in terms of several occupancy variables or in terms of the average number of bonds in system. The latter consideration yields a closer percolation thresholds in different lattices. In particular, it is ~2.09 and ~2.37 for site percolation and ~1.96 and ~2.00 for bond percolation on the honeycomb and square lattices, respectively, which are the most relevant to the case of adsorbed water [612]. Therefore, the water percolation threshold in various biosystems may be expected to be rather universal in terms of water–water H-bonds.

The average number of H-bonds n_H, which create each water molecule with its neighbors, constantly increases with increasing hydration. Two examples are shown in Fig. 108 for a single lysozyme and lysozyme powder. The dependence of the fractal dimension d_f on n_H for these systems is shown in Fig. 109. Below the percolation threshold d_f is essentially an effective fractal dimension because most of the largest water clusters are not true (infinite) fractal objects. Thus, the values of d_f noticeably depend on the system size and geometry at low hydration levels. In the system of the same size, such as water at the rigid and flexible lysozyme molecules, the structure of the largest water cluster described by d_f is practically identical at the same n_H. At the percolation threshold, the structure of the largest water cluster is close to a fractal and d_f approaches the threshold fractal dimension d_f^{2D} at $n_H \approx 2.31$ in all systems, including the lysozyme

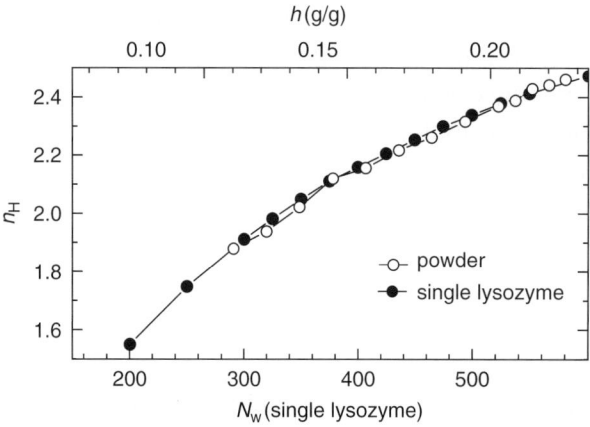

Figure 108: Average number n_H of water–water hydrogen bonds on the surface of a flexible lysozyme molecule and in the rigid lysozyme powder shown as functions of N_w (number of water molecules per lysozyme) and hydration level h (data from [630]).

Water in low-hydrated biosystems

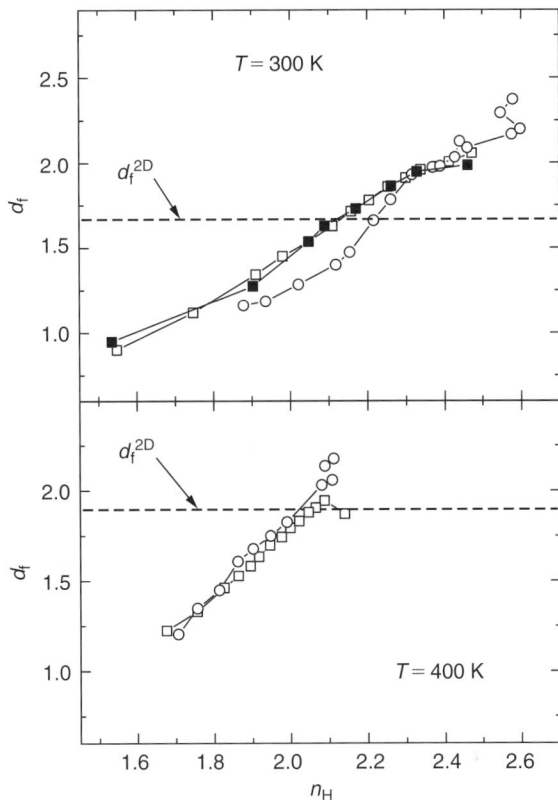

Figure 109: Fractal dimension of the largest cluster d_f as a function of the average number n_H of H-bonds between water molecules at the surface of rigid (open squares) and flexible (solid squares) lysozymes and in the hydrated lysozymes powder (open circles). Reprinted, with permission, from [631].

powder (Fig. 109, upper panel). An increase of temperature to $T = 400$ K reduces the threshold value of n_H to ≈ 2.03, i.e. by about 15%. This trend corresponds to the growing importance of the "bond percolation" relative to the "site percolation" with increasing temperature in site-bond percolation of water. Note that the reductions in n_H is accompanied by a general increase of the hydration level, where the percolation transition occurs in the studied systems.

Very similar conclusions were arrived at for water percolation threshold at the smooth planar surfaces [394, 631]. At $T = 425$ K, percolation threshold of hydration water occurs when $n_H \approx 2.22$. A formation of percolating water network at the curved spherical surfaces needs higher hydration level than at the planar surface and, accordingly, n_H at the percolation threshold becomes slightly lower ($\approx$ 2.15–2.11, depending on curvature). Increasing temperature decreases n_H at the percolation threshold, although less than at the surfaces of biomolecules [631].

About 2.0 water–water H-bonds are necessary to create a spanning network of hydration water around B- and A-DNA molecules at ambient temperature [621]. The lower value of n_H in comparison with lysozyme systems obviously reflects the trend of water clustering toward three dimensionality. Indeed, a 2D percolation threshold of water in binary mixtures close to ambient temperature occurs when $n_H \approx 1.80$ [100]. Note that 3D percolation transition in neat supercritical water occurs when $n_H \approx 1.78$ [24].

A universal structure of the largest water cluster close to the percolation threshold is one more phenomenon that originated from the universality of the percolation transitions in various systems. We have already shown that the largest cluster of hydration water at the percolation threshold may be described by a universal fractal dimension d_f^{2D} at the smooth planar and spherical surfaces, surfaces of rigid and flexible lysozyme molecules and in lysozyme powder. The local structure of the largest water cluster may be characterized by the oxygen–oxygen pair correlation function $g_{O-O}(r)$, calculated within the largest cluster. The maximum of $g_{O-O}(r)$ at about 5.4 Å, seen in all systems, indicates the domination of chain-like structure in the largest water cluster in a wide range of hydrations see Fig. 110). Note that peak at 5.4 Å, indicating the well-developed chain-like and polygon-like arrangement of water molecules, dominates in the surface water also at higher hydrations, when surface is covered by one complete monolayer, two monolayers, or many layers of water [208]. Approaching the percolation threshold, the largest water cluster appears as a rarefied network that grows via attachments of more and more molecules (or small clusters) without noticeable changes in the internal structure of the network. Such scenario was also observed in aqueous solutions and in supercritical water [24, 25].

Change in topology of the largest cluster near the percolation threshold may be studied through the evolution of the average number of hydrogen-bonded neighbors of water molecules calculated within the largest cluster only, n_H^{max}. A relatively slow increase in n_H^{max} upon hydration near planar surfaces and on the surface of flexible lysozyme is evident from Fig. 111.

Water in low-hydrated biosystems   193

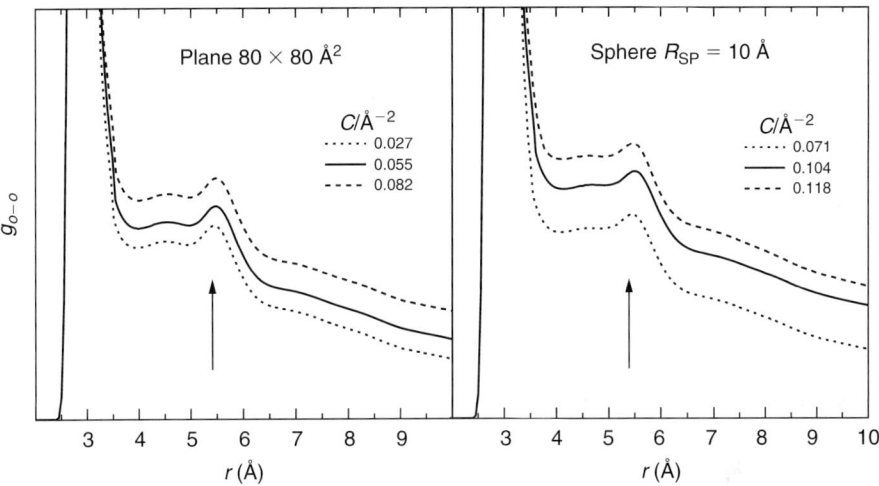

Figure 110: Water oxygen–oxygen pair correlation function $g_{O-O}(r)$ calculated for the members of the largest cluster of hydration water near planar and spherical surfaces. Surface coverage is shown in legends. The maxima at $r \approx 5.4$ Å, reflecting the chain-like structures, are indicated by arrows. Reprinted, with permission, from [398].

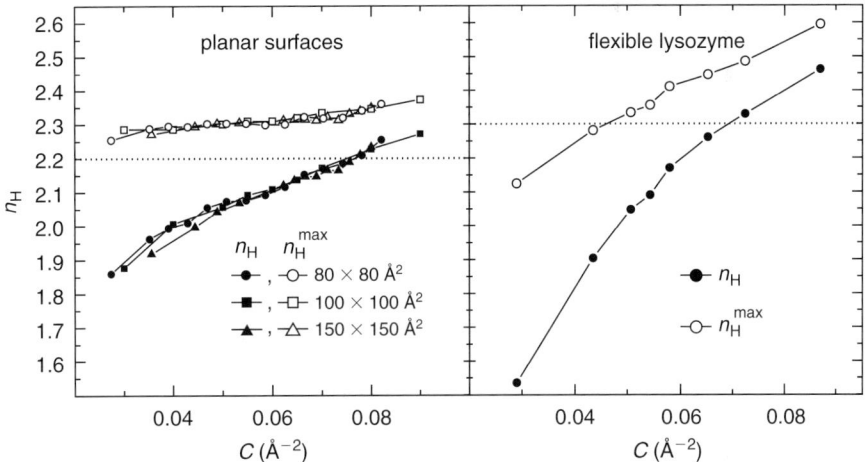

Figure 111: Average number of hydrogen-bonded neighbors of a water molecule calculated for all water molecules (n_H, filled symbols) and for molecules within the largest cluster only (n_H^{max}, open symbols) near planar surfaces (left panel) and on the surface of a flexible lysozyme molecule (right panel.) Values n_H at the percolation thresholds are indicated by horizontal dotted lines (data from [398]).

At the planar surface, n_H^{max} increases upon hydration by about factor of 4 faster than n_H. On the biological surfaces, the difference between n_H^{max} and n_H is less impressive but remains qualitatively the same (see Fig. 111 and [621] for the case of lysozyme and DNA, respectively).

So, experimental and simulation studies show the formation of a spanning network of hydration water in various biosystems with increasing water content via the percolation transition. Analysis of the various properties of water clusters allows localization of the percolation threshold and characterization of the properties of the largest (spanning) water cluster. In the next section, we consider how increasing hydration level and appearance of a spanning water network affect the properties of hydrated biosystems.

7.2 Effect of hydration on the properties of biosystems

Strong changes in the various properties of biosystems occur with increasing hydration, especially in the hydration range corresponding to the formation of a spanning water network. In this hydration range, biological function starts to develop (Section 6). To understand the role of a spanning network of hydration water in biological function, it is reasonable to analyze various physical properties of water and biomolecules below the percolation threshold and above the percolation threshold of hydration water. In this section, we analyze various properties of a hydrated biosystems at hydration levels in the vicinity of the percolation threshold of water.

Effect of hydration on the properties of biosystems was extensively studied both experimentally and by computer simulations. We have already considered how biological activity and conformational dynamics of hydrated biomolecules (Section 6) as well as conductivity of biosystems (Section 7.1) develop upon hydration. Now we analyze some other physical properties of hydrated biosystems (first, their dynamical properties) in relation to the percolation transition of water. Typical biomolecular surface is characterized by heterogeneity (presence of strongly hydrophilic and strongly hydrophobic groups), roughness, and finite size (closed surface of a single biomolecule). These features determine several steps in the process of hydration of biomolecules.

The properties of lysozyme at various hydrations and percolation transition of hydration water were studied in most details (see Section 6, 7.1 and [473, 508–510, 512–515, 544, 585, 590–592, 601, 632–634, 636–639]). At low hydrations and up to $h \approx 0.07$ g/g (gram of water per gram of protein), water molecules are adjusted mostly to the charge groups of lysozyme and most protein motions are frozen. Rotational dynamics of methyl groups is observed at very low hydration, and it seems to be rather insensitive to the hydration level and temperature. With increasing hydration, water molecules hydrate the polar groups and form larger water clusters. Formation of a spanning network of hydration water causes a rapid increase in the proton conductivity in agreement with percolation theory [592]. Light and neutron scattering experiments show a sharp stepwise increase of the fast relaxation process at hydration range h between 0.1 g/g and 0.15 g/g, which was attributed to the rattling of residues in the cages formed by their neighbors [512, 513]. It is not clear whether this effect is related to the water percolation transition at $h \approx 0.15$ g/g due to the large interval between the hydration levels studied. These experiments suggest that sharp increase in the fast conformational fluctuations activates large-scale slow protein motions, which correlate well with the enzymatic (catalytic) activity [473, 508, 510]. Experimental studies of hydrated lysozyme powder [508, 509] indicate another important hydration level corresponding to the complete monolayer coverage of each lysozyme molecule (at $h \sim 0.38$ g/g in lysozyme powders). Below this hydration level (at about $h \sim 0.25$ g/g), all lysozyme molecules are covered with water, but water shells are shared between two or more lysozymes. Above the one monolayer coverage, the full internal motions of protein are recovered, although the characteristic time scales are slower than at the infinite dilution. Note that qualitatively similar changes of lysozyme dynamics are observed when dehydration is achieved by substitution of water by cosolvents [514–516].

The detailed studies of the percolation transition of hydration water in model biosystems (Section 7.1) makes it possibile to consider various physical properties of these systems below and above the percolation threshold. The total MSD $(\langle r^2 \rangle)$ of water at the surfaces of rigid and flexible lysozyme molecules continuously increases upon hydration (Fig. 112). Similar behavior was observed in the simulation studies of water near the surface of differently hydrated plastocyanin [640, 641].

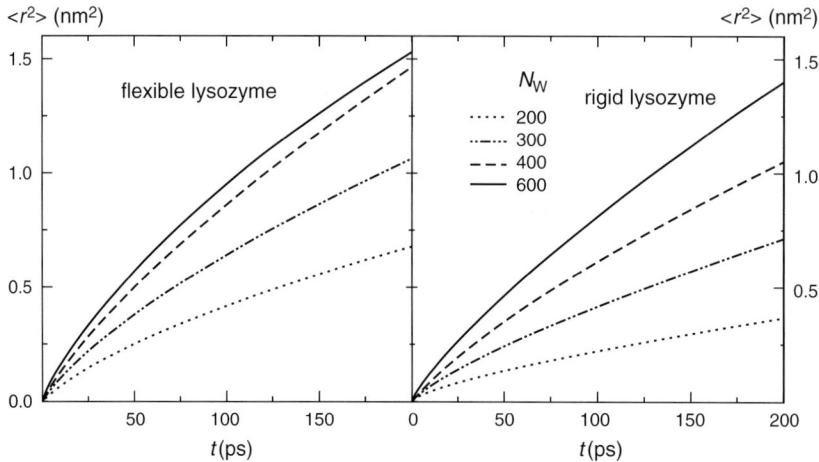

Figure 112: Total MSD $\langle r^2 \rangle$ of water molecules on the surfaces of the flexible and rigid lysozyme molecules at various hydrations N_w shown in legend (data from [630]).

Translational mobility of water on the surface of a flexible lysozyme is noticeably higher than on the surface of a rigid lysozyme. This difference is about a factor of two at low hydrations, and it progressively vanishes at higher hydration levels. Considerable enhancement of water translational motion at low hydrations is obviously caused by the motions of the surface groups of a flexible lysozyme molecule. This effect diminishes at higher hydrations when the role of water–water interactions in translational motion of water molecules becomes more important.

The time dependence of $\langle r^2 \rangle$ is essentially nonlinear at all hydration levels studied. Translational motion of water molecules in such complex system as low-hydrated biomolecules is determined by the following factors: restriction of the motions in the direction normal to the protein surface; restriction of the motion due to the finite size of a biomolecule; spatial disorder due to fractal-like structure of diffusion pathway; temporal disorder due to the presence of the strongly attractive sites on the surface. Relative importance of these factors depends on the time and length scales considered, on the properties of a biomolecule and on the hydration level. In pores, MSD of molecules normally to the pore wall (axis) nonlinearly increases at short times and achieves saturation at longer times. As a result, the time dependence of the total MSD is nonlinear at short times and becomes linear only when displacements essentially exceed pore width (diameter). Due to the same reason, total MSD of water molecules adsorbed on the surface of a single biomolecule (or the surface of any other finite object) cannot exceed some maximal value and achieves saturation at long times. At shorter times, when $\sqrt{\langle r^2 \rangle}$ is essentially smaller than the size of a biomolecule, $\langle r^2 \rangle$ varies with time t in accordance with a power law

$$\langle r^2 \rangle \sim t^\alpha, \qquad (25)$$

where the exponent $\alpha = 1$ for diffusion in homogeneous media (normal diffusion) and $\alpha < 1$ for diffusion in inhomogeneous media (anomalous diffusion). Anomalous diffusion of water molecules on the surfaces of various biomolecules originates from the strong spatial variations of the surface–water interaction, and it was seen in some experiments and simulations [640, 641, 643, 644].

The double-logarithmic plot of the time dependences of the total MSD $\langle r^2 \rangle$ for water molecules on the surfaces of flexible and rigid lysozyme molecules (Fig. 113) evidences that $\langle r^2 \rangle$ follows equation (25) in a wide range of hydrations. In the time interval 10 to 100 ps, the dependencies $\langle r^2 \rangle(t)$ for $N_w > 300$ may be well fitted to equation (25) with exponent $\alpha = 0.775 \pm 0.010$ for the flexible lysozyme and $\alpha = 0.793 \pm 0.010$ for the rigid lysozyme. The values of these exponents do not depend on the hydration level. Independence of the obtained values of the exponent α on hydration level indicates that the anomalous diffusion is caused mainly by the spatial disorder in the system.

Noticeable deviations of $\langle r^2 \rangle(t)$ from the equation (25) are seen at low hydrations, where effective value of α continuously decreases at $t < 10$ ps. These deviations should be attributed to the water molecules, which are strongly bound to lysozyme surfaces (there is about 36 water molecules, having two or more hydrogen bonds with lysozyme molecule [635, 636]). The total MSD of such water molecules quickly achieves saturation during their rather long residence times. So, the simulations indicate the presence of two main classes of water molecules with respect to the translational motion: molecules with short residence times, which show anomalous diffusion due to the spatial disorder already at the short times, and molecules with long residence times, which remain bound to some centers on lysozyme surface during hundreds of picoseconds.

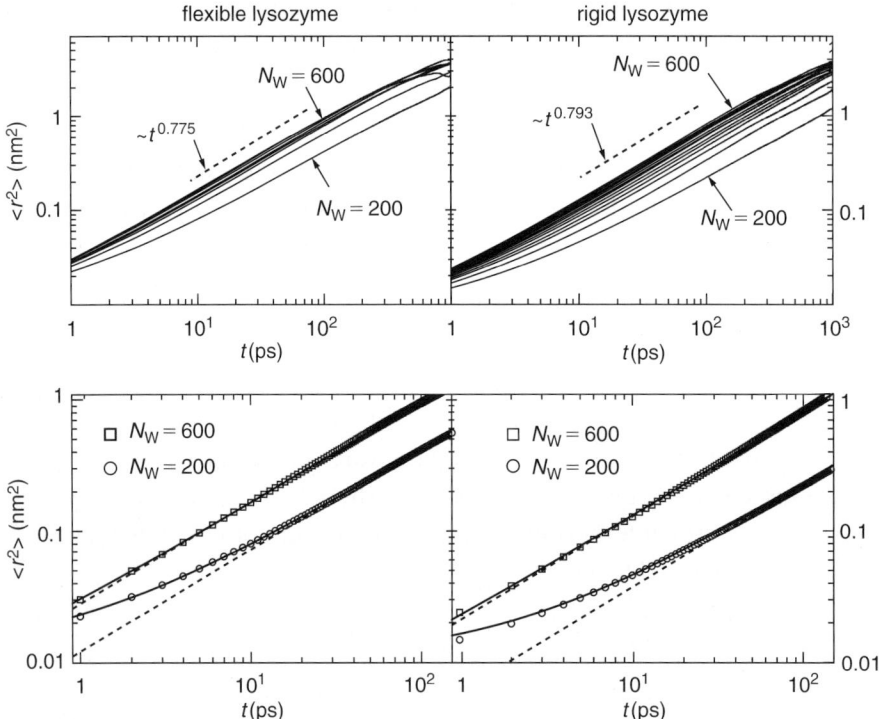

Figure 113: Time dependence of MSD $\langle r^2 \rangle$ of water molecules at surfaces of the flexible (left panels) and rigid (right panels) lysozyme molecules in double-logarithmic scale. Upper panels: simulated data (solid lines); power laws corresponding to the anomalous diffusion (equation (25)) with different values of α (dashed lines). Hydration increases from the bottom to the top. Lower panels: simulated data (open symbols); fits to equation (25) and to equation (26) are shown by dashed and solid lines, respectively (data from [630]).

Contribution of the latter molecules to the average MSD may be considered a constant in the first approximation, and the total MSD may be described by the equation

$$\langle r^2 \rangle \sim t^\alpha + \text{const.} \tag{26}$$

Fits of the time dependence of the total MSD to equations (25) and (26) for each hydration level indicate that equation (26) essentially better reproduces $\langle r^2 \rangle(t)$ at short t for low-hydrated systems (Fig. 113, lower panels). Relative contribution of the strongly bound water molecules to the total MSD decreases with increasing hydration level (compare the data for $N_w = 200$ and $N_w = 600$). When time t exceeds residence times of strongly bound water molecules, effect of spatial and temporal disorders cannot be distinguished, and equation (26) with presumably lower value of the exponent α should be valid. The diffusion of water is anomalous at the surfaces of both rigid and flexible lysozymes. This behavior is observed in time scale from few picoseconds up to 100–200 ps. At longer time scale, the finite size of a lysozyme molecule leads to the saturation of water MSD, which is typical for translational diffusion in confined systems [249, 645, 646]. The latter effect is noticeable when the total MSD exceeds ~1 nm^2, which corresponds to the linear displacement on about half of the linear size of lysozyme. The anomalous diffusion of water near lysozyme surface in a wide range of hydrations is well described by the power law with exponent α of about 0.78. A slightly lower value of α (~0.5) was obtained for translational motion of water near in low-hydrated silica pore [647]. Note that comparable values of the exponent α were obtained for hydration water in simulation studies of fully hydrated systems (see Section 9). Value of $\alpha \approx 0.6$ was observed in the first water layer near fully hydrated lysozyme [610]. Note that α in the surface water layer found from the simulations of fully hydrated systems strongly depends on the definition of hydration layer [640, 644] (see also Section 8). The value of α at about 0.40 was obtained for water diffusion at the surface of a low hydrated myoglobin molecule from experiment [643].

A qualitatively similar anomalous diffusion of water was observed in low-hydrated DNA systems [642]. MSD of water molecules evidences a power-law time dependence in the time interval up to 100 ps (Fig. 114). Fits of $\langle r^2 \rangle$ to equation (25) in this range is shown in Fig. 114. The exponent α of anomalous diffusion obtained in such a way is about 0.8, and it is almost independent on hydration Γ (Fig. 115, squares). Assuming the presence of strongly bound water molecules, which remain immobile during hundreds of picoseconds, and taking into account the cage effect, simulated data may be described by equation (26) with fitting values of α shown in Fig. 115, (open circles). Such approach yields slightly higher value of $\alpha \approx 0.835$ in the whole hydration range studied. Note that very similar α may be obtained from the fit of $\langle r^2 \rangle$ to equation (25) in shorter time interval: $3 \text{ ps} < t < 40 \text{ ps}$ (Fig. 115, solid circles).

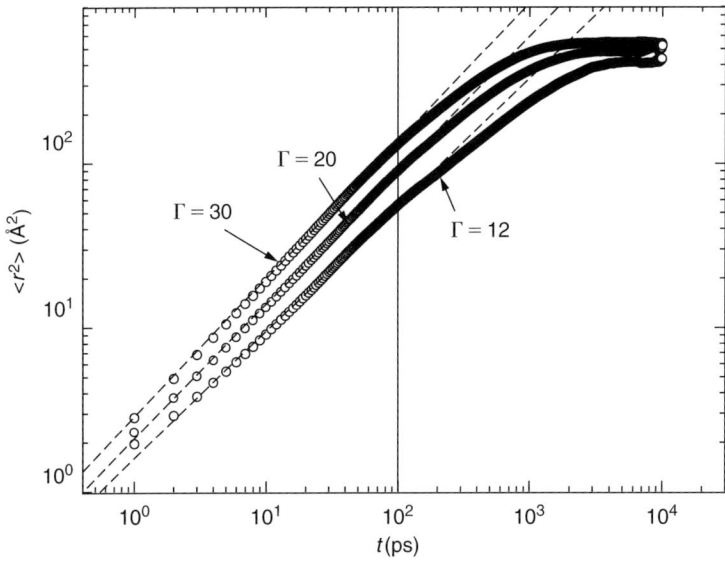

Figure 114: Time dependence of the MSD $\langle r^2 \rangle$ of water molecules on the surfaces of DNA molecule at various hydrations Γ shown in double-logarithmic scale (symbols). The time interval ($t < 100$ ps), where effect of the finite size of DNA molecule is negligible, is bounded by a vertical line. Fits to the equation (25) in this interval are shown by dashed lines (data from [642]).

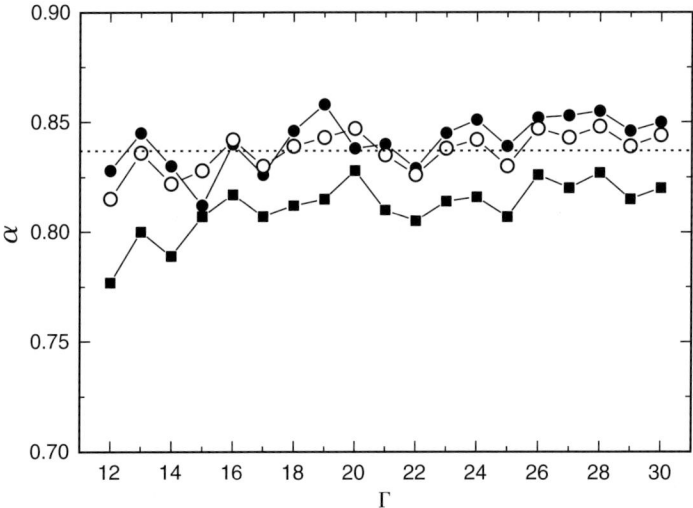

Figure 115: The exponent α of anomalous diffusion found from the fits of $\langle r^2 \rangle$ in the time interval $1 < t(\text{ps}) < 100$ to equation (25) (solid squares) and equation (26) (open circles), and in the interval $3 < t(\text{ps}) < 40$ to equation (25) (solid circles). Average value $\alpha = 0.837$ is shown by horizontal dotted line (data from [642]).

So, in partially hydrated biosystems (proteins and DNA), exponent α for water diffusion is almost independent of hydration level. This behavior can be hardly expected when anomalous diffusion is caused by the temporal disorder. Presence of a strongly adsorbing sites at the surface of a biomolecule, which provide long residence times for water molecules and facilitate the cage effect, affects time dependence of the MSD at short times only and decreases the effective value of α at this time scale. At low hydrations, it can be approximately taken into account by adding some constant to the time dependence of the MSD (equation (26)). This effect quickly disappears with increasing hydration level because the contribution to MSD of strongly bound molecules becomes progressively less important.

In case of anomalous diffusion, diffusion rates depend on time or spatial scale considered. A quantitative analysis of the hydration dependence of the water translational mobility may be done in different ways. One may compare the MSD $\langle r^2 \rangle$ at some chosen time t or, alternatively, compare the times t that yield the same value of $\langle r^2 \rangle$. Besides, one may characterize water mobility by the time-dependent effective diffusion coefficient $D_{\text{eff}}(t)$:

$$D_{\text{eff}} = \frac{\langle (r(t+\Delta t) - r(t))^2 \rangle}{2 d \Delta t}, \qquad (27)$$

where d is Euclidean dimension of a system ($d = 3$ for the bulk water) and $t \leq t + \Delta t$ is time interval used for estimation.

Short-time water diffusion may be characterized by the effective diffusion coefficients $D_{\text{eff}}^{\text{tot}}$ and $D_{\text{eff}}^{\parallel}$ calculated by equation (27) for the total displacement $\langle r^2 \rangle$ and for the displacement $\langle (xy)^2 \rangle$ parallel to the surface in the time interval 5 ps $< t <$ 15 ps, using $d = 3$ and $d = 2$, respectively. $D_{\text{eff}}^{\text{tot}}$ and $D_{\text{eff}}^{\parallel}$ gradually increase with hydration (Fig. 116). The values of $D_{\text{eff}}^{\text{tot}}$ are noticeably lower than $D_{\text{eff}}^{\parallel}$ due to the strong confining effect of a boundary on the displacement in z direction (normally to the surface). This effect is absent for the displacements parallel to the surface and $D_{\text{eff}}^{\parallel}$ may be compared with the self-diffusion coefficient of bulk water $D \approx 4.2 \cdot 10^{-9} m^2 s^{-1}$ for the water model studied [648] (see horizontal line in the left panel in Fig. 116). However, such comparison suffers from the fact that coefficient $D_{\text{eff}}^{\parallel}$ is obtained from equation (27), which assumes linear time dependence of MSD, whereas hydration water shows anomalous diffusion.

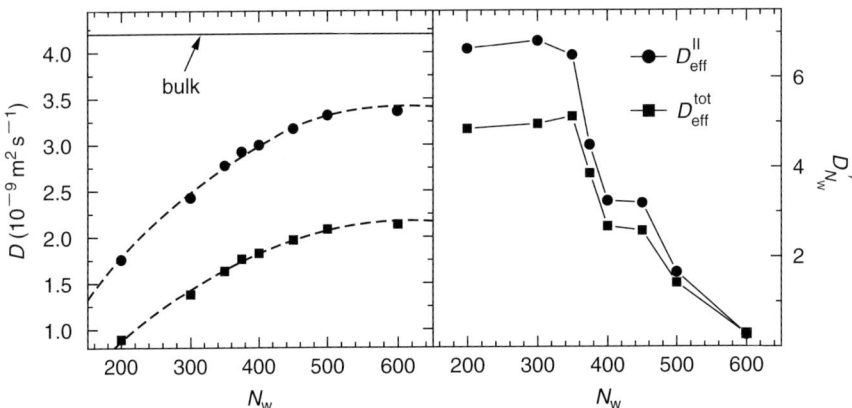

Figure 116: Effective diffusion coefficients $D_{\text{eff}}^{\text{tot}}$ and $D_{\text{eff}}^{\parallel}$ of water molecules at the surface of a flexible lysozyme molecule (left panel) and their derivatives with respect to N_w (right panel). Polynomial fits are shown by the dashed lines (data from [630]).

Dependence of water translational mobility on hydration may be further analyzed using the derivatives of MSD or $D_{\text{eff}}^{\text{tot}}$ and $D_{\text{eff}}^{\parallel}$ with respect to hydration level N_w (Fig. 116, right panel). They show a rapid drop in the hydration level $N_w \approx 380$ corresponding to the midpoint of the percolation transition and approaches zero above the percolation threshold ($N_w > 475$). Starting from the midpoint of the percolation transition, translational mobility of water correlates with the average size of water clusters, size of the maximal cluster, and spanning probability, which show roughly similar behavior in the considered range of hydrations. This correlation may be seen when water mobility is plotted as a function of the normalized average cluster size $S_{\text{av}}^* = S_{\text{av}}/N_w$. For water on the surface of a flexible lysozyme, there is a linear correlation between mobility and S_{av}^* in a wide hydration range. For example, this correlation is seen when water mobility is characterized by the inverse time t^{-1} corresponding to the total MSD $\langle r^2 \rangle = 0.1$ nm^2 and $\langle r^2 \rangle = 1$ nm^2 (Fig. 117, left panel). Note that in the case of a rigid lysozyme, trend of water mobility toward saturation with approaching percolation threshold is much less pronounced and its correlation with cluster properties, such as S_{av}^*, is questionable [630]. This evidences importance of protein motions for mobility of hydration water.

Obviously, the fast translational motions of water molecules should reflect mainly a *local* environment of water molecule. This local environment

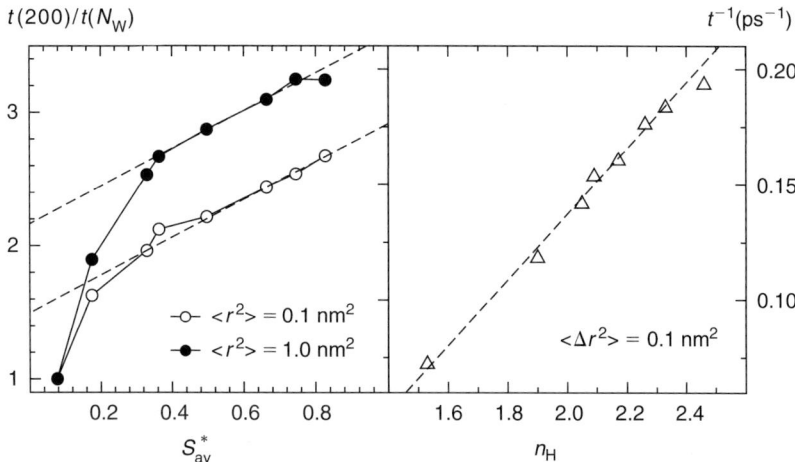

Figure 117: Water mobility on the surface of a flexible lysozyme. Left panel: normalized inverse time $t(200)/t(N_w)$ corresponding to the two total MSD $\langle r^2 \rangle = 0.1$ nm^2 and $\langle r^2 \rangle = 1.0$ nm^2 as function of the normalized average cluster size S^*_{av}. Linear dependences are shown by dashed lines. Right panel: inverse time t^{-1} corresponding to the total MSD $\langle r^2 \rangle = 0.1$ nm^2 as a function of the average number n_H of H-bonded neighbors. Linear fit is shown by dashed line (data from [630]).

may be divided into two parts: related to water–protein and water–water interactions. The latter part may be characterized, for example, by the average number n_H of H-bonded neighbors. Indeed, the short-time mobility of water on the surface of a flexible lysozyme estimated in different ways varies almost linearly with n_H in the whole hydration range studied (Fig. 117, right panel). Behavior of the short-range water mobility on the surface of a rigid lysozyme is qualitatively similar. However, contrary to the flexible lysozyme, the correlation between water mobility and n_H remains linear also for larger displacements.

Diffusion of water in the hydration shell of DNA molecule depends on hydration Γ, similarly as at the lysozyme surface. The effective diffusion coefficient calculated by equation (27) shows a gradual increase with hydration, which is followed by a weak trend toward saturation (Fig. 118, left panel). The short-range mobility and long-range mobility calculated in the time intervals 5 ps $< t <$ 15 ps and 100 ps $< t <$ 300 ps, respectively, depend on hydration in a rather similar way. However, the strong fluctuations of the short-range mobility are seen in the hydration range

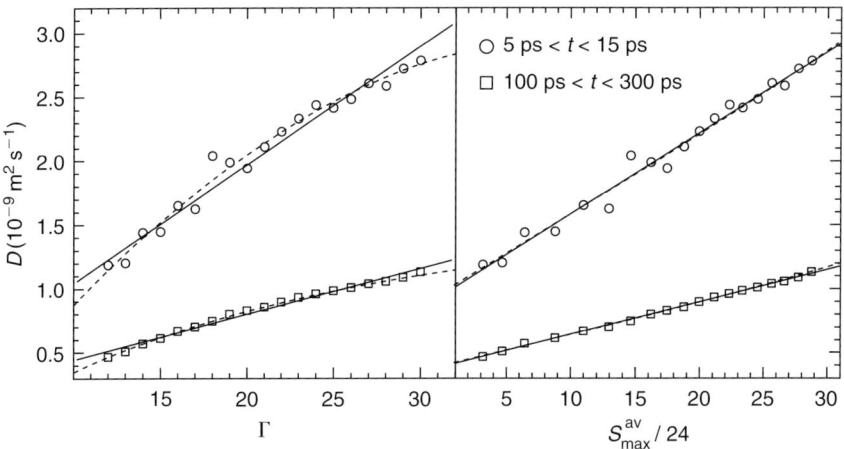

Figure 118: Effective diffusion coefficients of water on the surface of DNA shown as functions of hydration level Γ (left panel) and of average size S_{max}^{av} of the largest water clusters (right panel) estimated for short-time and long-time mobility in the time intervals shown in legend. Linear and second-order polynomial fits are shown in both panels by solid and dashed lines, respectively (data from [649]).

immediately above the percolation threshold ($\Gamma > 15.5$). A good linear correlation between water mobility and the average size of the largest water cluster S_{max}^{av} is observed for all ranges considered (Fig. 118, right panel).

Obviously, dynamics of a biomolecule and dynamics of its hydration water should be coupled. For example, pressure-induced dynamic transition in a crystalline Snase causes similar qualitative changes of pressure dependence of the MSD of both water molecules and hydrogens of Snase molecule [611]. So, comparison of water dynamics at rigid and flexible lysozymes may give insight into the dynamics of a lysozyme molecule. The coupling of water and lysozyme dynamics may be characterized by the difference $\Delta(t^{-1})$ between the mobilities of water on the surfaces of the flexible and rigid lysozyme molecules; in latter case this coupling is intrinsically absent. Variation of $\Delta(t^{-1})$ with hydration is not sensitive to the time/length scale considered [630]. $\Delta(t^{-1})$ rapidly increases up to hydration level $N_w = 350$–370 and rapidly falls at N_w of about 500. So, the hydration-induced enhancement of lysozyme dynamics is maximal in the hydration range, bounded by the appearance of a spanning water network from below and by the percolation threshold from above.

This picture agrees with the available experimental analysis of the effect of hydration on lysozyme dynamics [473, 508–510, 512, 513]. Namely, internal dynamics of lysozyme molecule is restored when it is covered by some minimal amount of hydration water. Note that correlation between the percolation transition of water and pressure-induced dynamic transition of protein molecules was also observed in simulations of crystalline Snase [612].

A strong retardation of the rotational motion of water is usually observed near surfaces of biomolecules. The rotational motion of water is usually analyzed through the reorientational dynamics of its electrical dipole defined as the vector pointing from the water oxygen to the middle point of the two hydrogen atoms. Two first- and second-rank autocorrelation functions Γ_1 and Γ_2 are the time average of the Legendre polynomials $P_n(\cos(\theta))$:

$$\Gamma_1 = \langle P_1(t) \rangle = \langle \cos \theta(t) \rangle, \tag{28}$$

$$\Gamma_2 = \langle P_2(t) \rangle = \langle 3\cos^2 \theta(t) - 1 \rangle, \tag{29}$$

where θ is the angle between dipole orientation at time t and its initial orientation. Both Γ_1 and Γ_2 decay faster toward zero at higher hydration levels.

At low hydrations, the decay of the autocorrelation function Γ_1 may be described by the Kohlrausch–Williams–Watts stretched exponential equation [650, 651]:

$$\Gamma_1 = \exp\left(-\left(\frac{t}{\tau_1}\right)^\beta\right). \tag{30}$$

For example, time dependence of Γ_1 for 200 water molecules on the surface of a flexible lysozyme may be well fitted by equation (30) with $\beta = 0.325$ and $\tau_1 \approx 48$ ps (see upper dashed line in Fig. 119). At higher hydration levels, decay of Γ_1 cannot be described by one-term equation (30) (see lower dashed line in Fig. 119 for $N_w = 600$). It was found at all hydrations studied that Γ_1 may be well described by two-term equation, which includes besides stretched exponential a simple Debye decay:

$$\Gamma_1 = (1-a)\exp\left(-\left(\frac{t}{\tau_1}\right)^\beta\right) + a\exp\left(-\frac{t}{\tau_2}\right). \tag{31}$$

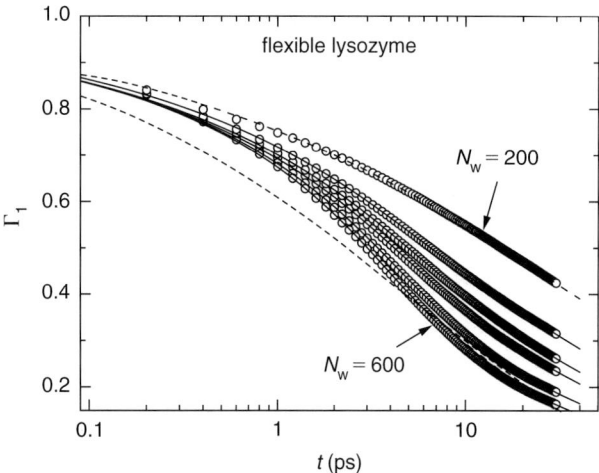

Figure 119: First-rank dipole–dipole autocorrelation function Γ_1 of water on the surface of a flexible lysozyme at selected hydration levels. The fits of stretched exponential equation (30) to Γ_1 for the lowest and highest hydrations studied are shown by dashed lines. The fits of the data to the two-term equation (31) are shown by solid lines (data from [630]).

Equation (31) assumes an existence of two kinds of molecules showing quite different reorientational dynamics in the considered time interval. The water molecules with strong retardation of rotational motion and with broad distribution of relaxation times (low value of β) should be considered *strongly bound*. Such kind of water molecules with similar low values of the stretching exponent were observed in neutron-scattering experiment for combined rotational–translational motion of water in hydrated myoglobin ($\beta \approx 0.3$) [643] and in simulations of water near mica surface ($\beta \approx 0.25$) [652]. Those water molecules, which show simple one-term exponential relaxation, like in the bulk, should be considered as *weakly bound*. Fractions of these two kinds of molecules ($(1-a)$ and a, respectively) should depend on the hydration level. Note, that weakly bound water molecules with Debye rotational relaxation were not distinguished in other simulation studies [610, 644].

The values of the relaxation time τ_1, obtained from the fitting of Γ_1 to equation (31) with the stretching exponent β fixed at 0.33, are shown in Fig. 120. τ_1 continuously decreases with hydration, and this decrease is much steeper at the surface of a rigid lysozyme because of much slower rotational relaxation of water at low hydrations.

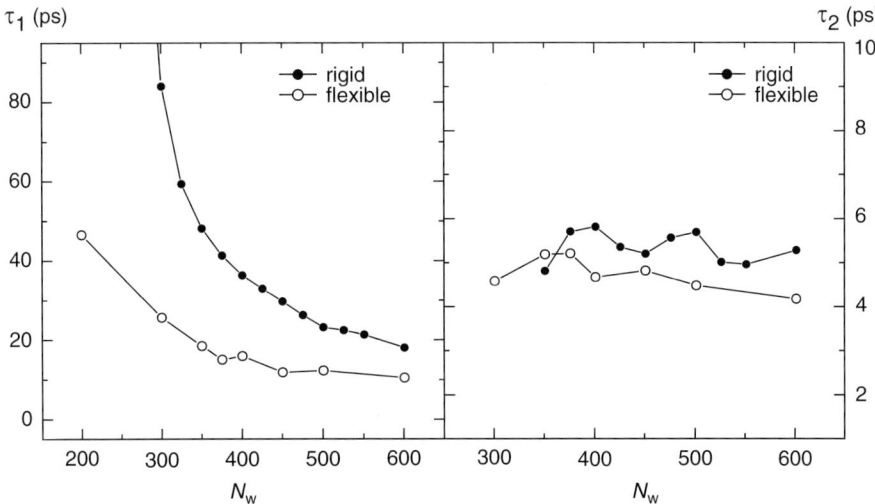

Figure 120: Relaxation times τ_1 and τ_2 found from the fits of equation (31) to the first-rank dipole–dipole autocorrelation function Γ_1 of water on the surface of flexible and rigid lysozyme molecules at various hydrations N_w (data from [630]).

The rotational relaxation may be also characterized by the second-rank correlation function Γ_2, which shows faster decay than Γ_1. Accordingly, the relaxation times τ_1 and τ_2 found from the fits of Γ_2 to equation (31) are lower at all hydration studied [630]. The values of stretching exponent β derived from the fits of the second-rank correlation function to equation (31) were found independent of hydration and equal to ≈ 0.245 for the rigid and ≈ 0.225 for the flexible lysozyme molecules.

In contrast to the stretched exponential relaxation time, the Debye relaxation time τ_2 does not vary with hydration and stays around $\tau_2 = 4.7 \pm 0.4$ ps and 2.5 ± 0.6 ps for Γ_1 and Γ_2, respectively (Fig. 120). These values are noticeably larger than the bulk values for the same water model (2.5 and 0.9 ps [648]). The parameter a in equation (31) reflects the fraction of these weakly bound water molecules with Debye rotational dynamics. The amplitude a increases with hydration level, as it is shown in Fig. 121. At low hydrations, a is negligibly small and therefore cannot be estimated from the fits with a reasonable accuracy. At the surface of a rigid lysozyme, we have detected the appearance of the water molecules with Debye-like rotational dynamics only when $N_w > 300$. On

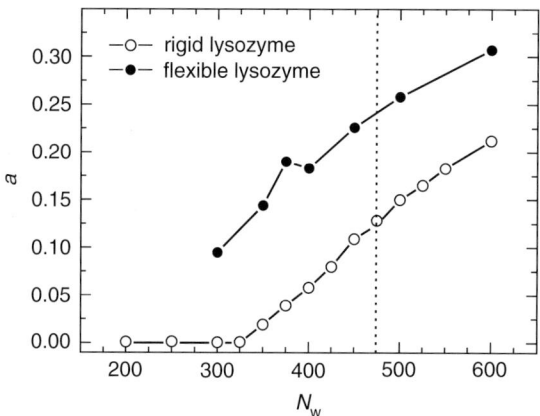

Figure 121: Amplitude a of the Debye term in equation (31), when it is fitted to the first-rank dipole–dipole autocorrelation functions Γ_1 of water on the surface of flexible and rigid lysozymes at various hydrations N_w. Water percolation threshold is shown by the vertical dotted line (data from [630]).

the flexible surface, such water molecules appear at much lower hydration (linear extrapolation of fraction a (N_w) yields $a = 0$ at $N_w \approx 130$), and their fraction is about 20 to 25% at the percolation threshold (see Fig. 121).

Rotational motion of water molecules determines the dielectric properties of hydrated biosystems, in general, and of a lyzozyme powder, in particular [633, 635, 653, 654]. The dielectric increment $\Delta\epsilon$ originates from the reorientation of water dipoles in the presence of external electric field. At a given frequency ν of external electric field, only dipoles with relaxation times $< 1/(2\pi\nu)$ contribute to $\Delta\epsilon$. For example, at the frequency $\nu = 10$ kHz, all water molecules (with an exception of irrotationally bound to protein) contribute to $\Delta\epsilon$. Whereas at high frequency (10 GHz), only a weakly bound water responds to the electric field and strongly bound water molecules may be considered being "frozen." A wide distribution of relaxation times of hydration water and protein and coupling of their dynamics make dielectric spectra of low-hydrated protein systems rather complicate. Nevertheless, a strong increase in a dielectric increment and dielectric constant is observed when water motion becomes strongly correlated upon formation of a large hydrogen-bonded network when approaching the water percolation threshold [630].

Water in low-hydrated biosystems

Diffusion of ions near biological surface has many common features with translational motion of hydration water. In general, the $\langle r^2 \rangle(t)$ traces of ions are qualitatively similar to water plots (see Fig. 122), with much lower absolute migration rates. MSD of ions are naturally limited by the size of hydration shell and with $t \to \infty$ plots $\langle r^2 \rangle(t)$ should reach some plateau at any finite hydration level. Due to the low mobility, ion displacements become comparable with the size of DNA molecule for much larger time intervals than simulation run, and in a large time interval, MSD time dependence is essentially nonlinear due to the temporal and/or spatial disorder. In this regime, referred to as anomalous diffusion, $\langle r^2 \rangle(t)$ shows a power-law behavior (equation (25)) with the exponent α smaller than 1. For diffusion of Na$^+$ ions in hydration shell of DNA molecule, anomalous diffusion is observed in the time interval from $\sim$ 10 ps to at least 1000 ps (Fig. 122). In the time interval less than 10 ps, notable deviations of MSD of ions from power law (25) (dashed lines in Fig. 122) may have the same nature, as in case of water diffusion. Accordingly, $\langle r^2 \rangle(t)$ of ions may be perfectly described by equation (26) (solid lines in Fig. 122).

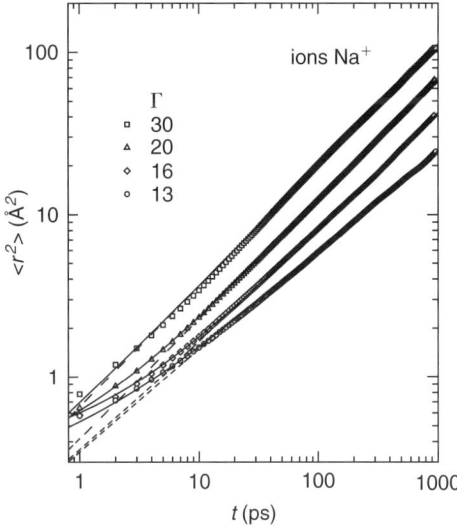

Figure 122: Time dependence of the MSD $\langle r^2 \rangle$ of Na$^+$ ions in the hydration shells of DNA in a double-logarithmic scale (symbols). Fits of the data in the range $1 < t(\text{ps}) < 1000$ to equations (25) and (26) are shown by dashed and solid lines, respectively (data from [642]).

Change of the fitting parameters in equation (26) with hydration is shown in Fig. 123. The exponent of α of anomalous diffusion of ions demonstrates sigmoid-like behavior with hydration. An inflection point of this dependence, which indicates transition between two regimes, is remarkably close to the percolation threshold of water at $\Gamma = 15.5$. Below and above this hydration level, different temporal/spatial disorder is probed by Na^+ ions in the hydration shell of DNA. When the hydration shell is dispersed in a large number of small clusters below the percolation threshold, α is low (at about 0.65). When the hydrogen-bonded network of water exists above the percolation threshold, α approaches 0.80. So, spanning water network "smooths" a DNA surface for translational motion of ions.

Effect of hydration on the ion mobility may be examined using effective diffusion coefficient or other related characteristics, such as a dimensionless relative mobility L obtained as an effective diffusion coefficient estimated in a given time interval and normalized by the corresponding value at $\Gamma = 12$ [200]. A linear growth of L with hydration was observed for short-range translations of ions Na^+. But for a long-range translations ($t \geq 200$ ps), ion mobility depends on hydration Γ on a very specific way (Fig. 124). From hydration $\Gamma = 12$ up to $\Gamma = 18$, L^{Na} exhibits a sigmoid

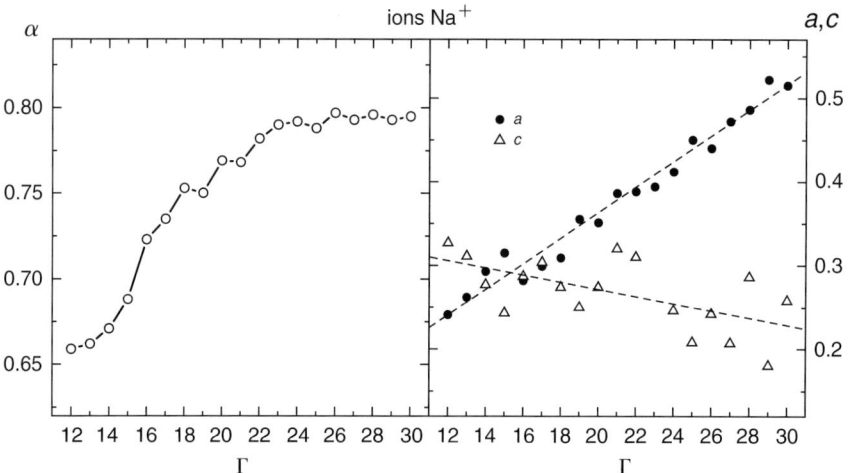

Figure 123: The exponent α and parameters a and c found from the fits of $\langle r^2 \rangle$ of Na^+ ions in the time interval $2 < t(ps) < 100$ to equation (26) (data from [642]).

Water in low-hydrated biosystems

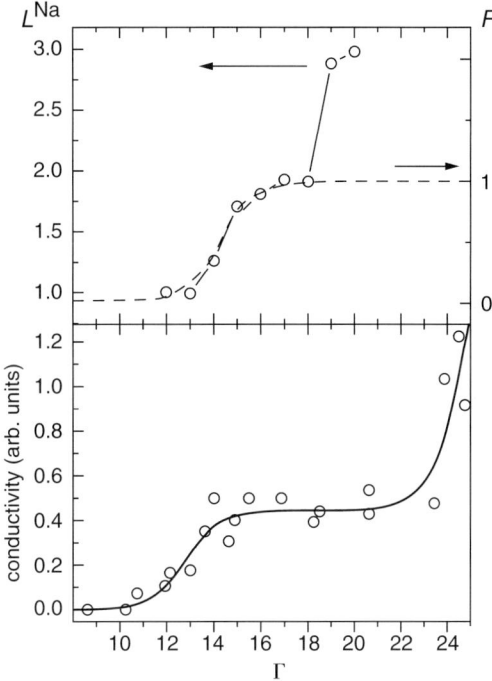

Figure 124: Upper panel: the long-range relative mobility L^{Na} of ions (open circles, left axis) and percolation probability R of hydration water (dashed line, right axis) as function of the hydration level Γ. Lower panel: dependence of the radiation-induced conductivity of hydrated DNA on the hydration level Γ [606]. Reprinted, with permission, from [200].

dependence closely following the spanning probability $R(\Gamma)$. So, the spanning H-bonded water network drastically increases the space around DNA accessible for Na^+, thus accelerating their long-range transport. Note that the magnitude of mobility acceleration shown in Fig. 124 is, in fact, a lower estimate because the hydration conditions in the minor DNA groove remain essentially unchanged and make a constant contribution to the average mobility of water and ions in the range of hydrations studied. The change of a long-range ion mobility with hydration obtained in simulations remarkably agrees with the experimentally observed dependence of the radiation-induced conductivity of hydrated DNA fibers [606] (compare lower and upper panels in Fig. 124). In the experimental curve, the sigmoid increase at about $\Gamma = 13$ is less steep because the DNA conformation changes with hydration, whereas a rigid DNA molecule was used in simulations. The importance of the hydration water network for ion transport is supported by the fact that the dynamic transition of hydrated DNA upon heating [526] is accompanied by a sharp increase in its conductivity [655].

Parameter a in equation (26) reflects effective mobility of ions which show anomalous diffusion and it increases strongly with hydration (Fig. 123, lower panel). Parameter c represents *cage effect* related to the trapping of ions in a potential well of DNA atoms. Caging of ions within a distance of about 0.5 Å (corresponding to $c \sim 0.25$ Å^2) is expected to be proper for all ions in the hydration shell. The ions strongly adsorbed at the DNA surface may stay within the cage during long time and do not leave it during hundreds of picoseconds. Therefore, cage effect is more important for overall ion mobility at low hydrations, when fraction of adsorbed ions is large.

Change of long-range mobility of ions with hydration demonstrates two steps at $\Gamma \approx 18.5$ and 23.5 (Fig. 125). An abrupt increase in the conductivity of DNA fibers at about $\Gamma = 24$ was also observed in experiment and was attributed to the charge transport via the second hydration shell [606]. In general, a mapping of the hydration levels studied in simulations of a single DNA fragment onto hydration of condensed DNA

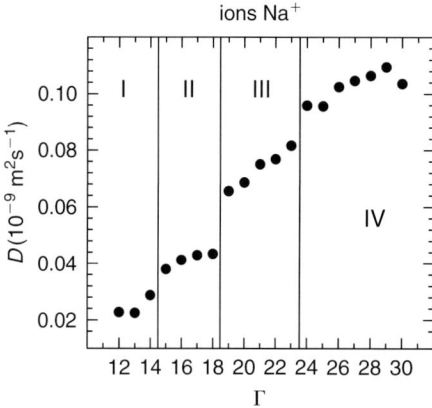

Figure 125: Long-range mobility of ions at various hydration levels characterized by the effective diffusion coefficient D calculated by equation (27) in the time interval 200 ps $< t <$ 500 ps. Vertical lines separate the regions with different behavior of D with hydration (data from [649]).

samples studied experimentally is not straighforward. It is, therefore, necessary to clarify the nature of these steps in further simulations.

An important information about the interplay between water structure and ion mobility may be extracted from the evolution of ion distributions around DNA. This distribution may be characterized by the radial distribution functions of Na^+ ions around heavy (nonhydrogen) DNA atoms. As the B-DNA surface is not uniform, so that solvation conditions of surface atoms in the minor groove change very little upon hydration, whereas the major groove hydration shell evolves. Here the spanning H-bonded water network is reversibly broken and formed via a percolation transition [10], therefore, the change of ions distribution was analyzed within the major groove of B-DNA [649]. Under hydration $\Gamma = 30$, the radial distribution function involves three peaks centered at 2.5, 4.2, and 6.5 Å. The first peak corresponds to Na^+ ions bound to negatively charged DNA atoms by strong electrostatic forces. The second peak corresponds to van der Waals contacts of Na^+ ions with neutral or weakly charged DNA atoms. The third broad peak appears under high hydrations, and it corresponds to fully hydrated Na^+ ions separated from DNA by a water molecule. Under low hydration ($\Gamma < 12$), ions are stuck to DNA and eventually form ion pairs. The relative weight of the second peak rapidly grows with $\Gamma \geq 12$ and plateaus at about $\Gamma = 15$, which is close to the percolation threshold. The drastic change in the ion radial distributions between $\Gamma = 12$ and 15 features the dissociation of ion pairs. Such dissociation may be explained by a rapid growth of the static dielectric constant of hydration water at about monolayer water coverage [632, 635, 656], or when a spanning H-bonded water network appears around a biomolecule [642]. The growth of the relative weight of the third peak of radial distributions essentially repeats that of the long-range ion mobility, shown in Fig. 125. Evidently, fully hydrated ions can move faster and cover much longer distances than ions blocked in the first DNA hydration layer. Such ions make a dominant contribution to the average long-range ion mobility.

Analysis of the properties of biosystems at various hydrations in the vicinity of the percolation transition of hydration water should clarify which particular property changes in a drastic way at the percolation threshold and makes possible the biological function. Rotational motions of water molecules change qualitatively near the percolation threshold, that affects dielectric properties of the system. This includes two factors: appearance of a water molecule with bulk-like dynamics and strong correlation of water motions upon formation of a spanning network. The latter factor causes noticeable increase in the dielectric constant of the system that provides screening of the charged groups of a biomolecule and, accordingly, promotes its dynamics. As we discussed in Section 7.1, conductivity of biosystems changes in a drastic way at the percolation threshold. This evidences that a spanning water network is an effective medium for the charge transfer along biosurfaces. Transfer of ions or protons may play some special role in biological function. Besides, a spanning water network may be an effective medium for the transfer of metabolites (see Section 6), which also makes its existence necessary for biological function. In this section, we have considered mainly properties of single water molecules and ions in hydration shells. Further studies are necessary to clarify the role of the specific properties of a spanning water *network* in a biological function.

8 States of interfacial water in fully hydrated biosystems

In dilute aqueous solutions, biomolecules are completely covered by water molecules. The structure of water near a boundary essentially differs from the structure of bulk water (see Sections 2 and 5). Specific water structure is seen in one or two water layers near hydrophilic surfaces, whereas the rest of liquid water is bulk-like. This is also the case for the surfaces of biomolecules, which allow consideration of hydration water as a separate subsystem. Conformational transitions and aggregation of biomolecules occur in dilute solution due to variations of temperature and/or pressure and due to additions of some cosolvents. It is natural to expect that these biologically important processes are related to the changes in the state of hydration water shell. First, we consider the effect of heating on the state of hydration water shell and on the properties of biomolecules. Then, we discuss the dynamic transition of biomolecules and pressure-induced denaturation in relation with the liquid–liquid transitions of hydration water.

Taking into account the presence of a spanning network of hydration water at relatively low hydration levels (Section 7.1), one may assume that such a network always spans the biomolecule in dilute aqueous solution. Most of the water molecules in the hydration shell of a biomolecule belong to the infinite H-bonded network of bulk liquid water. However, if we consider the network formed by the molecules in the first hydration shell only, this is not necessarily the case. First, in dilute solutions, water molecules from a complete second layer effectively reduce the direct interconnectivity between the molecules in the first layer of hydration water due to H-bonding between two water layers. Second, upon heating, the spanning network of hydration water will ultimately break up in some temperature interval as the number of water–water hydrogen bonds gradually decreases with increasing temperature. The spanning network of hydration water may break upon heating at some temperature or within some temperature interval. This break may affect the properties of biomolecules, and it is important to estimate the temperatures where it may be expected.

The connectivity and clustering of water molecules within the hydration shell may be analyzed in a similar way as in the case of a low-hydrated system. Such analysis requires the criteria for distinguishing water molecules in the hydration shell from the rest of the water. Various experiments yield estimations of a thickness of the hydration shell, which intrinsically depends on the properties considered. For example, terahertz spectroscopy measurements [657] evidence a hydration shell of about 5.1 Å at the surface of a lactose. In simulations, the width of a hydration shell may be estimated using water density profiles. Such profiles calculated based on the distribution of water oxygen and hydrogens around the atoms of elastin-like peptide (ELP) and Snase are shown in Fig. 126. For comparison, the liquid density profile of water near moderately hydrophilic smooth surface is also shown (Fig. 126, lower panel). It is reasonable to use the location of the first minimum of the density profile to define the shell width D. Note that the shallow minimum at $r \approx 3$ Å in the case of Snase (Fig. 126, middle panel) separates two contributions to the density profiles, originating from water molecules in the first hydration shell near polar (left peak) and nonpolar (right peak) atoms of Snase. For all systems presented, $D = 4.5$ Å seems to be an optimal choice for the width of the first hydration shell, which does not change noticeably with increasing temperature [566].

Water clustering may be studied by the methods applied for low-hydrated systems in Sections 5.1 and 7.1 with the only, but important, difference: we consider water clustering being exclusively established by direct H-bonding between molecules in the hydration shell. Probability distributions $P(S_{max})$ of the size S_{max} of the largest water cluster in the hydration shell ($D = 4.5$ Å) of the ELP at various temperatures are shown in Fig. 127. The evolution of $P(S_{max})$ with decreasing temperature is quite similar to the one observed for hydration water at various surfaces with increasing hydration level. In general, the probability distribution $P(S_{max})$ shows a two-peak structure with a left (low S_{max}) and right (high S_{max}) peaks corresponding to the nonspanning and spanning largest clusters, respectively. In the case of ELP, these two peaks are never clearly separated and, accordingly, a minimum of S_{max} is not observed. Obviously, this is caused by small size of ELP, whose hydration shell never contains more than ∼190 water molecules.

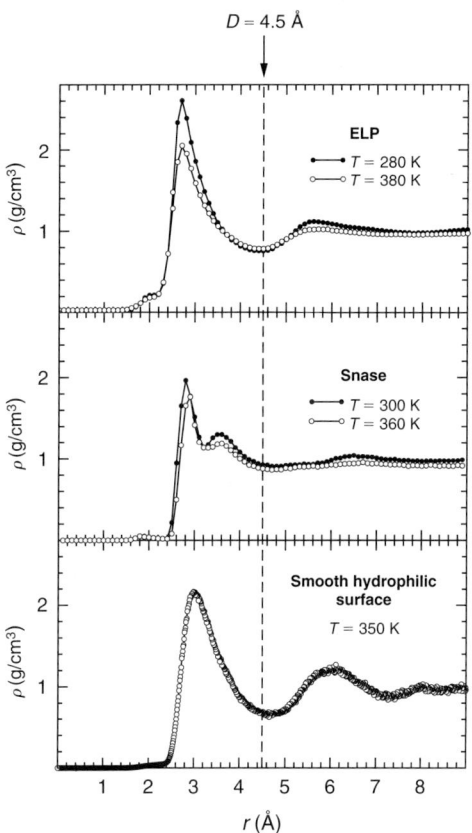

Figure 126: Water density profiles near the surface of ELP (upper panel), Snase (middle panel), and near a smooth hydrophilic surface (lower panel). The vertical dashed lines show the most realistic width of hydration shell: $D = 4.5$ Å. Reprinted, with permission, from [566].

Nevertheless, a two-peak distribution of S_{max} is manifested in pronounced shoulders or as an almost flat $P(S_{max})$ at $T = 320$ K. The latter temperature may be considered a midpoint of a temperature-induced percolation transition of hydration water. Distribution $P(S_{max})$ makes possible an estimation of some minimal size S_{max}^t required for the largest cluster to be spanning. One of the possible choice of S_{max}^t is being equally distant to both peaks of $P(S_{max})$. Integration of $P(S_{max})$ for $S_{max} > S_{max}^t$ yields an estimation of the spanning probability R. For a given temperature, the number of water molecules in the hydration shell fluctuates

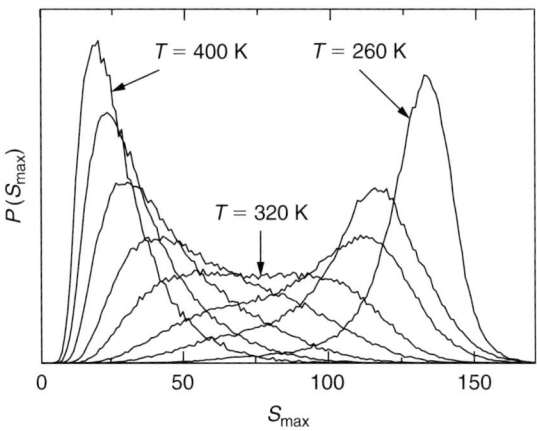

Figure 127: Probability distributions $P(S_{max})$ of the size S_{max} of the largest water cluster in the hydration shell of ELP at different temperatures.

around some most probable value. In the case of small and flexible biomolecules, these fluctuations may be relatively large. Under such circumstances, it is reasonable to analyze not S_{max} probability distribution but rather a distribution of S_{max} normalized by the current number of water molecules N_w in the hydration shell. The spanning probability calculated as an integral of the probability distribution for $S_{max}/N_w > 0.5$ is shown in Fig. 128.

Other properties of the largest water cluster within hydration shell also evidence a percolation transition. Probability distributions of the distance H between the center of mass of the largest water cluster and the center of mass of ELP indicate that spanning clusters practically never appear and nonspanning clusters (with large H) dominate at high temperatures $T = 380$ and 400 K (Fig. 129, left panel). With decreasing temperature, a peak of the probability distribution appears at $H \leq 1$ Å. This peak corresponds to the clusters that homogeneously envelope a biomolecule. Both peaks are comparable in the temperature range between 320 and 340 K. The same conclusion may be drawn from the temperature dependence of the probability distribution of the radius of gyration R_g of the largest water cluster (Fig. 129, right panel). So, all considered properties of the largest cluster of hydration water indicate a midpoint of the percolation transition at about 330 K.

The temperature evolution of the cluster size distribution n_S allows estimation of the "true" quasi-2D percolation transition of hydration water.

States of interfacial water in fully hydrated biosystems

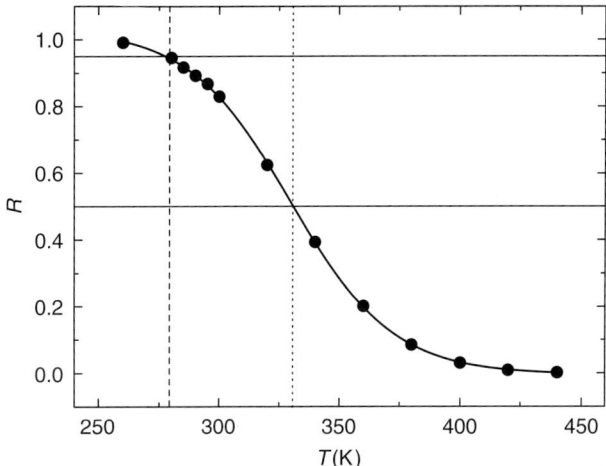

Figure 128: Spanning probability R for the largest water cluster in the hydration shell of ELP as a function of temperature (solid circles). Fit to the sigmoid function is shown by solid line. The midpoint of percolation transition where $R = 50\%$ is located at $T \approx 330$ K and denoted by a vertical dotted line. The percolation threshold, which corresponds to $R \approx 95\%$, is located at $T \approx 290$ K and denoted by a vertical dashed line.

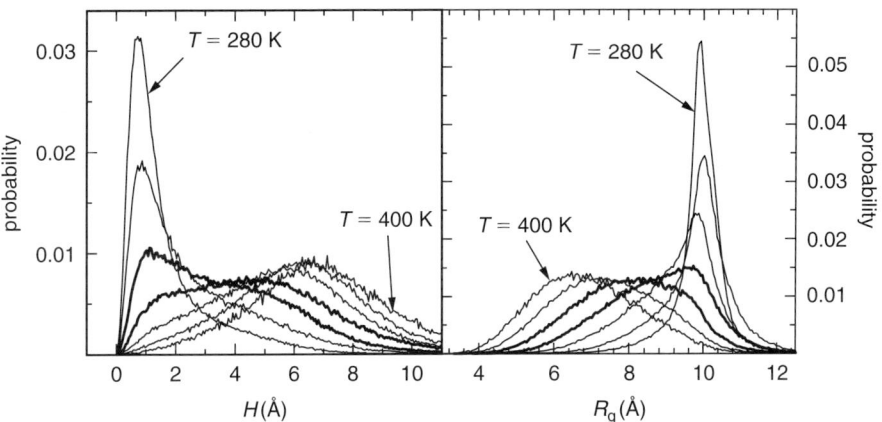

Figure 129: Probability distributions of the distance H between the center of mass of the largest water cluster and the center of mass of ELP (left panel) and of the radius of gyration R_g of the largest water cluster (right panel) at different temperatures. The distributions at $T = 320$ and 340 K closest to the midpoint of percolation transition are shown by thick lines.

A hump at large S, which reflects the truncation of the large clusters due to the finite size of the hydration shell, appears far below the percolation transition, when notable part of the largest water clusters becomes spanning. The percolation threshold may be located based on deviations of n_S from power law in the range of S before the hump. Fig. 130 evidences that at $T \approx 280$ K, n_S follows a power law for 2D percolation in the

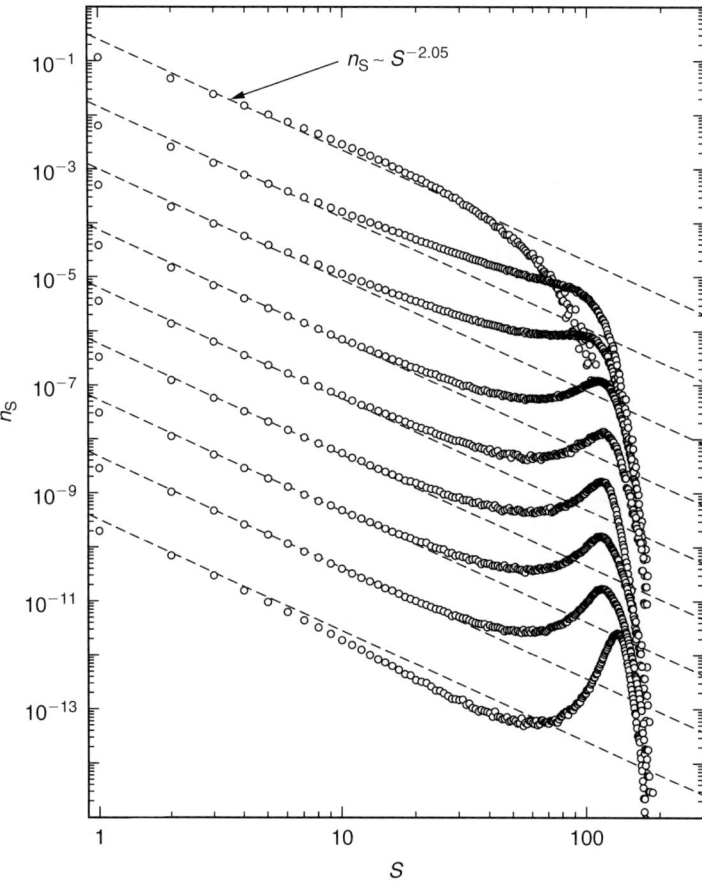

Figure 130: Size distribution n_S of water clusters in the hydration shell of an ELP at various temperatures (from bottom to top): $T = 260, 280, 285, 290, 295, 300, 320, 340,$ and 400 K. The distributions are shifted consecutively by one order of magnitude each, starting from the top. The power law expected at the percolation threshold is shown by dashed lines. Reprinted, with permission, from [398].

widest range of cluster size up to the hump. At this temperature, a spanning water network exists in the hydration shell with probability $\sim 95\%$ (see Fig. 128), which is in good agreement with the results obtained for low-hydrated systems. This indicates almost permanent existence of the spanning water network around ELP molecule at the temperatures below about 280 K [566].

A similar study carried out for the Snase molecule at full hydration has shown that the H-bonded water network envelopes Snase molecules permanently at temperatures below about 275 K [566]. A midpoint of the percolation transition was estimated at $T \approx 295$ K. So, the thermal break of a spanning water network occurs in a narrower temperature interval in the case of Snase molecule in comparison with ELP. The shrinkage of the temperature interval of the percolation transition should be attributed to the larger size of Snase molecule, which has about eight times more water molecules in the hydration shell than ELP.

Taking into account some ambiguity in the choice of the hydration shell width D, it is reasonable to estimate its effect on the temperature of the percolation transition. Such analysis was performed in Ref. [566] for various choices of D from 3.8 to 5.4 Å. Such variations of D were also useful for an accurate location of the percolation threshold at every temperature studied. Depending on the chosen value of D, the number N_w of water molecules in the hydration shell varied up to about a factor of two. Due to the increasing number of water molecules in the hydration shell, a percolation transition occurs at some value of D, particular for each temperature studied. Example of a percolation transition at constant temperature is shown in Fig. 131. With increase in the thickness of hydration shell, larger deviations from a strict 2D to 3D percolation transition may be expected. The respective power laws for n_S at the 2D and 3D percolation thresholds are shown in Fig. 131. Obviously, the behavior of n_S allows the location of the percolation threshold between $N_w = 147$ and $N_w = 153$ without any assumption about the dimensionality of the transition. This means, in particular, that at $T = 300$ K, water network around a peptide is spanning, if all water molecules within hydration shell of 4.75 Å width are considered as hydration water.

At any temperature, a true percolation transition of water upon increasing hydration shell may be located based on the cluster size distribution n_S, whereas a midpoint of this transition may be estimated based on

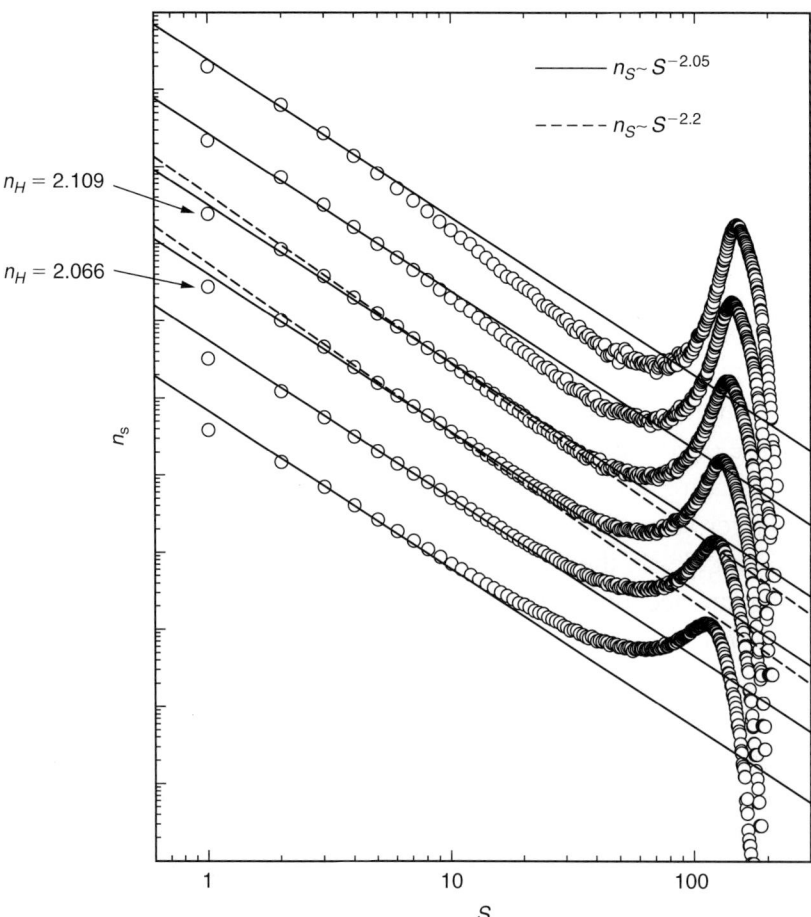

Figure 131: Cluster size distribution n_S in the hydration shell of an ELP at $T = 300$ K for several widths D of the hydration shell, which correspond to the following numbers N_w of water molecules in the shell: 135, 141, 147, 153, 159, and 164 (from bottom to top). The distributions are shifted consecutively by one order of magnitude each. The power laws for 2D and 3D percolation thresholds are shown by solid and dashed lines, respectively. Reprinted, with permission, from [631].

the spanning probability ($R = 50\%$) and using a maximum of the mean cluster size S_{mean}. The results of such studies for the hydration shell of ELP are summarized in Fig. 132. The thermal disruption of the hydration water shell occurs in a wide temperature range, which is about 50°C at the most reliable estimation of the thickness width of hydration shell ($D = 4.5$ Å) and may be slightly narrower (~40°C), if other

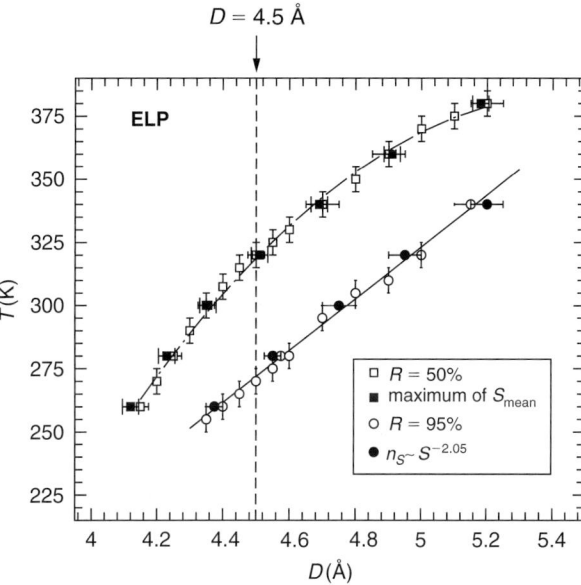

Figure 132: Temperature of the percolation threshold of water in the hydration shell of ELP as a function of the hydration shell width D. Percolation thresholds, estimated from the distributions $P(S_{max})$ of the largest cluster size (open circles), from the distributions n_S of the cluster size (closed circles) and linear fit of the joint data set (solid line). The shell widths, where the mean cluster size S_{mean} passes at a given temperature through a maximum, are shown by closed squares. The temperatures at which the spanning probability D, determined from the distribution $P(S_{max})$ at a given shell width, is about 50% are shown by open squares. Dot-dashed line is a guide for eyes only. Reprinted, with permission, from [566].

criterion for hydration shell is imposed. The temperature of the percolation threshold increases by ~10°C when the definition of hydration shell is increased by 0.1 Å. A similar estimation is valid for the hydration shell of Snase molecules, although the percolation transition here occurs in narrower temperature range. Interestingly, water clustering and percolation threshold in various hydration shells are highly universal in terms of water–water H-bond. In particular, a true percolation threshold occurs when the average number of H-bonded neigbors within hydration shell is ≈2.1. This value is close to but slightly lower than the corresponding value $n_H = 2.3$ in low-hydrated biosystems. This effect is obviously caused by the formation of H-bonds between water in hydration shell and surrounding water molecules.

The mass distribution $m(r)$ of the molecules within the largest cluster, calculated by using each of these molecules as an origin, yields the fractal dimension of the largest cluster d_f. The hydration shell of the ELP is too small for a meaningful estimation of the fractal dimension of the largest cluster: even spanning clusters do not show a mass distribution $m(r)$ that can be fitted to the power law. However, this can be done for larger protein molecules, such as Snase, which contains several times more water molecules in the hydration shell. The mass distribution displays a power law behavior up to the distance more than 25 Å and yields reliable estimation of d_f. The values of the fractal dimension of the largest cluster at the percolation threshold were found $d_f \approx 2.1$ at any reasonable choice of hydration shell and exceed the values $d_f^{2D} \approx 1.896$ expected theoretically and observed in low-hydrated systems. This can be attributed to the noticeable trend of the considered hydration shell toward three dimensionality due to rather large values of D or to the specific nonhomogeneous topology of the shell. The effective dimension d of the whole hydration shells of various widths, which takes into account all molecules in the shell (both bonded and nonbonded), is ≈ 2.22 [566]. This value practically does not depend on temperature and only slightly increases with increasing shell width D. Hence, the specific structure of the hydration shell is responsible for the fact that the effective dimension of the water shell d and the fractal dimension of the largest water cluster d_f at the threshold noticeably exceed 2. Interestingly, that the ratio $d_f/d = 2.10/2.22 \approx 0.946$ is extremely close to the value $1.896/2.00 \approx 0.948$ for fractals at the percolation threshold in 2D lattices.

Disruption of an ordered H-bonded water shell with temperature may provoke conformational changes of biomolecules. Typically, biomolecules undergo denaturation transition upon heating. This process is accompanied by phase separation into dilute (water rich) phase and organic-rich phase, which may appear as viscous liquid or amorphous solid. In real aqueous solutions of large elastin-based polymers, the phase separation into water-rich and organic-rich phases, accompanied by sharp conformational changes of the polymer (the so-called inverse temperature transition), occurs at about 300 K [558]. In solutions of small ELP, where the phase separation was not detected, pronounced conformational changes of biomolecules are still observed [659–661] and an inverse temperature transition occurs at ∼300–310 K [659]. Experimental studies evidence that even smaller ELP show a conformational transition at about 310–330 K [660, 661]. Hence, the experimentally measured inverse temperature transition of various ELP occurs in the temperature range where the spanning network of hydration water breaks into an ensemble of small water clusters in simulations.

Note that a correct comparison of the absolute values of the temperatures of the percolation transitions of water in the hydration shells of ELP and Snase, obtained in simulations, with the real temperature scale needs special consideration, as the phase diagrams of the available water models differ noticeably from the phase diagram of real water (see [5, 6] for a comparative analysis of the phase diagrams of various water models). There are two main characteristic temperatures that can be used for estimating the temperature shift of the phase diagram of model water with the behavior of real water: the critical temperature of the liquid–vapor phase transition and the temperature of the liquid density maximum. The latter temperature is the most important parameter for studies carried on close to ambient conditions. For example, the phase diagram of TIP3P water model is shifted downward by at least 35 K with respect to real water.

To clarify the effect of thermal breaking of hydration water shell on conformation of biomolecules and other properties of hydrated biosystems, all properties should be studied in the same model system. The temperature dependence of various conformational properties of the ELP obtained in simulations [658] shows two characteristic temperature intervals: at $T < 310$ K, a peptide is more compact and rigid, whereas at $T > 310$ K, it becomes much more flexible. For example, this temperature-induced conformational transition causes qualitative changes in the temperature dependence of the average value R_g of the radius of gyration of the ELP at $T \approx 310$ K (see Fig. 133, left panel). The average radius of gyration R_g of the H-bonded network of hydration water also shows two quite different temperature dependences: below about 310 K, R_g practically does not depend on temperature, whereas at higher temperatures, it almost linearly decreases with T (see Fig. 133, right panel).

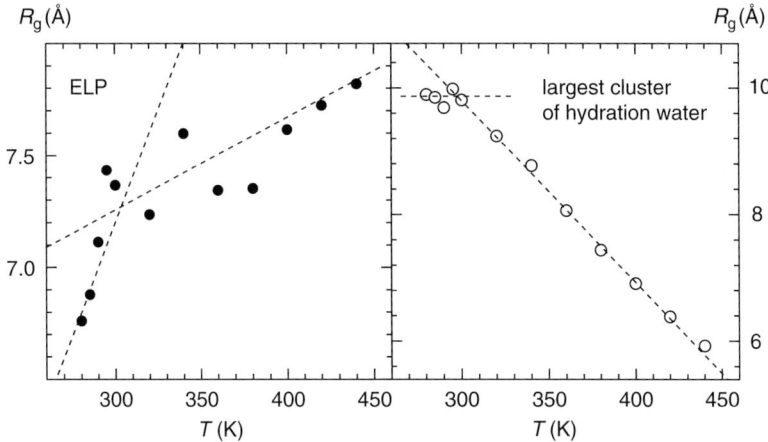

Figure 133: Average radius of gyration R_g of ELP (left panel) and average radius of gyration R_g of the largest hydrogen-bonded cluster in the hydration shell of ELP (right panel) (data from [658]).

By analysis of the probability distribution $P(R)$ of the end-to-end distance of macromolecule, it is possible to characterize its conformation and flexibility. At low temperatures, distribution $P(R)$ is rather narrow, whereas above $T = 320$ K, it may be successfully fitted to the equation of a random chain:

$$P(R) \sim (R - R_0)^2 \exp(-B_R(R - R_0)^2). \tag{32}$$

Simulation data and their fits are shown in Fig. 134. The ability of eq. (32) to describe adequately the probability distributions $P(R)$ is almost independent of T in the interval between 440 K and about 320 K. At lower temperatures (below 320 K), the shape of $P(R)$ changes qualitatively, and quality of fit drastically worsens. In this range, $P(R)$ strongly deviates from the distribution of the end-to-end distance for the random chain, indicating a growing fraction of the more rigid conformational state of ELP.

So, at high temperatures, the ELP is a highly flexible chain, which shows a random distribution of the end-to-end distance in agreement with the idea of Hoeve and Flory that the elasticity of elastin is rubber like [662, 663]. Irregular (or even random) location of the ordered

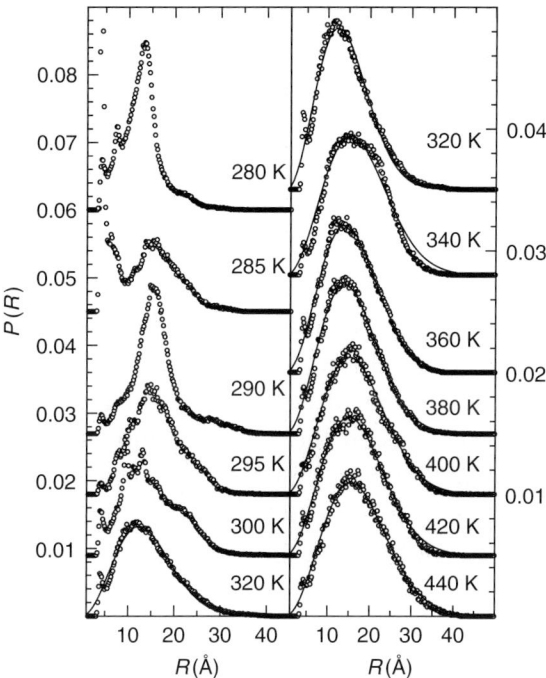

Figure 134: Probability distribution $P(R)$ of the end-to-end distance R of ELP at various temperatures (circles) and fits using eq. (32). The different distributions are shifted vertically to avoid overlapping. Reprinted, with permission, from [658].

structural elements along the chain together with a frequent interconversion between them provides a random distribution of the end-to-end distance of a chain. Obviously, this structure is enabled in the presence of disordered hydration water, which should strongly facilitate this rearrangement. Below 310 K, another more rigid conformational state of the ELP appears, and its fraction drastically increases upon cooling. The low-temperature rigid state differs from the high-temperature flexible conformational state by the presence of an irregular pattern of intramolecular H-bonds between amino acids. The state of the hydration water shell also changes drastically: a midpoint of percolation transition is located at $T \approx 330$ K, and below 310 K, most of the water molecules in the hydration shell form a spanning H-bonded network. So, percolation transition of water upon heating, which may be considered as a transition of the hydration water from a more ordered state (large spanning network) to a less ordered state (ensemble of small clusters), seems to be intrinsically related to temperature-induced transition of elastin to more flexible conformation.

The simulation studies of fully hydrated Snase molecule [566] predict a thermally induced break of hydration water network in *real* system in the temperature interval 310 K < T < 330 K, which includes $T \approx 325$ K, which is the unfolding temperature of Snase [664]. The fact that the spanning network of hydration breaks in the biologically relevant temperature interval seems to be not accidental and intrinsically related to the location of the critical point of quasi-2D hydration water [394]. There is a general relation between the critical temperatures of 2D and 3D systems consisting of the same interacting particles. The critical temperature of 2D water is about 50 to 60% of the critical temperature of a bulk 3D water [262]. Below this temperature, interaction of water molecules within hydration shell via H-bonded spanning network provides a specific collective properties of hydration water, which enable it to consider as an ordered subsystem. Above the critical temperature of 2D water, hydration water turns into disordered state. On the surface of a biomolecule, this transformation appears via a percolation transition of a quasi-2D hydration water. So, various biological processes occur in the temperature interval where the thermodynamic state of hydration water is not very far from 2D critical point of water. Further studies are needed to come to a definite conclusion on the effect of the surface properties on the temperature of the thermal break of H-bonded network of hydration water. In particular, effects of a surface hydrophobicity/hydrophilicity and heterogeneity remain unclear.

The studied reviewed above allow us to assume that the native conformation of a biomolecule and its function are possible when it is covered by a spanning H-bonded network of hydration water. Break of this network upon heating may be one of the decisive factor that determines the upper temperature limits for life (see Fig. 135, upper panel). Upon cooling, this network survives, but water may undergo liquid–liquid transition. As we have shown in Section 4.2, hydration water undergoes a phase transition from "normal" water to strongly tetrahedral water upon cooling. This phase transition is accompanied by the fragile to strong dynamic transition of water [243, 362]. The temperature of this transition is not very sensitive to the properties

States of interfacial water in fully hydrated biosystems

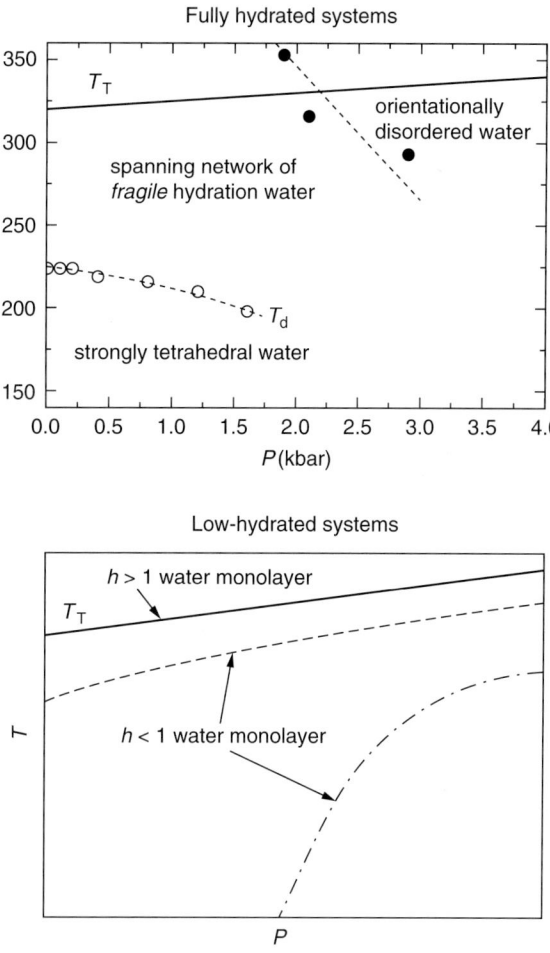

Figure 135: Phase diagram of surface water in fully hydrated systems (upper panel). Fragile to strong transition of the hydration water [243] and anomaly in thermophysical properties [108] that indicate a continuous transition from tetrahedrally ordered to orientationally disodered water [45] are shown by open and closed circles, respectively. The line of percolation transition of hydration water in the case of full hydration is shown schematically by solid lines based on the results of Ref. [566]. Location of the percolation transitions in low-hydrated systems is shown schematically by dashed and dot-dashed lines (lower panel). Reprinted, with permission, from [612].

of a surface and is very similar for silica surface and for biosurfaces [368, 369]. Approximately at the same temperature, dynamics of biomolecules diminishes in a drastic way, and they lose their function (see Section 6). So, at zero pressure, there is a temperature interval bound from above by the thermal breaking of the water shell (solid line) and from below by the liquid–liquid phase transition or frigile to strong transition of hydration water (open circles). Within this temperature interval, biomolecules are covered homogeneously by a H-bonded spanning network of "normal" (fragile) water and show biological function.

Change of pressure also can affect properties of hydration water in a drastic way. In low-hydrated system (Fig. 135, lower panel), an increasing pressure effectively improves hydration of protein molecules by filling the regions with depleted density near hydrophobic groups of proteins. In systems in which hydration level h is below the some critical value h_c, only finite (small) water clusters exist in the hydration shell of biomolecules at any temperature and pressure. If $h > h_c$, the largest cluster of hydration water, which increases upon pressurization, becomes spanning cluster at some pressure. This threshold pressure may be positive or negative, and its value should increase with temperature. At low temperatures, percolation transition of hydration water at low hydration levels is close to the conventional site percolation problem (dot-dashed line), whereas at high hydration levels, it looks similar to the bond percolation problem (solid and dashed lines).

In fully hydrated systems, the temperature of a thermal break of water network should increase with pressure due to densification of water in the hydration shell (solid line). However, parallel with densification, water progressively loses its tetrahedral ordering and transforms into a liquid with much weaker orientational ordering. This structural transformation is intrinsically related to the high-pressure liquid–liquid transition(s) of supercooled water, caused by separation of water phase enriched with water molecules showing highly isotropic arrangement of the neighbors in the first coordination shell [6, 10, 31, 45]. Above the critical temperature of the high-pressure liquid–liquid transition, the structural transformation occurs in a continuous way, and a line, which roughly separates normal water and orientationally disordered water, may be expected in $P - T$ plane. Water properties should change qualitatively with crossing this line (dotted line), and, indeed, these changes were observed experimentally (closed circles in Fig. 135) (see also Section 1). As the orientationally disordered hydration water noticeably differs from the normal (fragile) water, the pressure-induced denaturation transition of hydrated proteins (see [665–667] and Section 6) may be related to the disappearance of the spanning network of *fragile* hydration water. Summarizing, we suggest that fully hydrated biomolecules show their function only in a relatively narrow thermodynamic range, bounded by the transition to the strongly tetrahedral water upon cooling or upon applying negative pressure, by the transition to the orientationally disordered water upon applying positive pressure, and by the thermal disruption of a spanning water network upon heating.

9 Summary and outlook

We have analyzed the phase behavior and properties of interfacial and confined water, seen in experiments and simulations, from the point of view of statistical physics and based on the theory of phase transitions and critical phenomena. Such kind of analysis is obligatory, as in the phenomena considered above, we deal with macroscopic systems in various phase states. We show that the main regularities in the behavior of the confined water and water near surfaces are similar to those of other fluids. This conclusion is fully applicable to the surface critical behavior of water, the surface phase diagram of water, and the effect of confinement on the liquid–vapor phase transition. Comparison with other fluids allows distinguishing between the features typical for most fluids and features that are peculiar for water. Obviously, peculiar features of water are connected with hydrogen bonding, which dominates in the water–water interactions in an extremely wide range of thermodynamic conditions. Near any boundary, number of water–water hydrogen bonds unavoidably decreases, which is reflected in various surface properties of water. Besides, polyamorphism of bulk liquid water in supercooled region appears also in the behavior of interfacial and confined water.

Due to the strong water–water interactions, many surfaces on the earth are in fact hydrophobic, as they relatively weakly interact with water. Therefore, interfaces between liquid water and hydrophobic surface gain much importance in various biological, geological, and industrial processes. Accordingly, the depletion of liquid water density near surfaces is abundant, and solvophobic (hydrophobic) attraction between surfaces in liquid water is frequent phenomena. There is no increase in water ordering near hydrophobic surfaces. Similar to other fluids, weakly attractive surface has only disordering effect on a liquid water. For all one-component fluids, long-range interaction with a wall makes impossible a drying transition below the bulk liquid–vapor critical temperature. Accordingly, a vapor layer never appears between a liquid and a wall. Manifestations of a microscopic drying layer, which may be considered as a liquid–vapor interface completely attached to the wall, appear for strongly hydrophobic (paraffin-like) surfaces and at high temperatures only. In a wide range of thermodynamic conditions, depletion of water density near hydrophobic surface in both vapor and liquid phases or in supercritical states is caused by the missing neighbor effect, and it is governed by the bulk correlation length.

Wetting transition of water may occur in a rather narrow interval of the water–wall interactions. This feature may be common for all one-component fluids interacting with a wall via a long-range attractive potential. The first layering transition of water on the hydrophilic surface is located at temperatures essentially exceeding the bulk freezing temperature. Although water is not unique fluid that possesses such peculiarity, this feature may be of special significance as the critical temperatures of the first layering transition are close to ambient and biologically relevant temperatures. Hence, strong fluctuations of interfacial (hydration) water in biological systems due to the proximity of the 2D critical point should be expected. The temperature-induced break up of the hydrogen-bonded network of hydration water occurring via a percolation transition is intrinsically related to the critical point of quasi-2D water, and it also occurs at biologically relevant temperatures. Remarkably, the hydration water in thermodynamic conditions, typical for living organisms, is highly sensitive to external perturbations. This may be one of the factors that are responsible for the biological function of hydrated biomolecules.

A specific rearrangement of the hydrogen-bonded network of liquid water relative to the bulk case takes place near hydrophilic surfaces. This rearrangement is aimed at preserving more energetically favorable water–water hydrogen bonds to compensate the missing neighbor effect. Therefore, a strong orientational ordering of water molecules in the first surface layer near hydrophilic surfaces is seen even in the absence of water–surface hydrogen bonds. Much weaker signatures of orientational ordering are seen in the second surface water layer, and they are practically absent in the third layer. So, the second water layer connects orientationally ordered first layer with the third layer with almost isotropic orientation of water molecules. As a result, there are two specific water layers near hydrophilic surfaces, whose local structure strongly differs from the bulk liquid water, whereas local structure of water becomes essentially bulk-like starting from the third layer. This peculiar feature of water appears in various phenomena: there are only one or two layering transitions of water; "bound" water of about one to two layers thick is not freezable and shows properties different from those of the bulk ones; two-step sequential wetting scenario assumes condensation of two surface layers as a first step.

Ability of water molecules to form various kinds of local order in condensed state causes variety of its crystalline and amorphous phases at low temperatures. The transitions between liquid water phases with different local orders at low temperatures strongly affect the properties of water at ambient conditions. This effect is presumably responsible for various water properties, which makes water different from most other fluids and often called "anomalous" (liquid density maximum, heat capacity minimum, etc.). Naturally, the bulk polyamorphism appears also in water properties near surfaces. A transition of liquid water to strongly tetrahedrally ordered water upon cooling is the most important manifestation of this phenomenon as it occurs at ambient pressures. This transition is extremely difficult to detect in bulk water due to unavoidable crystallization. However, it is observed in many systems containing a confined water owing to the drastic change in various properties.

Despite the intensive studies of interfacial and confined water, many aspects of its behavior remain not well studied or even unclear. There are only a few studies of the phase diagram of confined water and of water adsorbed on the surface. Most of these studies are the simulations with very simple smooth surfaces. Clearly, experimental studies and simulations with more realistic surfaces are necessary. Repulsion between hydrophilic surfaces in liquid water gained much less attention than attraction between hydrophobic surfaces. However, this effect may be responsible, for example, for the destruction of some solids in environment with varying humidity. The liquid–liquid transitions of water, confined in various pores, should be studied because of their importance not only in understanding the properties of interfacial water but also aiming to locate these transitions in bulk water.

Further studies of interfacial water are necessary to clarify its role in functioning of living systems. This is a long-standing problem, and many efforts have been already made to understand the physical mechanisms behind the crucial role of water in biological function. Some aspects of this problem were clarified, but the self-consistent picture is far from being completed. There is an understanding that hydration water forms an environment for biomolecules, which enables their conformational dynamics and serves as a media for the mass and charge transfer along biosurfaces. There are numerous evidences that the changes of the state of the hydration water (formation of a spanning water shell via percolation transition upon hydration, its qualitative changes upon varying temperature or pressure or due to addition of cosolvents) are directly linked to the important biological processes. To understand the origin of the crucial role of water in biological function, it is necessary to relate particular states of hydration water to various physical properties of hydrated biosystems. We hope that knowledge of the water properties in various phase states and at various thermodynamic conditions summarized in our book will help in solving this problem.

References

[1] V. F. Petrenko, R. W. Whitworth, Physics of Ice, Oxford: Oxford University Press, 1999.

[2] J. M. H. Levelt-Sengers, J. Straub, K. Watanabe, P. G. Hill, Assessment of critical parameter values for H_2O and D_2O, J. Phys. Chem. Ref. Data 14 (1985) 193–207.

[3] W. Wagner, A. Pruss, The IAPWS formulation 1995 for the thermodynamic properties of ordinary water substance for general and scientific use, J. Phys. Chem. Ref. Data 31 (2002) 387–535.

[4] R. G. Fernandez, J. L. F. Abascal, C. Vega, The melting point of ice Ih for common water models calculated from direct coexistence of the solid-liquid interface, J. Chem. Phys. 124 (2006) 144506.

[5] B. Guillot, A reappraisal of what we have learnt during three decades of computer simulations on water, J. Mol. Liq. 101 (2002) 219–260.

[6] I. Brovchenko, A. Geiger, A. Oleinikova, Liquid-liquid phase transitions in supercooled water studied by computer simulations of various water models, J. Chem. Phys. 123 (2005) 044515.

[7] C. Vega, E. Sanz, J. Abascal, The melting temperature of the most common models of water, J. Chem. Phys. 122 (2005) 114507.

[8] C. Vega, J. L. F. Abascal, Relation between the melting temperature and the temperature of maximum density for the most common models of water, J. Chem. Phys. 123 (2005) 144504.

[9] C. Vega, J. L. F. Abascal, I. Nezbeda, Vapor-liquid equilibria from the triple point up to the critical point for the new generation of TIP4P-like models: TIP4P/Ew, TIP4P/2005, and TIP4P/ice, J. Chem. Phys. 125 (2006) 034503.

[10] I. Brovchenko, A. Oleinikova, Effect of confinement of the liquid-liquid phase transition of supercooled water, J. Chem. Phys. 126 (2007) 214701.

[11] Y. Guissani, B. Guillot, A computer-simulation study of the liquid-vapor coexistence curve of water, J. Chem. Phys. 98 (1993) 8221–8235.

[12] R. Guida, J. Zinn-Justin, Critical exponents of the N-vector model, J. Phys. A: Math. Gen. 31 (1998) 8103–8121.

[13] F. J. Wegner, Corrections to scaling laws, Phys. Rev. B 5 (1972) 4529–4536.

[14] B. Widom, J. S. Rowlinson, New model for the study of liquid-vapor phase transitions, J. Chem. Phys. 52 (1970) 1670–1684.

[15] M. E. Fisher, G. Orkoulas, The Yang-Yang anomaly in fluid criticality: Experiment and scaling theory, Phys. Rev. Lett. 85 (2000) 696–699.

[16] M. A. Anisimov, J. Wang, Nature of asymmetry in fluid criticality, Phys. Rev. Lett. 97 (2006) 025703.

[17] M. E. Fisher, The theory of equilibrium critical phenomena, Rep. Prog. Phys. 30 (1967) 615–730.

[18] H. E. Stanley, Phase Transitions and Critical Phenomena, Oxford: Clarendon Press, 1971.

[19] M. Bonetti, G. Romet-Lemonne, P. Calmettes, M.-C. Bellissent-Funel, Small-angle neutron scattering from heavy water in the vicinity of the critical point, J. Chem. Phys. 112 (2000) 268–274.

[20] C. Angell, M. Oguni, W. Sichina, Heat-capacity of water at extremes of supercooling and superheating, J. Phys. Chem. 86 (1982) 998–1002.

[21] J. V. Sengers, J. M. H. Levelt-Sengers, Thermodynamic behavior of fluids near the critical-point, Ann. Rev. Phys. Chem. 37 (1986) 189–222.

[22] N. Erokhin, B. Kalyanov, Extremal behavior of ultrasonic velocity and of some other quantities in the supercritical region of water, Therm. Eng. 27 (1980) 634–636.

[23] A. Coniglio, W. Klein, Clusters and Ising critical droplets—a renormalization group-approach, J. Phys. A: Math. Gen. 13 (1980) 2775–2780.

[24] L. B. Partay, P. Jedlovszky, I. Brovchenko, A. Oleinikova, Formation of mesoscopic water networks in aqueous systems, Phys. Chem. Chem. Phys. 9 (2007) 1341–1346.

[25] L. B. Partay, P. Jedlovszky, I. Brovchenko, A. Oleinikova, Percolation transition in supercritical water: A Monte Carlo simulation study, J. Phys. Chem. B 111 (2007) 7603–7609.

[26] A. Geiger, F. Stillinger, A. Rahman, Aspects of the percolation process for hydrogen-bond networks in water, J. Chem. Phys. 70 (1979) 4185–4193.

[27] H. Stanley, J. Teixeira, Interpretation of the unusual behavior of H_2O and D_2O at low temperatures: Tests of a percolation model, J. Chem. Phys. 73 (1980) 3404–3422.

[28] I. Brovchenko, A. Oleinikova, Molecular organization of gases and liquids at solid surfaces, in: M. Rieth, W. Schommers (Eds.), Handbook of Theoretical and Computational Nanotechnology, Volume 9, Chapter 3, Stevenson Ranch, California, American Scientific Publishers, 2006, pp. 109–206.

[29] I. Brovchenko, A. Geiger, A. Oleinikova, Surface critical behavior of fluids: Lennard-Jones fluid near a weakly attractive substrate, Eur. Phys. J. B 44 (2005) 345–358.

[30] I. Brovchenko, A. Geiger, A. Oleinikova, Phase equilibria of water in cylindrical nanopores, Phys. Chem. Chem. Phys. 3 (2001) 1567–1569.

[31] I. Brovchenko, A. Geiger, A. Oleinikova, Multiple liquid-liquid transitions in supercooled water, J. Chem. Phys. 118 (2003) 9473–9476.

[32] I. Brovchenko, A. Geiger, A. Oleinikova, Water in nanopores. I. Coexistence curves from Gibbs ensemble Monte Carlo simulations, J. Chem. Phys. 120 (2004) 1958–1972.

[33] I. Brovchenko, A. Geiger, A. Oleinikova, Multiple critical points of supercooled water, in: Water, steam and aqueous solutions for electric power: Advances in science and technology, M. Nakahara, N. Matubayasi, M. Ueno, K. Yasuoka, K. Watanabe (Eds.), Proceedings of 14th International Conference on the Properties of Water and Steam, Kyoto, 2004, eds. Kyoto: Maruzen Co., Ltd. 2005, pp. 194–199.

[34] P. H. Poole, I. Saika-Voivod, F. Sciortino, Density minimum and liquid-liquid phase transition, J. Phys.: Condens. Matt. 17 (2005) L431–L437.

[35] D. Liu, Y. Zhang, C.-C. Chen, C.-Y. Mou, P. H. Poole, S.-H. Chen, Observation of the density minimum in deeply supercooled confined water, Proc. Natl. Acad. Sci. U.S.A. 104 (2007) 9570–9574.

[36] P. G. Debenedetti, Supercooled and glassy water, J. Phys.: Condens. Matt. 15 (2003) R1669–R1726.

[37] C. Angell, Amorphous water, Ann. Rev. Phys. Chem. 55 (2004) 559–583.

[38] M. Chaplin, www.lsbu.ac.uk/water/.

[39] A. Voronel, Thermodynamic quantities near crystallization points of liquids, JETP Lett. (Russ.) 14 (1971) 174.

[40] C. Angell, J. Shuppert, J. Tucker, Anomalous properties of supercooled water—heat-capacity, expansivity, and proton magnetic-resonance chemical-shift from 0 to $-38°C$, J. Phys. Chem. 77 (1973) 3092–3099.

[41] R. Speedy, C. Angell, Isothermal compressibility of supercooled water and evidence for a thermodynamic singularity at $-45°C$, J. Chem. Phys. 65 (1976) 851–858.

[42] H. Kanno, C. A. Angell, Water: Anomalous compressibilities to 1.9 kbar and correlation with supercooling limits, J. Chem. Phys. 70 (1979) 4008–4016.

[43] N. N. Medvedev, Y. I. Naberukhin, Shape of the Delaunay simplices in dense random packings of hard and soft spheres, J. Non-Cryst. Solids 94 (1987) 402–406.

[44] I. Brovchenko, A. Oleinikova, Four phases of amorphous water: Simulations versus experiment, J. Chem. Phys. 124 (2006) 164505.

[45] A. Oleinikova, I. Brovchenko, Percolating networks and liquid-liquid transitions in supercooled water, J. Phys.: Condens. Matt. 18 (2006) S2247–S2259.

[46] I. Brovchenko, A. Oleinikova, Phases of supercooled liquid water, in: W. Kuhs (ed.) Proceedings of the 11th International Conference on the Physics and Chemistry of Ice, Cambridge, England: Cambridge: Royal Society of Chemistry, 2007, pp. 117–124.

[47] H. Whiting, A new theory of cohesion applied to the thermodynamics of liquids and solids, Harvard Physics PhD Thesis, 1884.

[48] W. Roentgen, Ueber die constitution des flussigen wassers, Annal. Phys. Chem. 45 (1892) 91.

[49] H. M. Chadwell, The molecular structure of water, Chem. Rev. 4 (1928) 375–398.

[50] C. A. Angell, Two-state thermodynamics and transport properties for water from "bond lattice" model, J. Phys. Chem. 75 (24) (1971) 3698–3705.

[51] F. H. Stillinger, Water revisited, Science 209 (1980) 451–457.

[52] F. Sciortino, A. Geiger, H. E. Stanley, Isochoric differential scattering functions in liquid water: The fifth neighbor as a network defect, Phys. Rev. Lett. 65 (1990) 3452–3455.

[53] M. Vedamuthu, S. Singh, G. W. Robinson, Properties of liquid water: Origin of the density anomalies, J. Phys. Chem. 98 (1994) 2222–2230.

[54] P. Poole, F. Sciortino, T. Grande, H. E. Stanley, C. Angell, Effect of hydrogen bonds on the thermodynamic behavior of liquid water, Phys. Rev. Lett. 73 (1994) 1632–1635.

[55] L. Bartell, On possible interpretations of the anomalous properties of supercooled water, J. Phys. Chem. B 101 (1997) 7573–7583.

[56] M.-C. Bellissent-Funel, Is there a liquid-liquid phase transition in supercooled water? Europhys. Lett. 42 (1998) 161–166.

[57] R. C. Dougherty, L. N. Howard, Equilibrium structural model of liquid water: Evidence from heat capacity, spectra, density, and other properties, J. Chem. Phys. 109 (1998) 7379–7393.

[58] H. Tanaka, Simple physical model of liquid water, J. Chem. Phys. 112 (2000) 799–809.

[59] A. K. Soper, M. A. Ricci, Structures of high-density and low-density water, Phys. Rev. Lett. 84 (2000) 2881–2884.

[60] H. Kanno, H. Yokoyama, Y. Yoshimura, A new interpretation of anomalous properties of water based on Stillinger's postulate, J. Phys. Chem. B 105 (2001) 2019–2026.

[61] T. Loerting, N. Giovambattista, Amorphous ices: experiments and numerical simulations, J. Phys.: Condens. Matt. 18 (2006) R919–R977.

[62] M. A. Anisimov, E. E. Gorodetskii, V. D. Kulikov, J. V. Sengers, Crossover between vapor-liquid and consolute critical phenomena, Phys. Rev. E 51 (1995) 1199–1215.

[63] T. Narayanan, A. Kumar, Reentrant phase transitions in multicomponent liquid mixtures, Phys. Rep. 249 (1994) 135–218.

[64] I. V. Brovchenko, A. V. Oleinikova, Structural changes of the molecular complexes of pyridines with water and demixing phenomena in aqueous solutions, J. Chem. Phys. 106 (1997) 7756–7765.

[65] C. M. Sorensen, G. A. Larsen, Light scattering and viscosity studies of a ternary mixture with a double critical point, J. Chem. Phys. 83 (1985) 1835–1842.

[66] S. S. Borick, P. G. Debenedetti, S. Sastry, A lattice model of network-forming fluids with orientation-dependent bonding—equilibrium, stability, and implications for the phase-behavior of supercooled water, J. Phys. Chem. 99 (1995) 3781–3792.

[67] C. J. Roberts, P. G. Debenedetti, Polyamorphism and density anomalies in network-forming fluids: Zeroth- and first-order approximations, J. Chem. Phys. 105 (1996) 658–672.

[68] S. Sastry, P. G. Debenedetti, F. Sciortino, H. E. Stanley, Singularity-free interpretation of the thermodynamics of supercooled water, Phys. Rev. E 53 (1996) 6144–6154.

[69] T. M. Truskett, P. G. Debenedetti, S. Sastry, A single-bond approach to orientation-dependent interactions and its implications for liquid water, J. Chem. Phys. 111 (1999) 2647–2656.

[70] O. Mishima, L. D. Calvert, E. Whalley, An apparently first-order transition between two amorphous phases of ice induced by pressure, Nature 314 (1985) 76–78.

[71] O. Mishima, Reversible first-order transition between two H_2O amorphs at 0.2 GPa and 135 K, J. Chem. Phys. 100 (1994) 5910–5912.

[72] O. Mishima, H. Stanley, Decompression-induced melting of ice IV and the liquid-liquid transition in water, Nature 392 (1998) 164–168.

[73] O. Mishima, Liquid-liquid critical point in heavy water, Phys. Rev. Lett. 85 (2000) 334–336.

[74] O. Mishima, Y. Suzuki, Propagation of the polyamorphic transition of ice and the liquid-liquid critical point, Nature 419 (2002) 599–603.

[75] S. Klotz, T. Strassle, R. Nelmes, J. Loveday, G. Hamel, G. Rousse, B. Canny, J. Chervin, A. Saitta, Nature of the polyamorphic transition in ice under pressure, Phys. Rev. Lett. 94 (2005) 025506.

[76] T. Loerting, W. Schustereder, K. Winkel, C. G. Salzmann, I. Kohl, E. Mayer, Amorphous ice: Stepwise formation of very-high-density amorphous ice from low-density amorphous ice at 125 K, Phys. Rev. Lett. 96 (2006) 025702.

[77] P. H. Poole, F. Sciortino, U. Essmann, H. E. Stanley, Phase-behavior of metastable water, Nature 360 (1992) 324–328.

[78] P. Poole, U. Essmann, F. Sciortino, H. E. Stanley, Phase diagram for amorphous solid water, Phys. Rev. E 48 (1993) 4605–4610.

[79] P. Poole, F. Sciortino, U. Essmann, H. E. Stanley, Spinodal of liquid water, Phys. Rev. E 48 (1993) 3799–3817.

[80] F. Sciortino, P. Poole, U. Essmann, H. E. Stanley, Line of compressibility maxima in the phase diagram of supercooled water, Phys. Rev. E 55 (1997) 727–737.

[81] S. Harrington, R. Zhang, P. Poole, F. Sciortino, H. E. Stanley, Liquid-liquid phase transition: Evidence from simulations, Phys. Rev. Lett. 78 (1997) 2409–2412.

[82] S. Harrington, P. Poole, F. Sciortino, H. E. Stanley, Equation of state of supercooled water simulated using the extended simple point charge intermolecular potential, J. Chem. Phys. 107 (1997) 7443–7450.

[83] A. Scala, F. Starr, E. LaNave, H. E. Stanley, F. Sciortino, Free energy surface of supercooled water, Phys. Rev. E 62 (2000) 8016–8020.

[84] M. Yamada, S. Mossa, H. E. Stanley, F. Sciortino, Interplay between time-temperature-transformation and the liquid–liquid phase transition in water, Phys. Rev. Lett. 88 (2002) 195701.

[85] M. Yamada, H. E. Stanley, F. Sciortino, Equation of state of supercooled water from the sedimentation profile, Phys. Rev. E 67 (2003) 010202.

[86] P. Jedlovszky, R. Vallauri, Liquid-vapor and liquid-liquid phase equilibria of the Brodholt-Sampoli-Vallauri polarizable water model, J. Chem. Phys. 122 (2005) 081101.

[87] P. Jedlovszky, L. B. Partay, A. P. Bartok, G. Garberoglio, R. Vallauri, Structure of coexisting liquid phases of supercooled water: Analogy with ice polymorphs, J. Chem. Phys. 126 (2007) 241103.

[88] S. Buldyrev, H. E. Stanley, A system with multiple liquid-liquid critical points, Physica A 330 (2003) 124–129.

[89] A. Skibinsky, S. Buldyrev, G. Franzese, G. Malescio, H. E. Stanley, Liquid-liquid phase transitions for soft-core attractive potentials, Phys. Rev. E 69 (2004) 061206.

[90] B. Pellicane, G. Pellicane, G. Malescio, Polymorphism in simple liquids: A Gibbs ensemble Monte Carlo study, J. Chem. Phys. 120 (2004) 8671–8675.

[91] L. Xu, P. Kumar, S. V. Buldyrev, S.-H. Chen, P. H. Poole, F. Sciortino, H. E. Stanley, Relation between the Widom line and the dynamic crossover in systems with a liquid-liquid phase transition, Proc. Natl. Acad. Sci. U.S.A. 102 (2005) 16558–16562.

[92] L. A. Cervantes, A. L. Banavides, Theoretical prediction of multiple fluid-fluid transitions in monocomponent fluids, J. Chem. Phys. 126 (2007) 084507.

[93] H. Schober, M. Koza, A.Toelle, F. Fujara, C. A. Angell, R. Boehmer, Amorphous polymorphism in ice investigated by inelastic neutron scattering, Physica B 241–243 (1998) 897–902.

[94] M. M. Koza, T. Hansen, R. P. May, H. Schober, Link between the diversity, heterogeneity and kinetic properties of amorphous ice structures, J. Non-Cryst. Solids 352 (2006) 4988–4993.

[95] M. M. Koza, R. P. May, H. Schober, On the heterogeneous character of water's amorphous polymorphism, J. Appl. Crystallogr. 40 (2007) S517–S521.

[96] H. Tanaka, General view of a liquid-liquid phase transition, Phys. Rev. E 62 (2000) 6968–6976.

[97] V. Brazhkin, R. Voloshin, A. Lyapin, S. Popova, Quasi-transitions in simple liquids at high pressures, Physics-Uspekhi 42 (1999) 1035–1039.

[98] S. Kiselev, J. Ely, Parametric crossover model and physical limit of stability in supercooled water, J. Chem. Phys. 116 (2002) 5657–5665.

[99] D. A. Fuentevilla, M. A. Anisimov, Scaled equation of state for supercooled water near the liquid-liquid critical point, Phys. Rev. Lett. 97 (2006) 195702.

[100] A. Oleinikova, I. Brovchenko, A. Geiger, B. Guillot, Percolation of water in aqueous solution and liquid–liquid immiscibility, J. Chem. Phys. 117 (2002) 3296–3304.

[101] H. E. Stanley, A polychromatic correlated-site percolation problem with possible relevance to the unusual behaviour of supercooled H_2O and D_2O, J. Phys. A: Math. Gen. 12 (1979) L329–L337.

[102] E. Lang, H.-D. Luedemann, Pressure and temperature dependence of the longitudinal deuterium relaxation times in supercooled heavy water to 300 MPa and 188 K, Ber. Bunsenges. Phys. Chem. 84 (1980) 462–470.

[103] F. X. Prielmeier, E. W. Lang, R. J. Speedy, H.-D. Luedemann, Diffusion in supercooled water to 300 MPa, Phys. Rev. Lett. 59 (1987) 1128–1131.

[104] F. X. Prielmeier, E. W. Lang, R. J. Speedy, H.-D. Luedemann, The pressure dependence of self diffusion in supercooled light and heavy water, Ber. Bunsenges. Phys. Chem. 92 (1988) 1111–1117.

[105] K. R. Harris, P. J. Newitt, Self-diffusion of water at low temperatures and high pressure, J. Chem. Eng. Data 42 (1997) 346–348.

[106] A. Cunsolo, A. Orecchini, C. Petrillo, F. Sacchetti, Quasielastic neutron scattering investigation of the pressure dependence of molecular motions in liquid water, J. Chem. Phys. 124 (2006) 084503.

[107] M. Krisch, P. Loubeyre, G. Ruocco, F. Sette, M. D'Astuto, R. L. Toulec, M. Lorenzen, A. Mermet, G. Monaco, R. Verbeni, Pressure evolution of the high-frequency sound velocity in liquid water, Phys. Rev. Lett. 89 (2002) 125502.

[108] F. Li, Q. Cui, Z. He, J. Zhang, Q. Zhou, G. Zou, S. Sasaki, High pressure-temperature Brillouin study of liquid water: Evidence of the structural transition from low-density water to high-density water, J. Chem. Phys. 123 (2005) 174511.

[109] T. Kawamoto, S. Ochiai, H. Kagi, Changes in the structure of water deduced from the pressure dependence of the Raman OH frequency, J. Chem. Phys. 120 (2004) 5867–5870.

[110] N. K. Alphonse, S. R. Dillon, R. C. Dougherty, D. K. Galligan, L. N. Howard, Direct Raman evidence for a weak continuous phase transition in liquid water, J. Phys. Chem. A 110 (2006) 7577–7580.

[111] T. Young, An essay on the cohesion of fluids, Philos. Trans. R. Soc. London 95 (1805) 65–87.

[112] J. W. Cahn, Critical point wetting, J. Chem. Phys. 66 (1977) 3667–3672.

[113] H. Nakanishi, M. E. Fisher, Multicriticality of wetting, prewetting, and surface transitions, Phys. Rev. Lett. 49 (1982) 1565–1568.

[114] K. Binder, D. P. Landau, Wetting and layering in the nearest-neighbor simple-cubic Ising lattice: A Monte Carlo investigation, Phys. Rev. B 37 (1988) 1745–1765.

[115] K. Binder, D. P. Landau, S. Wansleben, Wetting transitions near the bulk critical point: Monte Carlo simulations for the Ising model, Phys. Rev. B 40 (1989) 6971–6979.

[116] K. Binder, D. P. Landau, Wetting versus layering near the roughening transition in the three-dimensional Ising model, Phys. Rev. B 46 (1992) 4844–4854.

[117] C. Ebner, W. F. Saam, New reentrant wetting phenomena and critical behavior near bulk critical points, Phys. Rev. Lett. 58 (1987) 587–590.

[118] C. Ebner, W. F. Saam, Effect of long-range forces on wetting near bulk critical temperatures: An Ising-model study, Phys. Rev. B 35 (1987) 1822–1834.

[119] G. Forgacs, R. Lipowsky, T. M. Nieuwenhuizen, The behaviour of interfaces in ordered and disordered systems, in: C. Domb, J. L. Lebowitz (Eds.), Phase Transitions and Critical Phenomena, Vol. 14, London: Academic Press, 1991, pp. 135–363.

[120] S. Dietrich, Wetting phenomena, in: C. Domb, J. L. Lebowitz (Eds.), Phase Transitions and Critical Phenomena, Vol. 12, London: Academic Press, 1988, pp. 1–218.

[121] K. Binder, D. Landau, M. Mueller, Monte Carlo studies of wetting, interface localization and capillary condensation, J. Stat. Phys. 110 (2003) 1411–1514.

[122] M. P. Nightingale, W. F. Saam, M. Schick, Wetting and growth behaviors in adsorbed systems with long-range forces, Phys. Rev. B 30 (1984) 3830–3840.

[123] S. Dietrich, M. Schick, Critical wetting of surfaces in systems with long-range forces, Phys. Rev. B 31 (1985) 4718–4720.

[124] M. P. Nightingale, J. O. Indekeu, Examination of the necessity of complete wetting near critical points in systems with long-range forces, Phys. Rev. B 32 (1985) 3364–3366.

[125] M. J. P. Nijmeijer, C. Bruin, A. F. Bakker, J. M. J. van Leeuwen, Molecular dynamics of the wetting and drying of a wall with a long-ranged wall-fluid interaction, J. Phys.: Condens. Matt. 4 (1992) 15–31.

[126] A. Maciolek, R. Evans, N. B. Wilding, Effects of weak surface fields on the density profiles and adsorption of a confined fluid near bulk criticality, J. Chem. Phys. 119 (2003) 8663–8675.

[127] A. Oleinikova, I. Brovchenko, Effect of a fluid-wall interaction on a drying layer, Phys. Rev. E 76 (2007) 041603.

[128] J. E. Rutledge, P. Taborek, Prewetting phase diagram of ^{4}He on cesium, Phys. Rev. Lett. 69 (1992) 937–940.

[129] M. Yao, F. Hensel, Wetting of mercury on sapphire, J. Phys.: Condens. Matt. 8 (1996) 9547–9551.

[130] D. Ross, D. Bonn, J. Meunier, Wetting of methanol on the n-alkanes: Observation of short-range critical wetting, J. Chem. Phys. 114 (2001) 2784–2792.

[131] K. Ragil, J. Meunier, D. Broseta, J. O. Indekeu, D. Bonn, Experimantal observation of critical wetting, Phys. Rev. Lett. 77 (1996) 1532–1536.

[132] E. Bertrand, H. Dobbs, D. Broseta, J. Indekeu, D. Bonn, J. Meunier, First-order and critical wetting of alkanes on water, Phys. Rev. Lett. 85 (2000) 1282–1285.

[133] N. Shahidzadeh, D. Bonn, K. Ragil, D. Broseta, J. Meunier, Sequence of two wetting transitions induced by tuning the Hamaker constant, Phys. Rev. Lett. 80 (1998) 3992–3995.

[134] D. Bonn, D. Ross, Wetting transitions, Rep. Prog. Phys. 64 (2001) 1085–1163.

[135] B. M. Law, Wetting, adsorption and surface critical phenomena, Prog. Surf. Sci. 66 (2001) 159–216.

[136] G. B. Hess, M. J. Sabatini, M. H. W. Chan, Nonwetting of cesium by neon near its critical point, Phys. Rev. Lett. 78 (1997) 1739–1742.

[137] F. Ancilotto, S. Curtarolo, F. Toigo, M. W. Cole, Evidence concerning drying behavior of Ne near a Cs surface, Phys. Rev. Lett. 87 (2001) 206103.

[138] R. Evans, A. O. Parry, Liquids at interfaces: what can a theorist contribute, J. Phys.: Condens. Matt. 2 (1990) SA15–SA32.

[139] D. Nicolaides, R. Evans, Nature of the prewetting critical point, Phys. Rev. Lett. 63 (1989) 778–781.

[140] V. F. Kozhevnikov, D. I. Arnold, S. P. Naurzakov, M. E. Fisher, Prewetting transitions in a near-critical metallic vapor, Phys. Rev. Lett. 78 (1997) 1735–1738.

[141] A. Oleinikova, I. Brovchenko, A. Geiger, Drying layer near a weakly attractive surface, J. Phys.: Condens. Matt. 17 (2005) 7845–7866.

[142] P. G. de Gennes, Wetting: Statics and dynamics, Rev. Mod. Phys. 57 (1985) 827–863.

[143] A. Doerr, M. Tolan, T. Seydel, W. Press, The interface structure of thin liquid hexane films, Physica B 248 (1998) 263–268.

[144] A. K. Doerr, M. Tolan, J.-P. Schlomka, W. Press, Evidence for density anomalies of liquids at the solid/liquid interface, Europhys. Lett. 52 (2000) 330–336.

[145] R. Steitz, T. Gutberlet, T. Hauss, B. Klosgen, R. Krastev, S. Schemmel, A. C. Simonsen, G. H. Findenegg, Nanobubbles and their precursor layer at the interface of water against a hydrophobic substrate, Langmuir 19 (2003) 2409–2418.

[146] R. Steitz, S. Schemmel, H. Shi, G. H. Findenegg, Boundary layers of aqueous surfactant and block copolymer solutions against hydrophobic and hydrophilic solid surfaces, J. Phys.: Condens. Matt. 17 (2005) S665–S683.

[147] V. Weiss, J. Indekeu, Contact angle at the first-order transition in sequential wetting, Physica A 301 (2001) 37–51.

[148] N. S. Desai, S. Peach, C. Franck, Critical adsorption in the undersaturated regime, Phys. Rev. E 52 (1995) 4129–4133.

[149] J. Bowers, A. Zarbakhsh, A. Querol, H. K. Chistenson, I. A. McLur, R. Cubitt, Adsorption from alkane plus perfluoroalkane mixtures at fluorophobic and fluorophilic surfaces. II. Crossover from critical adsorption to complete wetting, J. Chem. Phys. 121 (2004) 9058–9065.

[150] H. Wu, G. B. Hess, Multilayer adsorption of deuterium hydride on graphite, Phys. Rev. B 57 (1998) 6720–6730.

[151] F. Millot, Y. Larher, C. Tessier, Critical temperatures of two-dimensional condensation in monolayers of Ar, Kr, or Xe adsorbed on lamellar halides, J. Chem. Phys. 76 (1982) 3327–3335.

[152] A. Z. Panagiotopoulos, Molecular simulation of phase coexistence: Finite-size effects and determination of critical parameters for two- and three- dimensional Lennard-Jones fluids, Int. J. Thermophys. 15 (1994) 1057–1072.

[153] H. Mannebach, U. G. Volkmann, J. Faul, K. Knorr, Order-parameter kinetics in the liquid-gas coexistence region of Ar monolayers physisorbed on graphite, Phys. Rev. Lett. 67 (1991) 1566–1569.

[154] H. K. Kim, M. H. W. Chan, Experimental determination of a two-dimensional liquid-vapor critical-point exponent, Phys. Rev. Lett. 53 (1984) 170–173.

[155] Q. M. Zhang, Y. P. Feng, H. K. Kim, M. H. W. Chan, Layering and layer-critical-point transitions of ethylene on graphite, Phys. Rev. Lett. 57 (1986) 1456–1459.

[156] Y. Larher, The critical exponent β associated with the two-dimensional condensation in the second adlayer of argon on the cleavage face of cadmium chloride, Mol. Phys. 38 (1979) 789–795.

[157] W. Gac, M. Kruk, A. Patrykiejew, S. Sokolowski, Effects of random quenched impurities on layering transitions: A Monte Carlo study, Langmuir 12 (1996) 159–169.

[158] P. A. Thiel, T. E. Madey, The interaction of water with solid surfaces: Fundamental aspects, Surf. Sci. Rep. 7 (1987) 211–385.

[159] K. Morishige, S. Kittaka, T. Morimoto, Studies of two-dimensional condensation of water on hydroxylated ZnO, SnO_2 and Cr_2O_3—determination of two-dimensional critical-temperature, Surf. Sci. 109 (1981) 291–300.

[160] T. Ishikawa, N. Kodaira, K. Kandori, Step-like adsorption isotherms of molecules on γ-FeOOH and the surface homogeneity of γ-FeOOH, J. Chem. Soc., Faraday Trans. 88 (1992) 719–722.

[161] D. R. Stull, Vapor pressure of pure substances organic compounds, Ind. Eng. Chem. 39 (1947) 517–540.

[162] K. Morishige, S. Kittaka, T. Morimoto, Two-dimensional condensation of water and alcohols on NaF, Surf. Sci. 120 (1982) 223–238.

[163] Y. Kuroda, Effect of chemisorbed water on the two-dimensional condensation of water and argon on CaF_2, J. Chem. Soc., Faraday Trans. 1 81 (1985) 757–768.

[164] Y. Kuroda, S. Kittaka, K. Miura, T. Morimoto, Effect of chemisorbed water on the two-dimensional condensation of water and argon on strontium fluoride, Langmuir 4 (1988) 210–215.

[165] S. Folsch, A. Stock, M. Henzler, two-dimensional water condensation on the NaCl(100) surface, Surf. Sci. 264 (1992) 65–72.

[166] L. W. Bruch, A. Glebov, J. P. Toennies, H. Weiss, A helium atom scattering study of water adsorption on the NaCl(100) single crystal surface, J. Chem. Phys. 103 (1995) 5109–5120.

[167] T. Morimoto, T. Kadota, Y. Kuroda, Adsorption of water on CaF_2: Two-dimensional condensation of water, J. Colloid Interface Sci. 106 (1985) 104–109.

[168] Y. Kuroda, Y. Yoshikawa, Y. Yokota, T. Morimoto, Effect of changing exposed surfaces of strontium fluoride crystal on the two-dimensional condensation of water and krypton, Langmuir 6 (1990) 1544–1548.

[169] Y. Kuroda, T. Matsuda, M. Nagao, Heat of adsorption of water on SrF_2: Relation to two-dimensional condensation of water adsorbed on SrF_2, J. Chem. Soc., Faraday Trans. 89 (1993) 2041–2048.

[170] Y. Kuroda, Y. Yoshikawa, T. Morimoto, M. Nagao, Dielectric behavior in the SrF_2-H_2O system. 1. Measurement at room temperature, Langmuir 11 (1995) 259–264.

[171] M. Nagao, R. Kumashiro, T. Matsuda, Y. Kuroda, Calorimetric study of water two-dimensionally condensed on the homogeneous surface of a solid, Thermochim. Acta 253 (1995) 221–233.

[172] T. Morimoto, M. Nagao, Adsorption anomaly in the system zinc oxide-water, J. Phys. Chem. 78 (1974) 1116–1120.

[173] S. Kittaka, S. Kanemoto, T. Morimoto, Interaction of water molecules with the surface of tin(IV) oxide, J. Chem. Soc., Faraday Trans. 1 74 (1978) 676–685.

[174] T. Morimoto, Y. Yokota, S. Kittaka, Adsorption anomaly in the system tin(IV) oxide-water, J. Phys. Chem. 82 (1978) 1996–1999.

[175] S. Kittaka, J. Nishiyama, K. Morishige, T. Morimoto, Two-dimensional condensation of water on the surface of Cr_2O_3, Colloids Surf. 3 (1981) 51–60.

[176] S. Kittaka, K. Morishige, J. Nishiyama, T. Morimoto, The effect of surface hydroxyls of Cr_2O_3 on the adsorption of N_2, Ar, Kr, and H_2O in connection with the two-dimensional condensation, J. Colloid Interface Sci. 91 (1983) 117–124.

[177] S. Kittaka, T. Sasaki, N. Fukuhara, H. Kato, Fourier-transform infrared spectroscopy of H_2O molecules on the Cr_2O_3 surface, Surf. Sci. 282 (1993) 255–261.

[178] Y. Kuroda, S. Kittaka, S. Takahara, T. Yamaguchi, M.-C. Bellissent-Funel, Characterization of the state of two-dimensionally condensed water on hydroxylated chromium(III) oxide surface through FT-IR, quasielastic neutron scattering, and dielectric relaxation measurements, J. Phys. Chem. B 103 (1999) 11064–11073.

[179] T. Miyazaki, Y. Kuroda, K. Morishige, S. Kittaka, J. Umemura, T. Takenaka, T. Morimoto, Interaction of the surface of BeO with water: In connection with the two-dimensional condensation of water, J. Colloid Interface Sci. 106 (1985) 154–160.

[180] D. Ferry, A. Glebov, V. Senz, J. Suzanne, J. P. Toennies, H. Weiss, Observation of the second ordered phase of water on the MgO(100) surface: Low energy electron diffraction and helium atom scattering studies, J. Chem. Phys. 105 (1996) 1697–1701.

[181] B. Demirdjian, J. Suzanne, D. Ferry, J. P. Coulomb, L. Giordano, Neutron diffraction investigation of water on MgO(001) surfaces, from monolayer to bulk condensation, Surf. Sci. 462 (2000) L581–L586.

[182] S. Peters, G. Ewing, Water on salt: An infrared study of adsorbed H_2O on NaCl(100) under ambient conditions, J. Phys. Chem. B 101 (1997) 10880–10886.

[183] S. Takahara, S. Kittaka, T. Mori, Y. Kuroda, T. Yamaguchi, K. Shibata, Neutron scattering study on the dynamics of water molecules adsorbed on SrF_2 and ZnO surfaces, J. Phys. Chem. B 106 (2002) 5689–5694.

[184] M. C. Foster, G. E. Ewing, Adsorption of water on the NaCl(001) surface. II. An infrared study at ambient temperatures, J. Chem. Phys. 112 (2000) 6817–6826.

[185] M. Nagao, Physisorption of water on zinc oxide surface, J. Phys. Chem. 75 (1971) 3822–3828.

[186] A. Z. Panagiotopoulos, Direct determination of phase coexistence properties of fluids by Monte Carlo simulation in a new ensemble, Mol. Phys. 61 (1987) 813–826.

[187] A. Z. Panagiotopoulos, Adsorption and capillary condensation of fluids in cylindrical pores by Monte Carlo simulation in the Gibbs ensemble, Mol. Phys. 62 (1987) 701–719.

[188] A. Patrykiejew, D. P. Landau, K. Binder, Lattice gas models for multilayer adsorption: variation of phase diagrams with the strength of the substrate potential, Surf. Sci. 238 (1990) 317–329.

[189] M. Kruk, A. Patrykiejew, S. Sokolowski, The crossover from strong to intermediate substrate regimes in multilayer adsorption, Thin Sol. Films 238 (1994) 302–311.

[190] S. Sokolowski, A. Patrykiejew, Monte-Carlo study of physical adsorption: Comparison of the critical properties for two- and three-dimensional models of adsorption, Thin Sol. Films 128 (1985) 171–180.

[191] Y. C. Kim, M. E. Fisher, G. Orkoulas, Asymmetric fluid criticality. I. Scaling with pressure mixing, Phys. Rev. E 67 (2003) 061506.

[192] K. Shirono, H. Daiguji, Molecular simulation of the phase behavior of water confined in silica nanopores, J. Phys. Chem. C 111 (2007) 7938–7946.

[193] M. Drir, H. S. Nham, G. B. Hess, Multilayer adsorption and wetting: Ethylene on graphite, Phys. Rev. B 33 (1986) 5145–5148.

[194] H. K. Kim, Y. P. Feng, Q. M. Zhang, M. H. W. Chan, Phase transitions of ethylene on graphite, Phys. Rev. B 37 (1986) 3511–3523.

[195] H. S. Nham, G. B. Hess, Layer critical points of multilayer ethane adsorbed on graphite, Phys. Rev. B 38 (1988) 5166–5169.

[196] X. Zhao, S. Kwon, R. D. Vidic, E. Borquet, J. K. Johnson, Layering and orientational ordering of propane on graphite: An

experimental and simulation study, J. Chem. Phys. 117 (2002) 7719–7731.

[197] M. Drir, G. B. Hess, Multilayer adsorption of oxygen on graphite near the triple point, Phys. Rev. B 33 (1986) 4758–4761.

[198] K. Morishige, K. Hayashi, K. Izawa, I. Ohfuzi, Y. Okuda, Formation of a bimolecular first layer in tert-butanol on graphite, Phys. Rev. Lett. 68 (1992) 2196–2199.

[199] G. T. Gao, X. C. Zeng, H. Tanaka, The melting temperature of proton-disordered hexagonal ice: A computer simulation of 4-site transferable intermolecular potential model of water, J. Chem. Phys. 112 (2000) 8534–8538.

[200] I. Brovchenko, A. Krukau, A. Oleinikova, A. K. Mazur, Water percolation governs polymorphic transitions and conductivity of DNA, Phys. Rev. Lett. 97 (2006) 137801.

[201] A. Huerta, O. Pizio, S. Sokolowski, Phase transitions in an associating, network-forming, Lennard-Jones fluid in slit-like pores. II. Extension of the density functional method, J. Chem. Phys. 112 (2000) 4286–4295.

[202] B. M. Malo, L. Salazar, S. Sokolowski, O. Pizio, Application of the density functional method to study adsorption and phase transitions in two-site associating, Lennard-Jones fluids in cylindrical pores, J. Phys.: Condens. Matt. 12 (2000) 8785–8800.

[203] A. Huerta, O. Pizio, P. Bryk, S. Sokolowski, Application of the density functional method to study phase transitions in an associating Lennard-Jones fluid adsorbed in energetically heterogeneous slit-like pores, Mol. Phys. 98 (2000) 1859–1869.

[204] I. Brovchenko, A. Geiger, A. Oleinikova, Clustering of water molecules in aqueous solutions: Effect of water-solute interaction, Phys. Chem. Chem. Phys. 6 (2004) 1982–1987.

[205] I. Brovchenko, A. Geiger, A. Oleinikova, Liquid-vapor phase diagrams of water in nanopores, in: V. V. Brazhkin, S. V. Buldyrev,

V. N. Rhyzhov, H. E. Stanley (Eds.), New Kinds of Phase Transitions: Transformations in Disordered Substances, Proceedings of NATO Advanced Research Workshop, Volga River, Kluver, Dordrecht, 2002, pp. 367–380.

[206] I. Brovchenko, D. Paschek, A. Geiger, Gibbs ensemble simulation of water in spherical cavities, J. Chem. Phys. 113 (2000) 5026–5036.

[207] I. Brovchenko, A. Geiger, D. Paschek, Simulation of confined water in equilibrium with a bulk reservoir, Fluid Phase Equilib. 183 (2001) 331–339.

[208] I. Brovchenko, A. Geiger, Water in nanopores in equilibrium with a bulk reservoir: Gibbs ensemble Monte Carlo simulations, J. Mol. Liq. 96 (2002) 195–206.

[209] D. Nicolaides, R. Evans, Monte Carlo study of phase transitions in a confined lattice gas, Phys. Rev. B 39 (1989) 9336–9342.

[210] J. O. Indekeu, Thin-thick adsorption phase-transitions and competing short-range forces, Europhys. Lett. 10 (1989) 165–170.

[211] J. O. Indekeu, Must thin-thick transitions precede long-range critical wetting? Phys. Rev. Lett. 85 (2000) 4188.

[212] H. K. Christenson, P. M. Claesson, Direct measurements of the force between hydrophobic surfaces in water, Adv. Colloid Interface Sci. 91 (2001) 391–436.

[213] O. I. Vinogradova, Slippage of water over hydrophobic surfaces, Int. J. Miner. Process. 56 (1999) 31–60.

[214] P. Attard, Nanobubbles and the hydrophobic attraction, Adv. Colloid Interface Sci. 104 (2003) 75–91.

[215] J. W. G. Tyrrell, P. Attard, Images of nanobubbles on hydrophobic surfaces and their interactions, Phys. Rev. Lett. 87 (2001) 176104.

[216] N. Ishida, M. Sakamoto, M. Miyahara, K. Higashitani, Attraction between hydrophobic surfaces with and without gas phase, Langmuir 16 (2000) 5681–5687.

[217] G. E. Yakubov, H.-J. Butt, O. I. Vinogradova, Interaction forces between hydrophobic surfaces. Attractive jump as an indication of formation of "stable" submicrocavities, J. Phys. Chem. 104 (2000) 3407–3410.

[218] A. C. Simonsen, P. L. Hansen, B. Klosgen, Nanobubbles give evidence of incomplete wetting at a hydrophobic interface, J. Colloid Interface Sci. 273 (2004) 291–299.

[219] U. K. Sur, V. Lakshminarayanan, Existence of a hydrophobic gap at the alkanethiol SAM-water interface: An interfacial capacitance study, J. Colloid Interface Sci. 254 (2002) 410–413.

[220] V. Lakshminarayanan, U. K. Sur, Hydrophobicity-induced drying transition in alkanethiol self-assembled monolayer–water interface, Pramana J. Phys. 61 (2003) 361–371.

[221] U. K. Sur, V. Lakshminarayanan, A study of the hydrophobic properties of alkanethiol self-assembled monolayers prepared in different solvents, J. Electroanal. Chem. 565 (2004) 343–350.

[222] S. M. Dammer, D. Lohse, Gas enrichment at liquid-wall interfaces, Phys. Rev. Lett. 96 (2006) 206101.

[223] M. Mao, J. Zhang, R.-H. Yoon, W. Ducker, Is there a thin film of air at the interface between water and smooth hydrophobic solids? Langmuir 20 (2004) 1843–1849.

[224] Y. Takata, J.-H. Cho, B. Law, M. Aratono, Ellipsometric search for vapor layers at liquid-hydrophobic solid surfaces, Langmuir 22 (2006) 1715–1721.

[225] Y.-S. Seo, S. Satija, No intrinsic depletion layer on a polystyrene thin film at a water interface, Langmuir 22 (2006) 7113–7116.

[226] D. Schwendel, T. Hayashi, R. Dahint, A. Pertsin, M. Grunze, R. Streitz, F. Schreiber, Interaction of water with self-assembled monolayers: Neutron reflectivity measurements of the water density in the interface region, Langmuir 19 (2003) 2284–2293.

[227] D. A. Doshi, E. B. Watkins, J. N. Israelachvili, J. Majewski, Reduced water density at hydrophobic surfaces: Effect of dissolved gases, Proc. Natl. Acad. Sci. U.S.A. 102 (2005) 9458–9462.

[228] M. Maccarini, R. Steitz, M. Himmelhaus, J. Fick, S. Tatur, M. Wolff, M. Grunze, J. Janecek, R. Netz, Density depletion at solid-liquid interfaces: A neutron reflectivity study, Langmuir 23 (2007) 598–608.

[229] T. R. Jensen, M. O. Jensen, N. Reitzel, K. Balashev, G. H. Peters, K. Kjaer, T. Bjornholm, Water in contact with extended hydrophobic surfaces: Direct evidence of weak dewetting, Phys. Rev. Lett. 90 (2003) 086101.

[230] A. Poynor, L. Hong, I. K. Robinson, S. Granick, Z. Zhang, P. A. Fenter, How water meets a hydrophobic surface, Phys. Rev. Lett. 97 (2006) 266101.

[231] M. Mezger, H. Reichert, S. Schoder, J. Okasinski, H. Schroder, H. Dosch, D. Palms, J. Ralston, V. Honkimaki, High-resolution in situ X-ray study of the hydrophobic gap at the water-octadecyltrichlorosilane interface, Proc. Natl. Acad. Sci. U.S.A. 103 (2006) 18401–18404.

[232] L. Castro, A. Almeida, D. Petri, The effect of water or salt solution on thin hydrophobic films, Langmuir 20 (2004) 7610–7615.

[233] Z. Ge, D. G. Cahill, P. V. Braun, Thermal conductance of hydrophilic and hydrophobic interfaces, Phys. Rev. Lett. 96 (2006) 186101.

[234] R. Helmy, Y. Kazakevich, C. Ni, A. Fadeev, Wetting in hydrophobic nanochannels: A challenge of classical capillarity, J. Am. Chem. Soc. 127 (2005) 12446–12447.

[235] A. Nakajima, K. Hashimoto, T. Watanabe, Recent studies on super-hydrophobic films, Monatsh. Chem. 132 (2001) 31–41.

[236] A. Wallqvist, B. J. Berne, Computer simulation of hydrophobic hydration forces on stacked plates at short range, J. Phys. Chem. 99 (1995) 2893–2899.

[237] K. Lum, D. Chandler, J. D. Weeks, Hydrophobicity at small and large length scales, J. Phys. Chem. B 103 (1999) 4570–4577.

[238] D. M. Huang, D. Chandler, Cavity formation and the drying transition in the Lennard-Jones fluid, Phys. Rev. E 61 (2000) 1501–1506.

[239] X. Huang, C. J. Margulis, B. J. Berne, Dewetting-induced collapse of hydrophobic particles, Proc. Natl. Acad. Sci. U.S.A. 100 (2003) 11953–11958.

[240] K. Leung, A. Luzar, D. Bratko, Dynamics of capillary drying in water, Phys. Rev. Lett. 90 (2003) 065502.

[241] Q. Huang, S. Ding, C.-Y. Hua, H.-C. Yang, C.-L. Chen, A computer simulation study of water drying at the interface of protein chains, J. Chem. Phys. 121 (2004) 1969–1977.

[242] J. H. Walther, R. L. Jaffe, E. M. Kotsalis, T. Werder, T. Halicioglu, P. Koumoutsakos, Hydrophobic hydration of C60 and carbon nanotubes in water, Carbon 42 (2004) 1185–1194.

[243] L. Liu, S.-H. Chen, A. Faraone, C.-W. Yen, C.-Y. Mou, Pressure dependence of fragile-to-strong transition and a possible second critical point in supercooled confined water, Phys. Rev. Lett. 95 (2005) 117802.

[244] X. Huang, R. Zhou, B. Berne, Drying and hydrophobic collapse of paraffin plates, J. Phys. Chem. B 109 (8) (2005) 3546–3552.

[245] S. Singh, J. Houston, F. van Swol, C. J. Brinker, Superhydrophobicity: Drying transition of confined water, Nature 442 (2006) 526.

[246] F. H. Stillinger, Structure in aqueous solutions of nonpolar solutes from the standpoint of scaled-particle theory, J. Solution Chem. 2 (1973) 141–158.

[247] D. Chandler, Interfaces and the driving force of hydrophobic assembly, Nature 437 (2005) 640–647.

[248] D. Chandler, Oil on troubled waters, Nature 445 (2007) 831–832.

[249] I. Brovchenko, A. Geiger, A. Oleinikova, D. Paschek, Phase coexistence and dynamic properties of water in nanopores, Eur. Phys. J. E 12 (2003) 69–76.

[250] I. Brovchenko, A. Geiger, A. Oleinikova, Water in nanopores: II. The liquid-vapour phase transition near hydrophobic surfaces, J. Phys.: Condens. Matt. 16 (2004) S5345–S5370.

[251] T. Werder, J. Walther, R. Jaffe, T. Halicioglu, P. Koumoutsakos, On the water-carbon interaction for use in molecular dynamics simulations of graphite and carbon nanotubes, J. Phys. Chem. B 107 (2003) 1345–1352.

[252] T. Werder, J. Walther, R. Jaffe, P. Koumoutsakos, Water-carbon interactions: Potential energy calibration using experimental data, Nanotechnology 3 (2003) 546–548.

[253] Y. Ikezoe, N. Hirota, J. Nakagawa, K. Kitazawa, Making water levitate, Nature 393 (1998) 749–750.

[254] K. Binder, Critical behaviour at surfaces, in: C. Domb, J. L. Lebowitz (Eds.), Phase Transitions and Critical Phenomena, London: Academic Press, 1983, pp. 1–144.

[255] M. E. Fisher, P.-G. de Gennes, Phénomènes aux parois dans un mélange binaire critique : physique des colloïdes, C. R. Acad. Sc. B (Paris) 287 (1978) 207–208.

[256] A. Bray, M. Moor, Critical behavior of semi-infinite systems, J. Phys. A: Math. Gen. 10 (1977) 1927–1962.

[257] H. W. Diehl, Critical adsorption of fluids and the equivalence of extraordinary to normal surface transitions, Ber. Bunsenges. Phys. Chem. 98 (1994) 466–471.

[258] T. W. Burkhardt, H. W. Diehl, Ordinary, extraordinary, and normal surface transitions: Extraordinary-normal equivalence and simple explanation of $|t - t_c|^{2-\alpha}$ singularities, Phys. Rev. B 50 (1994) 3894–3898.

[259] U. Ritschel, P. Czerner, Near-surface long-range order at the ordinary transition, Phys. Rev. Lett. 77 (1996) 3645–3648.

[260] P. Czerner, U. Ritschel, Near-surface long-range order at the ordinary transition: Scaling analysis and Monte Carlo results, Physica A 237 (1997) 240–256.

[261] P. Czerner, U. Ritschel, Magnetization profile in the D=2 semi-infinite Ising model and crossover between ordinary and normal transition, Int. J. Mod. Phys. B 11 (1997) 2075–2091.

[262] I. Brovchenko, A. Oleinikova, Surface critical behaviour of fluids and magnets, Mol. Phys. 104 (2006) 3535–3549.

[263] J. Specovius, G. H. Findenegg, Study of a fluid/solid interface over a wide density range including the critical region. 1. Surface exccess of ethylene/graphite, Ber. Bunsenges. Phys. Chem. 84 (1980) 690–696.

[264] R. Garcia, S. Scheidemantel, K. Knorr, M. H. W. Chan, Critical adsorption in a well-defined geometry, Phys. Rev. E 68 (2003) 056111.

[265] S. Blumel, G. H. Findenegg, Critical adsorption of a pure fluid on a graphite substrate, Phys. Rev. Lett. 54 (1985) 447–450.

[266] M. Thommes, G. Findenegg, H. Lewandowski, Critical adsorption of SF_6 on a finely divided graphite substrate, Ber. Bunsenges. Phys. Chem. 98 (1994) 477–481.

[267] M. Thommes, G. H. Findenegg, M. Schoen, Critical depletion of a pure fluid in controlled-pore glass. Experimental results and grand canonical ensemble Monte Carlo simulation, Langmuir 11 (1995) 2137–2142.

[268] A. Rajendran, T. Hocker, O. Di Giovanni, M. Mazzotti, Experimental observation of critical depletion: Nitrous oxide adsorption on silica gel, Langmuir 18 (2002) 9726–9734.

[269] J. Schulz, J. Bowers, G. H. Findenegg, Crossover from preferential adsorption to depletion: Aqueous systems of short-chain C_nE_m amphiphiles at liquid/liquid interfaces, J. Phys. Chem. 105 (2001) 6956–6964.

[270] L. Sigl, W. Fenzl, Order-parameter exponent β_1 of a binary liquid mixture at a boundary, Phys. Rev. Lett. 57 (1986) 2191–2194.

[271] D. Durian, C. Franck, Wetting phenomena of binary liquid mixtures on chemically altered substrates, Phys. Rev. Lett. 59 (1987) 555–558.

[272] W. Fenzl, Scaling properties of the surface tension of isobutyric-acid-water mixtures, Europhys. Lett. 24 (1993) 557–562.

[273] J.-H. Cho, B. Law, Ellipsometric study of undersaturated critical adsorption, Phys. Rev. E 65 (2002) 011601.

[274] A. Oleinikova, I. Brovchenko, A. Geiger, Behavior of a wetting phase near a solid boundary: vapor near a weakly attractive surface, Eur. Phys. J. B 52 (2006) 507519.

[275] H. W. Diehl, Field-theoretical approach to critical behaviour at surfaces, in: C. Domb, J. L. Lebowitz (Eds.), Phase Transitions and Critical Phenomena, New York: Academic Press, 1986, pp. 75–267.

[276] H. K. Christenson, Confinement effects on freezing and melting, J. Phys.: Condens. Matt. 13 (2001) R95–R133.

[277] C. Alba-Simionesco, B. Coasne, G. Dosseh, G. Dudziak, K. E. Gubbins, R. Radhakrishnan, M. Sliwinska-Bartkowiak, Effects of confinement on freezing and melting, J. Phys.: Condens. Matt. 18 (2006) R15–R68.

[278] L. Gelb, K. Gubbins, R. Radhakrishnan, M. Sliwinska-Bartkowiak, Phase separation in confined systems, Rep. Prog. Phys. 62 (1999) 1573–1659.

[279] S. Gross, G. H. Findenegg, Pore condensation in novel highly ordered mesoporous silica, Ber. Bunsenges. Phys. Chem. 101 (1991) 1726–1730.

[280] A. Schreiber, H. Bock, M. Schoen, G. H. Findenegg, Effect of surface modification on the pore condensation of fluids: Experimental results and density functional theory, Mol. Phys. 100 (2002) 2097–2107.

[281] K. Morishige, H. Fujii, M. Uga, D. Kinakawa, Capillary critical point of argon, nitrogen, oxygen, ethylene, and carbon dioxide in MCM-41, Langmuir 13 (1997) 3494–3498.

[282] K. Morishige, M. Shikimi, Adsorption hysteresis and pore critical temperature in a single cylindrical pore, J. Chem. Phys. 108 (1998) 7821–7824.

[283] K. Morishige, M. Ito, Capillary condensation of nitrogen in MCM-41 and SBA-15, J. Chem. Phys. 117 (2002) 8036–8041.

[284] W. Machin, Capillary condensation of hexafluoroethane in an ordered mesoporous silica, Phys. Chem. Chem. Phys. 5 (2003) 203–207.

[285] C. Sonwane, S. Bhatia, N. Calos, Experimental and theoretical investigations of adsorption hysteresis and criticality in MCM-41: Studies with O_2, Ar, and CO_2, Ind. Eng. Chem. Res. 47 (1998) 2271–2283.

[286] M. Kruk, M. Jaroniec, A. Sayari, Adsorption study of surface and structural properties of MCM-41 materials of different pore sizes, J. Phys. Chem. B 101 (1997) 583–589.

[287] J. Rathousky, A. Zukal, O. Franke, G. Schulz-Ekloff, Adsorption on MCM-41 mesoporous molecular sieves. 2. Cyclopentane isotherms and their temperature dependence, J. Chem. Soc. Faraday Trans. 91 (1995) 937–940.

[288] C. G. V. Burgess, D. Everett, S. Nutall, Adsorption hysteresis in porous materials, Pure Appl. Chem. 61 (1989) 1845–1852.

[289] A. de Keizer, T. Michalski, G. H. Findenegg, Fluids in pores: experimental and computer-simulation studies of multilayer adsorption, pore condensation and critical-point shifts, Pure Appl. Chem. 63 (1991) 1495–1502.

[290] W. Machin, Temperature dependence of hysteresis and the pore size distributions of two mesoporous adsorbents, Langmuir 10 (1994) 1235–1240.

[291] W. Machin, Properties of three capillary fluids in the critical region, Langmuir 15 (1999) 169–173.

[292] M. Thommes, G. H. Findenegg, Pore condensation and critical-point shift of a fluid in controlled-pore glass, Langmuir 10 (1994) 4270–4277.

[293] N. Wilkinson, M. Alam, J. Clayton, R. Evans, H. Fretwell, S. Usmar, Positron annihilation study of capillary condensation of nitrogen gas in a mesoporous solid, Phys. Rev. Lett. 69 (1992) 3535–3538.

[294] A. Wong, M. Chan, Liquid-vapor critical point of ^{4}He in aerogel, Phys. Rev. Lett. 65 (1990) 2567–2570.

[295] M. Chan, Phase transitions of helium in aerogel, Chech. J. Phys. 46, Suppl.S6 (1996) 2915–2922.

[296] A. Wong, S. Kim, W. Goldburg, M. Chan, Phase separation, density fluctuation, and critical dynamics of N_2 in aerogel, Phys. Rev. Lett. 70 (1993) 954–957.

[297] T. Herman, J. Beamish, Acoustic studies of liquid-vapor critical behavior of neon and helium in aerogels, J. Low Temp. Phys. 261 (2002) 661–666.

[298] S. Inoue, N. Ichikuni, T. Suzuki, T. Uematsu, K. Kaneko, Capillary condensation of N_2 on multiwall carbon nanotubes, J. Phys. Chem. B 102 (1998) 4689–4692.

[299] E. Kierlik, M. Rosinberg, G. Tarjus, P. Viot, Equilibrium and out-of-equilibrium (hysteretic) behavior of fluids in disordered porous materials: Theoretical predictions, Phys. Chem. Chem. Phys. 3 (2001) 1201–1206.

[300] R. Evans, Fluids adsorbed in narrow pores: phase equilibria and structure, J. Phys.: Condens. Matt. 46 (1990) 8989–9007.

[301] V. Privman, L. Schulman, Analytic properties of thermodynamic functions at 1st-order phase transitions, J. Phys. A: Math. Gen. 15 (1982) L231–L238.

[302] B. K. Peterson, K. E. Gubbins, G. S. Heffelfinger, U. M. B. Marconi, F. Swol, Lennard-Jones fluids in cylindrical pores: Non-local theory and computer simulation, J. Chem. Phys. 88 (1988) 6487–6500.

[303] G. Heffelfinger, F. Swol, K. Gubbins, Liquid-vapor coexistence in a cylindrical pore, Mol. Phys. 60 (1987) 1381–1390.

[304] G. Heffelfinger, F. Swol, K. Gubbins, Adsorption hysteresis in narrow pores, J. Chem. Phys. 89 (1988) 5202–5205.

[305] C. Bichara, J. Raty, R.-M. Pellenq, Adsorption of selenium wires in silicalite-1 zeolite: A first order transition in a microporous system, Phys. Rev. Lett. 89 (2002) 016101.

[306] V. Privman, M. Fisher, Finite-size effects at 1st-order transitions, J. Stat. Phys. 33 (1983) 385–417.

[307] R. Evans, U. M. B. Marconi, P. Tarazona, Capillary condensation and adsorption in cylindrical and slit-like pores, J. Chem. Soc. Faraday. Trans. 2 82 (1986) 1763–1787.

[308] L. Gelb, K. Gubbins, Kinetics of liquid-liquid phase separation of a binary mixture in cylindrical pores, Phys. Rev. E 55 (1997) R1290–R1293.

[309] L. Gelb, K. Gubbins, Liquid-liquid phase separation in cylindrical pores: Quench molecular dynamics and Monte Carlo simulations, Phys. Rev. E 56 (1997) 3185–3196.

[310] Z. Zhang, A. Chakrabarti, Phase separation of binary fluids confined in a cylindrical pore: A molecular dynamics study, Phys. Rev. E 50 (1994) R4290–R4293.

[311] Y. Melnichenko, G. Wignall, D. Core, H. Frielinghaus, Density fluctuations near the liquid-gas critical point of a confined fluid, Phys. Rev. E 69 (2004) 057102.

[312] E. Kierlik, P. Monson, M. Rosinberg, L. Sarkisov, G. Tarjus, Capillary condensation in disordered porous materials: Hysteresis versus equilibrium behavior, Phys. Rev. Lett. 87 (2001) 055701.

[313] F. Detcheverry, E. Kierlik, M. Rosinberg, G. Tarjus, Local mean-field study of capillary condensation in silica aerogels, Phys. Rev. E 68 (2003) 061504.

[314] F. Detcheverry, E. Kierlik, M. Rosinberg, G. Tarjus, Hysteresis in capillary condensation of gases in disordered porous solids, Physica B 343 (2004) 303–307.

[315] L. Sarkisov, P. Monson, Lattice model of adsorption in disordered porous materials: Mean-field density functional theory and Monte Carlo simulations, Phys. Rev. E 65 (2001) 011202.

[316] F. Brochard, P. de Gennes, Phase transitions of binary mixtures in random media, J. Phys. Lett. (Paris) 44 (1983) L785–L791.

[317] P. de Gennes, Liquid-liquid demixing inside a rigid network— qualitative features, J. Phys. Chem. 88 (1984) 6469–6472.

[318] R. L. C. Vink, K. Binder, H. Löwen, Critical behavior of colloid-polymer mixtures in random porous media, Phys. Rev. Lett. 97 (2006) 230603.

[319] W. Thomson, On the equilibrium of a vapour at a curved surface of liquid, Philos. Mag. 42 (1871) 448–452.

[320] M. Fisher, A. Berker, Scaling for first-order phase transitions in thermodynamic and finite systems, Phys. Rev. 26 (1982) 2507–2513.

[321] R. Evans, U. M. B. Marconi, P. Tarazona, Fluids in narrow pores: Adsorption, capillary condensation, and critical points, J. Chem. Phys. 84 (1986) 2376–2399.

[322] M. E. Fisher, H. Nakanishi, Scaling theory for the criticality of fluids between plates, J. Chem. Phys. 75 (1981) 5857–5863.

[323] H. Nakanishi, M. E. Fisher, Critical point shifts in films, J. Chem. Phys. 78 (1983) 3279–3293.

[324] R. J.-M. Pellenq, B. Rousseau, P. E. Levitz, A grand canonical Monte Carlo study of argon adsorption/condensation in mesoporous silica glasses, Phys. Chem. Chem. Phys. 3 (2001) 1207–1212.

[325] A. Vishnyakov, E. Piotrovskaya, E. Brodskaya, E. Votyakov, Y. Tovbin, The critical properties of a Lennard-Jones fluid in narrow slit-like pores, Russ. J. Phys. Chem. 74 (2000) 162–167.

[326] A. Vishnyakov, E. Piotrovskaya, E. Brodskaya, E. Votyakov, Y. Tovbin, Critical properties of Lennard-Jones fluids in narrow slit-shaped pores, Langmuir 17 (2001) 4451–4458.

[327] Y. Hiejima, M. Yao, Phase behaviour of water confined in Vycor glass at high temperatures and pressures, J. Phys.: Condens. Matt. 16 (2004) 7903–7908.

[328] A. Striolo, K. Gubbins, M. Gruszkiewicz, D. Cole, J. Simonson, A. Chialvo, P. Cummings, T. Burchell, K. More, Effect of temperature on the adsorption of water in porous carbons, Langmuir 21 (2005) 9457–9467.

[329] J. Rivera, C. McCabe, P. T. Cummings, Layering behavior and axial phase equilibria of pure water and water plus carbon dioxide inside single wall carbon nanotubes, Nano Lett. 2 (2002) 1427–1431.

[330] M. Jorge, N. A. Seaton, Molecular simulation of phase coexistence in adsorption in porous solids, Mol. Phys. 100 (2002) 3803–3815.

[331] H. Furukawa, K. Binder, Two-phase equilibria and nucleation barriers near a critical point, Phys. Rev. A 26 (1982) 556–566.

[332] L. G. MacDowell, P. Virnau, M. Muller, K. Binder, The evaporation/condensation transition of liquid droplets, J. Chem. Phys. 120 (2004) 5293–5308.

[333] M. E. Fisher, Proceedings of International School of Physics "Enrico Fermi" on Critical Phenomena Course 51, New York: Academic Press.

[334] K. Binder, Monte-Carlo study of thin magnetic Ising films, Thin Sol. Films 20 (1974) 367–381.

[335] P. Schilbe, S. Siebentritt, K.-H. Rieder, Monte Carlo calculations on the dimensional crossover of thin Ising films, Phys. Lett. A 216 (1996) 20–25.

[336] O. Dillmann, W. Janke, M. Mueller, K. Binder, A Monte Carlo test of the Fisher-Nakanishi-scaling theory for the capillary condensation critical point, J. Chem. Phys. 114 (2001) 5853–5862.

[337] A. Schreiber, I. Ketelsen, G. H. Findenegg, Melting and freezing of water in ordered mesoporous silica materials, Phys. Chem. Chem. Phys. 3 (2001) 1185–1195.

[338] K. Morishige, K. Kawano, Freezing and melting of water in a single cylindrical pore: The pore-size dependence of freezing and melting behavior, J. Chem. Phys. 110 (1999) 4867–4872.

[339] D. C. Steytler, J. C. Dore, C. J. Wright, Neutron diffraction study of cubic ice nucleation in a porous silica network, J. Phys. Chem. B 87 (1983) 2458–2459.

[340] D. C. Steytler, J. C. Dore, Neutron diffraction studies of water in porous silica, Mol. Phys. 56 (1985) 1001–1015.

[341] Y. P. Handa, M. Zakrzewski, C. Fairbridge, Effect of restricted geometries on the structure and thermodynamic properties of ice, J. Phys. Chem. B 96 (1992) 8594–8599.

[342] M.-C. Bellissent-Funel, J. Lal, L. Bosio, Structural study of water confined in porous glass by neutron scattering, J. Chem. Phys. 98 (1993) 4246–4252.

[343] J. Baker, J. Dore, P. Behrens, Nucleation of ice in confined geometry, J. Phys. Chem. B 101 (1997) 6226–6229.

[344] K. Overloop, L. Vangerven, Freezing phenomena in adsorbed water as studied by NMR, J. Magn. Reson., Ser. A 101 (1993) 179–187.

[345] R. Schmidt, E. W. Hansen, M. Stoecker, D. Akporiaye, O. H. Ellestad, Pore size determination of MCM-51 mesoporous materials by means of 1H NMR spectroscopy, N_2 adsorption, and HREM. A preliminary study, J. Am. Chem. Soc. 117 (1995) 4049–4056.

[346] Y. Hirama, T. Takahashi, M. Hino, T. Sato, Studies of water adsorbed in porous Vycor glass, J. Colloid Interface Sci. 184 (1996) 349–359.

[347] C. Faivre, D. Bellet, G. Dolino, Phase transitions of fluids confined in porous silicon: A differential calorimetry investigation, Eur. Phys. J. B 7 (1999) 19–36.

[348] Y. F. T. Ishizaki, M. Maruyama, J. G. Dash, Premelting of ice in porous silica glass, J. Cryst. Growth 163 (1996) 455–460.

[349] K. Ishikiriyama, M. Todoki, Evaluation of water in silica pores using differential scanning calorimetry, Thermochim. Acta 256 (1995) 213–226.

[350] E. Hansen, M. Stocker, R. Schmidt, Low-temperature phase transition of water confined in mesopores probed by NMR. Influence on pore size distribution, J. Phys. Chem. B 100 (1996) 2195–2200.

[351] E. Hansen, E. Tangstad, E. Myrvold, T. Myrstad, Pore structure characterization of mesoporous/microporous materials by 1H NMR using water as a probe molecule, J. Phys. Chem. B 101 (1997) 10709–10714.

[352] T. S. T. Iiyama, K. Nishikawa, K. Kaneko, Study of the structure of a water molecular assembly in a hydrophobic nanospace at low temperature with in situ X-ray diffraction, Chem. Phys. Lett. 274 (1997) 152–158.

[353] K. Koga, X. C. Zeng, H. Tanaka, Freezing of confined water: A bilayer ice phase in hydrophobic nanopores, Phys. Rev. Lett. 79 (1997) 5262–5265.

[354] J. Slovák, K. Koga, H. Tanaka, X. C. Zeng, Confined water in hydrophobic nanopores: Dynamics of freezing into bilayer ice, Phys. Rev. E 60 (1999) 5833–5840.

[355] R. Zangi, A. E. Mark, Monolayer ice, Phys. Rev. Lett. 91 (2003) 025502.

[356] R. Zangi, Water confined to a slab geometry: A review of recent computer simulation studies, J. Phys.: Condens. Matt. 16 (2004) S5371–S5388.

[357] P. Kumar, S. V. Buldyrev, F. W. Starr, N. Giovambattista, H. E. Stanley, Thermodynamics, structure and dynamics of water confined between hydrophobic plates, Phys. Rev. E 72 (2005) 051503.

[358] K. Koga, H. Tanaka, Phase diagram of water between hydrophobic surfaces, J. Chem. Phys. 122 (2005) 104711.

[359] T. Takamuku, M. Yamagami, H. Wakita, Y. Masuda, T. Yamaguchi, Thermal property, structure, and dynamics of

supercooled water in porous silica by calorimetry, neutron scattering, and NMR relaxation, J. Phys. Chem. B 101 (1997) 5730–5739.

[360] J.-M. Zanotti, M.-C. Bellissent-Funel, S.-H. Chen, Experimental evidence of a liquid-liquid transition in interfacial water, Europhys. Lett. 75 (2005) 91–97.

[361] J.-M. Zanotti, M. C. Bellissent-Funel, S.-H. Chen, A. I. Kolesnikov, Further evidence of a liquid-liquid transition in interfacial water, J. Phys.: Condens. Matt. 18 (2006) S2299–S2304.

[362] A. Faraone, L. Liu, C.-Y. Mou, C.-W. Yen, S.-H. Chen, Fragile-to-strong liquid transition in deeply supercooled confined water, J. Chem. Phys 121 (2004) 10843–10846.

[363] L. Liu, S.-H. Chen, A. Faraone, C.-W. Yen, C.-Y. Mou, A. I. Kolesnikov, E. Mamontov, J. Leno, Quasielastic and inelastic neutron scattering investigation of fragile-to-strong crossover in deeply supercooled water confined in nanoporous silica matrices, J. Phys.: Condens. Matt. 18 (2006) S2261–S2284.

[364] F. Mallamace, M. Broccio, C. Corsaro, A. Faraone, L. Liu, C.-Y. Mou, S.-H. Chen, Dynamical properties of confined supercooled water: An NMR study, J. Phys.: Condens. Matt. 18 (2006) S2285–S2297.

[365] F. Mallamace, M. Broccio, C. Corsaro, A. Faraone, U. Wanderlingt, L. Liu, C.-Y. Mou, S.-H. Chen, The fragile-to-strong dynamic crossover transition in confined water: Nuclear magnetic resonance results, J. Chem. Phys. 124 (2006) 161102.

[366] D. W. Hwang, C.-C. Chu, A. K. Sinha, L.-P. Hwang, Dynamics of supercooled water in various mesopore sizes, J. Chem. Phys. 126 (2007) 044702.

[367] E. Mamontov, C. J. Burnham, S.-H. Chen, A. P. Moravsky, C.-K. Loong, N. R. de Souza, A. I. Kolesnikov, Dynamics of water confined in single- and double-wall carbon nanotubes, J. Chem. Phys. 124 (2006) 194703.

[368] S.-H. Chen, L. Liu, X. Chu, Y. Zhang, E. Fratini, P. Baglioni, A. Faraone, E. Mamontov, Experimental evidence of fragile-to-strong dynamic crossover in DNA hydration water, J. Chem. Phys. 125 (2006) 171103.

[369] S.-H. Chen, L. Liu, E. Fratini, P. Baglioni, A. Faraone, E. Mamontov, Observation of fragile-to-strong dynamic crossover in protein hydration water, Proc. Natl. Acad. Sci. U.S.A. 103 (2006) 9012–9016.

[370] T. Truskett, P. Debenedetti, S. Torquato, Thermodynamic implications of confinement for a waterlike fluid, J. Chem. Phys. 114 (2001) 2401–2418.

[371] K. Koga, X. C. Zeng, H. Tanaka, Effects of confinement on the phase behavior of supercooled water, Chem. Phys. Lett. 285 (1998) 278–283.

[372] M. Meyer, H. E. Stanley, Liquid-liquid phase transition in confined water: A Monte Carlo study, J. Phys. Chem. B 103 (1999) 9728–9730.

[373] A. Koch, S. Siegesmund, The combined effect of moisture and temperature on the anomalous expansion behaviour of marble, Environ. Geol. 46 (2004) 350–363.

[374] H. Dominguez, M. P. Allen, R. Evans, Monte Carlo studies of the freezing and condensation transitions of confined fluids, Mol. Phys. 96 (1999) 209–229.

[375] B. Lefevre, A. Saugey, J. L. Barrat, L. Bocquet, E. Charlaix, P. F. Gobin, G. Vigier, Intrusion and extrusion of water in hydrophobic mesopores, J. Chem. Phys. 120 (2004) 4927–4938.

[376] N. Desbiens, I. Demachy, A. Fuchs, H. Kirsch-Rodeschini, M. Soulard, J. Patarin, Water condensation in hydrophobic nanopores, Angew. Chem. 117 (2005) 5444–5447.

[377] T. Hayashi, A. J. Pertsin, M. Grunze, Grand canonical Monte Carlo simulation of hydration forces between nonorienting and orienting structureless walls, J. Chem. Phys. 117 (2002) 6271–6280.

[378] N. Giovambattista, P. J. Rossky, P. G. Debenedetti, Effect of pressure on the phase behavior and structure of water confined between nanoscale hydrophobic and hydrophilic plates, Phys. Rev. E 73 (2006) 041604.

[379] Y. C. Kong, D. Nicholson, N. G. Parsonage, L. Thomson, Monte Carlo simulations of a polyoxyethylene $C_{12}E_2$ lamellar bilayer in water, Mol. Phys. 89 (1996) 835–865.

[380] S. C. McGrother, K. E. Gubbins, Constant pressure Gibbs ensemble Monte Carlo simulations of adsorption into narrow pores, Mol. Phys. 97 (1999) 955–965.

[381] D. R. Berard, P. Attard, G. N. Pattey, Cavitation of a Lennard-Jones fluid between hard walls, and the possible relevance to the attraction measured between hydrophobic surfaces, J. Chem. Phys. 98 (1993) 7236–7244.

[382] J. P. Noworyta, D. Henderson, S. Sokolowski, Density profiles and solvation force for a liquid in a slit, Mol. Phys. 96 (1999) 1139–1143.

[383] V. Yaminsky, S. Ohnishi, Physics of hydrophobic cavities, Langmuir 19 (2003) 1970–1976.

[384] G. Hummer, J. C. Rasaiah, J. P. Noworyta, Water conduction through the hydrophobic channel of a carbon nanotube, Nature 414 (2001) 188–190.

[385] R. Allen, J.-P. Hansen, S. Melchionna, Molecular dynamics investigation of water permeation through nanopores, J. Chem. Phys. 119 (2003) 3905–3919.

[386] O. Beckstein, M. Sansom, Liquid-vapor oscillations of water in hydrophobic nanopores, Proc. Natl. Acad. Sci. U.S.A. 100 (2003) 7063–7068.

[387] P. Attard, C. P. Ursenbach, G. N. Pattey, Long-range attractions between solutes in near-critical fluids, Phys. Rev. A 45 (1992) 7621–7623.

[388] J. Forsman, B. Jonsson, C. Woodward, H. Wennerstrom, Attractive surface forces due to liquid density depression, J. Phys. Chem. B 101 (1997) 4253–4259.

[389] J. Forsman, C. E. Woodward, B. Johnsson, Repulsive hydration forces and attractive hydrophobic forces in a unified picture, J. Colloid Interface Sci. 195 (1997) 264–266.

[390] B. Derjaguin, E. Obuchov, Anomalien Dunner Flussigkeitsschichten III, Ultramikrometrische Untersuchungen der Solvathullen und Des elementaren Quellungsaktes, Acta Physicochimica U.R.S.S. 5 (1936) 1–22.

[391] I. Langmuir, The role of attractive and repulsive forces in the formation of tactoids, thixotropic gels, protein crystals and coacervates, J. Chem. Phys. 6 (1938) 873–896.

[392] J. N. Israelachvili, H. Wennerstroem, Hydration or steric forces between amphiphilic surfaces? Langmuir 6 (1990) 873–876.

[393] M. Fisher, The theory of condensation and critical point, Physics 3 (1967) 255–283.

[394] A. Oleinikova, I. Brovchenko, A. Geiger, Percolation transition of hydration water at hydrophilic surfaces, Physica A 364 (2006) 1–12.

[395] A. Geiger, H. E. Stanley, Tests of universality of percolation exponents for a three-dimensional continuum system of interacting waterlike particles, Phys. Rev. Lett. 49 (1982) 1895–1898.

[396] D. Stauffer, Introduction to Percolation Theory, London and Philadelphia; Taylor and Francis, 1985.

[397] N. Jan, Large lattice random site percolation, Physica A 266 (1999) 72–75.

[398] A. Oleinikova, I. Brovchenko, Percolation transition of hydration water in biosystems, Mol. Phys. 104 (2006) 3841–3855.

[399] J.-P. Hovi, A. Aharony, Scaling and universality in the spanning probability for percolation, Phys. Rev. E 53 (1996) 235–253.

[400] P. Sen, Nature of the largest cluster size distribution at the percolation threshold, J. Phys. A: Math. Gen. 34 (2001) 8477–8483.

[401] A. Oleinikova, N. Smolin, I. Brovchenko, A. Geiger, R. Winter, Formation of spanning water networks on protein surfaces via 2D percolation transition, J. Phys. Chem. B 109 (2005) 1988–1998.

[402] N. Smolin, A. Oleinikova, I. Brovchenko, A. Geiger, R. Winter, Properties of spanning water networks at protein surfaces, J. Phys. Chem. B 109 (2005) 10995–11005.

[403] M. E. J. Newman, R. M. Ziff, Efficient Monte Carlo algorithm and high-precision results for percolation, Phys. Rev. Lett. 85 (2000) 4104–4107.

[404] E. J. Murphy, A. C. Walker, Electrical conduction in textiles. I. The dependence of the resistivity of cotton, silk and wool on relative humidity and moisture content, J. Phys. Chem. 32 (1928) 1761–1786.

[405] E. J. Murphy, The dependence of the conductivity of cellulose, silk and wool on their water content, J. Phys. Chem. Solids 16 (1960) 115–122.

[406] E. J. Murphy, Ionic conduction in keratin (wool), J. Colloid Interface Sci. 54 (1976) 400–408.

[407] S. Kurosaki, The dielectric behavior of sorbed water on silica gel, J. Phys. Chem. 58 (1954) 320–324.

[408] S. M. Nelson, A. C. D. Newman, T. E. Tomlinson, L. E. Sutton, A dielectric study of the adsorption of water by magnesium hydroxide, Trans. Faraday Soc. 55 (1959) 2186–2202.

[409] M. G. Baldwin, J. C. Morrow, Dielectric behavior of water adsorbed on alumina, J. Chem. Phys. 36 (1962) 1591–1593.

[410] A. Kyritsis, M. Siakantari, A. Vassilikou-Dova, P. Pissis, P. Varotsos, Dielectric and electrical properties of polycrystalline rocks at various hydration levels, IEEE Transactions on Dielectrics and Electrical Insulation 7 (2000) 493–497.

[411] W. Y. Hsu, J. R. Barkley, P. Meakin, Ion percolation and insulator-to-conductor transition in nafion perfluorosulfonic acid membranes, Macromolecules 13 (1980) 198–200.

[412] R. Wodzki, A. Nareogonbska, W. K. Nioch, Percolation conductivity in nafion membranes, J. Appl. Polym. Sci. 30 (1985) 769–780.

[413] X. Tongwen, Y. Weihua, H. Binglin, Ionic conductivity threshold in sulfonated poly (phenylene oxide) matrices: A combination of three-phase model and percolation theory, Chem. Eng. Sci. 56 (2001) 5343–5350.

[414] C. A. Edmondson, J. J. Fontanella, S. H. Chung, S. G. Greenbaum, G. E. Wnek, Complex impedance studies of S-SEBS block polymer proton-conducting membranes, Electrochim. Acta 46 (2001) 1623–1628.

[415] C. A. Edmondson, J. J. Fontanella, Free volume and percolation in S-SEBS and fluorocarbon proton conducting membranes, Solid State Ionics 152-153 (2002) 355–361.

[416] D. Stauffer, A. Aharony, Introduction to Percolation Theory, London: Taylor and Francis, 1992.

[417] C.-J. Yu, A. Richter, A. Datta, M. Durbin, P. Dutta, Observation of molecular layering in thin liquid films using X-ray reflectivity, Phys. Rev. Lett. 82 (1999) 2326–2329.

[418] C.-J. Yu, A. Richter, J. Kmetko, A. Datta, P. Dutta, X-ray diffraction evidence of ordering in a normal liquid near the solid-liquid interface, Europhys. Lett. 50 (2000) 487–493.

[419] G. Evmenenko, S. Dugan, J. Kmetko, P. Dutta, Molecular ordering in thin liquid films of polydimethylsiloxanes, Langmuir 17 (2001) 4021–4024.

[420] C. Lee, J. McCammon, P. Rossky, The structure of liquid water at an extended hydrophobic surface, J. Chem. Phys. 80 (1984) 4448–4455.

[421] A. Belch, M. Berkowitz, Molecular-dynamics simulations of TIPS2 water restricted by a spherical hydrophobic boundary, Chem. Phys. Lett. 113 (1985) 278–282.

[422] J. Valleau, A. Gardner, Water-like particles at surfaces. I. The uncharged, unpolarized surface, J. Chem. Phys. 86 (1987) 4162–4170.

[423] L. Zhang, H. T. Devis, D. M. Kroll, H. S. White, Molecular-dynamics simulations of water in a spherical cavity, J. Phys. Chem. 99 (1995) 2878–2884.

[424] J. C. Shelley, G. N. Patey, Modeling and structure of mercury-water interfaces, J. Chem. Phys. 107 (1997) 2122–2141.

[425] C. H. Bridgeman, N. T. Skipper, A Monte Carlo study of water at an uncharged clay surface, J. Phys.: Condens. Matt. 9 (1997) 4081–4087.

[426] E. Spohr, K. Heinzinger, Molecular dynamics simulation of a water metal interface, Chem. Phys. Lett. 123 (1986) 218–221.

[427] I.-C. Yeh, M. Berkowitz, Aqueous solution near charged Ag(111) surfaces: comparison between a computer simulation and experiment, Chem. Phys. Lett. 301 (1999) 81–86.

[428] I.-C. Yeh, M. Berkowitz, Effects of the polarizability and water density constraint on the structure of water near charged surfaces: Molecular dynamics simulations, J. Chem. Phys. 112 (2000) 10491–10495.

[429] P. Gallo, M. A. Ricci, M. Rovere, Layer analysis of the structure of water confined in vycor glass, J. Chem. Phys. 116 (2002) 342–346.

[430] J. Puibasset, R. J.-M. Pellenq, Grand canonical Monte Carlo simulation study of water structure on hydrophilic mesoporous and plane silica substrates, J. Chem. Phys. 119 (2003) 9226–9232.

[431] J. Puibasset, R. J.-M. Pellenq, A comparison of water adsorption on ordered and disordered silica substrates, Phys. Chem. Chem. Phys. 6 (2004) 1933–1937.

[432] D. Ferry, A. Glebov, V. Senz, J. Suzanne, J. Toennies, H. Weiss, The properties of a two-dimensional water layer on MgO(001), Surf. Sci. 377–379 (1997) 634–638.

[433] K. Jug, G. Geudtner, Quantum chemical study of water adsorption at the NaCl(100) surface, Surf. Sci. 371 (1997) 95–99.

[434] D. P. Taylor, W. P. Hess, M. I. McCarthy, Structure and energetics of the water/NaCl(100) interface, J. Phys. Chem. 101 (1997) 7455–7463.

[435] A. Marmier, P. Hoang, S. Picaud, C. Girardet, R. M. Lynden-Bell, A molecular dynamics study of the structure of water layers adsorbed on MgO(100), J. Chem. Phys. 109 (1998) 3245–3254.

[436] L. Giordano, J. Goniakowski, J. Suzanne, Partial dissociation of water molecules in the (3×2) water monolayer deposited on the MgO (100) surface, Phys. Rev. Lett. 81 (1998) 1271–1273.

[437] O. Engkvist, A. J. Stone, Adsorption of water on NaCl(001). I. Intermolecular potentials and low temperature structures, J. Chem. Phys. 110 (1999) 12089–12096.

[438] M. Odelius, Mixed molecular and dissociative water adsorption on MgO[100], Phys. Rev. Lett. 82 (1999) 3919–3922.

[439] A. Rahman, F. H. Stillinger, Hydrogen-bond patterns in liquid water, J. Am. Chem. Soc. 95 (1973) 7943–7948.

[440] P. Mausbach, J. Schnitker, A. Geiger, Hydrogen bond ring structures in liquid water. A molecular dynamics study, J. Tech. Phys. 28 (1987) 67–76.

[441] P. Fenter, N. C. Sturchio, Mineral-water interfacial structures revealed by synchrotron X-ray scattering, Prog. Surf. Sci. 77 (2004) 171–258.

[442] M. Chaplin, Do we underestimate the importance of water in cell biology? Nat. Rev. Mol. Cell Biol. 7 (2006) 861–866.

[443] Y. Levy, J. N. Onuchic, Water mediation in protein folding and molecular recognition, Annu. Rev. Biophys. Biomol. Struct. 35 (2006) 389–415.

[444] G. Singh, I. Brovchenko, A. Oleinikova, R. Winter, Aggregation of fragments of the islet amyloid polypeptide as a phase transition: A cluster analysis, in: U. H. E. Hansmann, J. Meinke, S. Mohanty, O. Zimmermann (Eds.), Proceedings of the NIC Workshop From Computational Biophysics to Systems Biology, NIC Series, Jülich: John von Neumann Institute for Computing, 2007, pp. 275–278.

[445] D. Keilin, The Leeuwenhoek lecture: The problem of anabiosis or latent life: History and current concept, Proc. R. Soc. London, Ser. B 150 (1959) 149–191.

[446] J. H. Crowe, J. S. Clegg (Eds.), Dry Biological Systems, New York: Academic Press, 1978.

[447] A. C. Leopold (Ed.), Membranes, Metabolism, and Dry Organisms, Ithaca: Cornell University Press, 1986.

[448] J. H. Crowe, F. A. Hoekstra, L. M. Crowe, Anhydrobiosis, Ann. Rev. Physiol. 54 (1992) 579–599.

[449] J. C. Wright, Cryptobiosis 300 years on from van Leuwenhoek: What have we learned about tardigrades?, Zoologischer Anzeiger 240 (2001) 563–582.

[450] A. Tunnacliffe, J. Lapinski, Resurrecting van Leeuwenhoek's rotifers: A reappraisal of the role of disaccharides in anhydrobiosis, Philos. Trans. R. Soc. London, Ser. B 358 (2003) 1755–1771.

[451] M. Watanabe, Anhydrobiosis in invertebrates, Appl. Entomol. Zool. 41 (2006) 15–31.

[452] G. V. Rijnberk, L. C. Palm (Eds.), The Collected Letters of Antoni van Leeuwenhoek, vol. 14, Lisse, The Netherlands: Swets and Zeitlinger, 1999.

[453] J. S. Clegg, Hydration-dependent metabolic transitions and the state of cellular water in Artemia cysts, in: J. H. Crowe, J. S. Clegg (Eds.), Dry Biological Systems, New York: Academic Press, 1978, pp. 117–154.

[454] J. S. Clegg, Interrelationships between water and metabolism in Artemia salina cysts: Hydration-dehydration from the liquid and vapour phases, J. Exp. Biol. 61 (1974) 291–308.

[455] J. S. Clegg, Interrelationships between water and metabolism in Artemia cysts. III. Respiration, Comp. Biochem. Physiol. 53a (1976) 89–92.

[456] A. Pigon, B. Weglarska, Rate of metabolism in Tardigrades during active life and anabiosis, Nature 176 (1955) 121–122.

[457] J. S. Clegg, Interrelationships between water and metabolism in Artemia cysts. II. Carbohydrates, Comp. Biochem. Physiol. 53a (1976) 83–87.

[458] J. S. Clegg, J. Cavagnaro, Interrelationships between water and metabolism in Artemia cysts. IV. Adenosine 5-triphosphate and cyst hydration, J. Cell. Physiol. 88 (1976) 159–166.

[459] J. S. Clegg, Interrelationships between water and metabolism in Artemia cysts. V. CO_2 incorporation, J. Cell. Physiol. 89 (1976) 369–380.

[460] J. S. Clegg, Interrelationships between water and metabolism in Artemia cysts. VI. RNA and protein synthesis, J. Cell. Physiol. 91 (1977) 143–154.

[461] J. S. Clegg, J. Lovallo, Interrelationships between water and metabolism in Artemia cysts. VII. Free amino acids, J. Cell. Physiol. 93 (1977) 161–168.

[462] J. S. Clegg, Interrelationships between water and metabolism in Artemia cysts. VIII. Sorption isotherms and derive a thermodynamic quantities, J. Cell. Physiol. 94 (1978) 123–137.

[463] S. Koga, A. Echigo, K. Nunomura, Physical properties of cell water in partially dried Saccharomyces cerevisiae, Biophys. J. 6 (1966) 665–674.

[464] D. A. Cowan, T. G. Green, A. T. Wilson, Lichen metabolism. 1. The use of tritium labelled water in studies of anhydrobiotic metabolism in Ramalina celastri and Peltigera polydactyla, New Phytol. 82 (1979) 489–503.

[465] C. W. Vertucci, A. C. Leopold, Bound water in soybean seed and its relation to respiration and imbibitional damage, Plant Physiol. 75 (1984) 114–117.

[466] C. W. Vertucci, J. L. Ellenson, A. C. Leopold, Chlorophyll fluorescence characteristics associated with hydration level in pea cotyledons, Plant Physiol. 79 (1985) 248–252.

[467] C. W. Vertucci, A. C. Leopold, Physiological activities associated with hydration level in seeds, in: A. C. Leopold (Ed.), Membranes, Metabolism, and Dry Organisms, Ithaca, Cornell University Press, 1986, pp. 35–49.

[468] C. W. Vertucci, A. C. Leopold, Oxidative processes in soybean and pea seeds, Plant Physiol. 84 (1987) 1038–1043.

[469] J. J. Skujins, A. D. McLaren, Enzyme reaction rates at limited water activities, Science 158 (1967) 1569–1570.

[470] Y. I. Khurgin, N. V. Medvedeva, V. Y. Rosliakov, Solid-state enzymatic reactions. II. Chymotrypsin hydrolysis of N-succinyl-L-phenylalanine n-nitroanilide, Biofizika 22 (1977) 1010–1014.

[471] E. Stevens, L. Stevens, The effect of restricted hydration on the rate of reaction of glucose 6-phosphate dehydrogenase, phosphoglucose isomerase, hexokinase and fumarase, Biochem. J. 179 (1979) 161–167.

[472] J. A. Rupley, P.-H. Yang, G. Tollin, Thermodynamic and related studies of water interacting with proteins, in: S. P. Rouland (Ed.), Water in Polymers. ACS Symposium Series. v.127, Washington, DC: American Chemical Society, 1980, pp. 111–132.

[473] J. A. Rupley, E. Gratton, G. Careri, Water and globular proteins, Trends Biochem. Sci. 8 (1983) 18–22.

[474] F. Yang, A. J. Russel, The role of hydration in enzyme activity and stability: 2. Alcohol dehydrogenase activity and stability in a continuous gas phase reactor, Biotechnol. Bioeng. 49 (2000) 709–716.

[475] P. A. Lind, R. M. Daniel, C. Monk, R. V. Dunn, Esterase catalysis of substrate vapour: Enzyme activity occurs at very low hydration, Biochim. Biophys. Acta 1702 (2004) 103–110.

[476] R. V. Dunn, R. M. Daniel, The use of gas-phase substrates to study enzyme catalysis at low hydration, Philos. Trans. R. Soc. London, Ser. B 339 (2004) 1309–1320.

[477] J. Partridge, P. R. Dennison, B. D. Moore, P. J. Halling, Activity and mobility of subtilisin in low water organic media: Hydration is more important than solvent dielectric, Biochim. Biophys. Acta 1386 (1998) 79–89.

[478] P. J. Halling, What can we learn by studying enzymes in non-aqueous media? Philos. Trans. R. Soc. London, Ser. B 359 (2004) 1287–1297.

[479] A. Zaks, A. M. Klibanov, The effect of water on enzyme action in organic media, J. Biol. Chem. 263 (1988) 8017–8021.

[480] R. Korenstein, B. Hess, Hydration effects on *cis-trans* isomerization of bacteriorhodopsin, FEBS Lett. 82 (1977) 7–11.

[481] G. Varo, L. Keszthelyi, Photoelectric signals from dried oriented purple membranes of Halobacterium halobium, Biophys. J. 43 (1983) 47–51.

[482] H. Sass, I. Schachowa, G. Rapp, M. Koch, D. Oesterhelt, N. Dencher, G. Büldt, The tertiary structural changes in bacteriorhodopsin occur between m states: X-ray diffraction and fourier transform infrared spectroscopy, EMBO J. 16 (1997) 1484–1491.

[483] J. Fitter, S. A. W. Verclas, R. E. Lechner, N. A. Dencher, Function and picosecond dynamics of bacteriorhodopsin in purple membrane at different lipidation and hydration, FEBS Lett. 433 (1998) 321–325.

[484] J. Fitter, R. Lechner, N. Dencher, Interactions of hydration water and biological membranes studied by neutron scattering, J. Phys. Chem. B 103 (1999) 8036–8050.

[485] G. Thiedemann, J. Heberle, N. A. Dencher, Bacteriorhodopsin pump activity at reduced humidity, in: J. L. Rigaud (Ed.), Structures and Functions of Retinal Proteins, Vol. 221 of Colloque INSERM, Paris, John Libbey Eurotext Ltd., 1992, pp. 217–220.

[486] U. Lehnert, V. Reat, M. Weik, G. Zaccai, C. Pfister, Thermal motion of bacteriorhodopsin at different hydration levels, Biophys. J. 75 (1998) 1945–1952.

[487] W. Saenger, Principles of Nucleic Acid Structure, New York: Springer-Verlag, 1984.

[488] A. G. W. Leslie, S. Arnott, R. Chandrasekaran, R. L. Ratliff, Polymorphism of DNA double helices, J. Mol. Biol. 143 (1980) 49–72.

[489] W. Saenger, W. N. Hunter, O. Kennard, DNA conformation is determined by economics in the hydration of phosphate groups, Nature 324 (1986) 385–388.

[490] J. Texter, Nucleic acid – water interactions, Prog. Biophys. Molec. Biol. 33 (1979) 83–97.

[491] L. van Dam, N. Korolev, L. Nordenskild, Polyamine-nucleic acid interactions and the effects on structure in oriented DNA fibers, Nucleic Acids Res. 30 (2002) 419–428.

[492] G. Malenkov, L. Minchenkova, E. Minyat, A. Schyolkina, V. Ivanov, The nature of the B-A transition of DNA in solution, FEBS Lett. 51 (1975) 38–42.

[493] V. I. Ivanov, L. E. Minchenkova, G. Burckhardt, E. Birch-Hirschfeld, H. Fritzsche, C. Zimmer, The detection of B-form/A-form junction in a deoxyribonucleotide duplex, Biophys. J. 71 (1996) 3344–3349.

[494] I. D. Kuntz, W. Kauzmann, Hydration of proteins and polypeptides, Adv. Protein Chem. 28 (1974) 239–345.

[495] J. L. Finney, P. L. Poole, Solvent effects on the structure, dynamics and activity of lysozyme, Proc. Int. Symp. Biomol. Struct. Interactions, Suppl. J. Biosci. 8 (1985) 25–35.

[496] P. L. Poole, The role of hydration in lysozyme structure and activity: Relevance in protein engineering and design, J. Food Eng. 22 (1994) 349–365.

[497] R. B. Gregory, M. Gangoda, R. K. Gilpin, W. Su, The influence of hydration on the conformation of lysozyme studied by solid-state ^{13}C-NMR spectroscopy, Biopolymers 33 (1993) 513–519.

[498] S. J. Prestrelski, N. Tedeschi, T. Arakawa, J. F. Carpenter, Dehydration-induced conformational transitions in proteins and their inhibition by stabilizers, Biophys. J. 65 (1993) 661–671.

[499] R. Affleck, Z.-F. Xu, V. Suzawa, K. Focht, D. S. Clark, J. S. Dordick, Enzymatic catalysis and dynamics in low-water environements, Proc. Natl. Acad. Sci. U.S.A. 89 (1992) 1100–1104.

[500] N. O. R. Martin Neto L, Tabak M, Effect of hydration in metHb: Reversible changes monitored by ESR of iron, J. Inorg. Biochem. 40 (1990) 309–321.

[501] A. G. Salvay, M. F. Colombo, J. R. Grigera, Hydration effects on the structural properties and haemhaem interaction in haemoglobin, Phys. Chem. Chem. Phys. 5 (2003) 192–197.

[502] E. W. Simon, Phospholipids and plant membrane permeability, New Phytologist 73 (1974) 377–420.

[503] M. J. Janiak, D. M. Small, G. G. Shipley, Temperature and compositional dependence of the structure of hydrated dimyristoyl lecithin, J. Biol. Chem. 254 (1979) 6068–6078.

[504] V. Seewaldt, D. A. Priestley, A. C. Leopold, G. W. Feigenson, F. Goodsaid-Zalduondo, Membrane organization in soybean seeds during hydration, Planta 152 (1981) 19–23.

[505] J. H. Crowe, F. A. Hoekstra, L. M. Crowe, Membrane phase transitions are responsible for imbibitional damage in dry pollen, Proc. Natl. Acad. Sci. U.S.A. 86 (1989) 520–523.

[506] J. Fitter, R. E. Lechner, G. Buldt, N. A. Dencher, Internal molecular motions of bacteriorhodopsin: Hydration-induced flexibility studied by quasielastic incoherent neutron scattering using oriented purple membranes, Proc. Natl. Acad. Sci. U.S.A. 93 (1996) 7600–7605.

[507] J. Fitter, R. E. Lechner, N. A. Dencher, Picosecond molecular motions in bacteriorhodopsin from neutron scattering, Biophys. J. 73 (1997) 2126–2137.

[508] J. A. Rupley, G. Careri, Protein hydration and function, Adv. Protein Chem. 41 (1991) 37–172.

[509] J. E. Schinkel, N. W. Downer, J. A. Rupley, Hydrogen exchange of lysozyme powders. Hydration dependence of internal motions, Biochemistry 24 (1985) 352–366.

[510] P.-H. Yang, J. A. Rupley, Protein-water imteractions. Heat capacity of the lysozyme-water system, Biochemistry 18 (1979) 2654–2661.

[511] O. V. Belonogova, E. N. Frolov, S. A. Krasnopol'skaya, B. P. Atanasov, B. K. Gins, Effect of the degree of hydration on the mobility of Mossbauer atoms in the active centers of metalloenzymes and carriers, Dokl. Akad. Nauk SSSR 241 (1978) 219–222.

[512] J. H. Roh, V. N. Novikov, R. B. Gregory, J. E. Curtis, Z. Chowdhuri, A. P. Sokolov, Onsets of anharmonicity in protein dynamics, Phys. Rev. Lett. 95 (2005) 038101.

[513] J. H. Roh, J. E. Curtis, S. Azzam, V. N. Novikov, I. Peral, Z. Chowdhuri, R. B. Gregory, A. P. Sokolov, Influence of hydration on the dynamics of lysozyme, Biophys. J. 91 (2006) 2573–2588.

[514] A. Paciaroni, S. Cinelli, G. Onori, Effect of the environment on the protein dynamical transition: A neutron scattering study, Biophys. J. 83 (2002) 1157–1164.

[515] S. Cinelli, A. D. Francesco, G. Onori, A. Paciaroni, Thermal stability and internal dynamics of lysozyme as affected by hydration, Phys. Chem. Chem. Phys. 6 (2004) 3591–3595.

[516] S. Cinelli, M. Freda, G. Onori, A. Paciaroni, A. Santucci, Hydration-dependent internal dynamics of macromolecules: A neutron scattering study, J. Mol. Liq. 117 (2005) 99–105.

[517] S. R. Kakivaya, C. A. J. Hoeve, The glass point of elastin, Proc. Natl. Acad. Sci. U.S.A. 72 (1975) 3505–3507.

[518] X. L. Yao, V. P. Conticello, M. Hong, Investigation of the dynamics of an elastin-mimetic polypeptide using solid-state NMR, Magn. Reson. Chem. 42 (2004) 267–275.

[519] R. F. Tilton, J. C. Dewan, G. A. Petsko, Effects of temperature on protein structure and dynamics: X-ray crystallographic studies of

the protein ribonuclease-A at nine different temperatures from 98 to 320 K, Biochemistry 31 (1992) 2469–2481.

[520] B. F. Rasmussen, A. M. Stock, D. Ringe, G. A. Petsco, Crystalline ribonuclease A loses function below the dynamic transition at 220 K, Nature 357 (1992) 423–424.

[521] X. Ding, B. F. Rasmussen, G. A. Petsko, D. Ringe, Direct structural observation of an acyl-enzyme intermediate in the hydrolysis of an ester substrate by elastase, Biochemistry 33 (1994) 9285–9293.

[522] R. H. Austin, K. W. Beeson, L. Eisenstein, H. Frauenfelder, I. C. Gunsalus, Dynamics of ligand binding to myoglobin, Biochemistry 14 (1975) 5355–5373.

[523] F. Parak, E. N. Frolov, A. A. Kononenko, R. L. Mssbauer, V. I. Goldanskii, A. B. Rubin, Evidence for a correlation between the photoinduced electron transfer and dynamic properties of the chromatophore membranes from Rhodospirillum rubrum, FEBS Lett. 117 (1980) 368–372.

[524] H. Keller, P. G. Debrunner, Evidence for conformational and diffusional mean square displacements in frozen aqueous solution of oxymyoglobin, Phys. Rev. Lett. 45 (1980) 68–71.

[525] W. Doster, S. Cusack, W. Petry, Dynamical transition of myoglobin revealed by inelastic neutron scattering, Nature 337 (1989) 754–756.

[526] A. P. Sokolov, H. Grimm, R. Kahn, Glassy dynamics in DNA: Ruled by water of hydration? J. Chem. Phys. 110 (1999) 7053–7057.

[527] H. Grimm, A. Sokolov, A. Dianoux, Relaxational dynamics in dry and humid DNA, Appl. Phys. A 74 (2002) S1248–S1250.

[528] G. Caliskan, R. Briber, D. Thirumalai, V. Garcia-Sakai, S. Woodson, A. Sokolov, Dynamic transition in tRNA is solvent induced, J. Am. Chem. Soc. 128 (2006) 32–33.

[529] H. Frauenfelder, G. A. Pestko, D. Tsernoglou, Temperature-dependent X-ray diffraction as a probe of protein structural dynamics, Nature 280 (1979) 558–563.

[530] W. Doster, A. Bachleitner, R. Dunau, M. Hiebl, E. Luscher, Thermal properties of water in myoglobin crystals and solutions at subzero temperatures, Biophys. J. 50 (1986) 213–219.

[531] R. J. Loncharich, B. R. Brooks, Temperature dependence of dynamics of hydrated myoglobin: Comparison of force field calculations with neutron scattering data, J. Mol. Biol. 215 (1990) 439–455.

[532] J. Smith, K. Kuczera, M. Karplus, Dynamics of myoglobin: Comparison of simulation results with neutron scattering spectra, Proc. Natl. Acad. Sci. U.S.A. 87 (1990) 1601–1605.

[533] P. Steinbach, B. Brooks, Protein hydration elucidated by molecular dynamics simulation, Proc. Natl. Acad. Sci. 90 (1993) 9135–9139.

[534] P. J. Steinbach, B. R. Brooks, Hydrated myoglobin's anharmonic fluctuations are not primarily due to dihedral transitions, Proc. Natl. Acad. Sci. U.S.A. 93 (1996) 55–59.

[535] J. Norberg, L. Nilsson, Glass transition in DNA from molecular dynamics simulations, Proc. Natl. Acad. Sci. U.S.A. 93 (1996) 10173–10176.

[536] A. L. Tournier, J. C. Smith, Principal components of the protein dynamical transition, Phys. Rev. Lett. 91 (2003) 208106.

[537] A. L. Tournier, J. Xu, J. C. Smith, Translational hydration water dynamics drives the protein glass transition, Biophys. J. 85 (2003) 1871–1875.

[538] M. Ferrand, A. J. Dianoux, W. Petry, G. Zaccai, Thermal motions and function of bacteriorhodopsin in purple membranes: Effect of temperature and hydration studied by neutron scattering, Proc. Natl. Acad. Sci. U.S.A. 90 (1993) 9668–9672.

[539] A. M. Tsai, D. A. Neumann, L. N. Bell, Molecular dynamics of solid-state lysozyme as affected by glycerol and water: A neutron scattering study, Biophys. J. 79 (2000) 2728–2732.

[540] M. Di Bari, F. Cavatorta, A. Deriu, G. Albanese, Mean square fluctuations of hydrogen atoms and water-biopolymer interactions in hydrated saccharides, Biophys. J. 81 (2001) 1190–1194.

[541] J. Fitter, The temperature dependence of internal molecular motion in hydrated and dry α-amylase: The role of hydration water in the dynamical transition of proteins, Biophys. J. 76 (1999) 1034–1042.

[542] V. Kurkal, R. M. Daniel, J. L. Finney, M. Tehei, R. V. Dunn, J. C. Smith, Enzyme activity and flexibility at very low hydration, Biophys. J. 89 (2) (2005) 1282–1287.

[543] P. Pissis, Dielectric studies of protein hydration, J. Mol. Liq. 41 (1989) 271–289.

[544] P. Pissis, A. Anagnostopoulou-Konsta, Protonic percolation on hydrated lysozyme powders studied by the method of thermally stimulated depolarization currents, J. Phys. D: Appl. Phys. 23 (1990) 932–939.

[545] R. B. Gregory, Protein hydration and glass transitions, in: D. Reid (Ed.), The Properties of Water in Foods: Isopow 6: Kluwer Academic 1997.

[546] Y. Fujita, Y. Noda, Effect of hydration on the thermal denaturation of lysozyme as measured by differential scanning calorimetry, Bull. Chem. Soc. Jpn. 51 (1978) 1567–1568.

[547] K. Kashefi, D. R. Lovley, Extending the upper temperature limit for life, Science 301 (2003) 934.

[548] R. M. Daniel, M. Dines, H. H. Petach, The denaturation and degradation of stable enzymes at high temperatures, Biochem. J. 317 (1996) 111.

[549] P. F. Mullaney, Dry thermal inactivation of trypsin and ribonuclease, Nature 210 (1966) 953.

[550] A. M. Klibanov, Enzymatic catalysis in anhydrous organic solvents, Trends Biochem. Sci. 14 (1989) 141–144.

[551] V. Mozhaev, K. Poltevsky, V. Slepnev, G. Badun, A. Levashov, Homogeneous solutions of hydrophilic enzymes in nonpolar organic solvents. New systems for fundamental studies and biocatalytic transformations, FEBS Lett. 292 (1991) 159–161.

[552] Y. Shen, C. R. Safinya, K. S. Liang, A. F. Ruppert, K. J. Rothschild, Stabilization of the membrane protein bacteriorhodopsin to 140 °C in two-dimensional films, Nature 366 (1993) 48–50.

[553] Y. Fujita, Y. Noda, Effect of hydration on the thermal stability of protein as measured by differential scanning calorimetry. Lysozyme – D_2O system, Bull. Chem. Soc. Jpn. 52 (1979) 2349–2352.

[554] L. N. Bell, M. J. Hageman, L. M. Muraoka, Thermally induced denaturation of lyophilized bovine somatotropin and lysozyme as impacted by moisture and excipients, J. Pharm. Sci. 84 (1995) 707–712.

[555] Y. Fujita, Y. Noda, The effect of hydration on the thermal stability of ovalbumin as measured by means of differential scanning calorimetry, Bull. Chem. Soc. Jpn. 54 (1981) 3233–3234.

[556] H. Batzer, U. T. Kreibich, Influence of water on thermal transitions in natural polymers and synthetic polyamides, Polym. Bull. 5 (1981) 585–590.

[557] G. Nimtz, A. Enders, B. Binggeli, Hydration dependence of the head group mobility in phospholipid (DMPC) membranes, Ber. Bunsenges. Phys. Chem. 89 (1985) 842–845.

[558] D. Urry, Physical chemistry of biological free energy transduction as demonstrated by elastic protein-based polymers, J. Phys. Chem. B 101 (1997) 11007–11028.

[559] F. M. Winnik, Fluorescence studies of aqueous solutions of poly(N-isopropylacrylamide) below and above their LCST, Macromolecules 23 (1990) 233–242.

[560] G. Luna-Barcenas, J. C. Meredith, I. C. Sanchez, K. P. Johnston, D. G. Gromov, J. J. de Pablo, Relationship between polymer chain conformation and phase boundaries in a supercritical fluid, J. Chem. Phys. 107 (1997) 10782–10792.

[561] Y. Maeda, T. Nakamura, I. Ikeda, Changes in the hydration states of poly(N-alkylacrylamide)s during their phase transitions in water observed by FTIR spectroscopy, Macromolecules 34 (2001) 1391–1399.

[562] N. Muller, Search for a realistic view of hydrophobic effects, Acc. Chem. Res. 23 (1990) 23–28.

[563] B. Lee, G. Graziano, A two-state model of hydrophobic hydration that produces compensating enthalpy and entropy changes, J. Am. Chem. Soc. 118 (1996) 5163–5168.

[564] S. Moelbert, P. DeLosRios, Hydrophobic interaction model for upper and lower critical solution temperatures, Macromolecules 36 (2003) 5845–5853.

[565] A. Shiryayev, D. L. Pagan, J. D. Gunton, D. S. Rhen, A. Saxena, T. Lookman, Role of solvent for globular proteins in solution, J. Chem. Phys. 122 (2005) 234911.

[566] I. Brovchenko, A. Krukau, N. Smolin, A. Oleinikova, A. Geiger, R. Winter, Thermal breaking of spanning water networks in the hydration shell of proteins, J. Chem. Phys. 123 (2005) 224905.

[567] P. Mentre, G. H. B. Hoa, Effects of high hydrostatic pressures on living cells: A consequence of the properties of macromolecules and macromolecule-associated water, Int. Rev. Cytology 201 (2001) 1–84.

[568] W. Doster, J. Friedrich, Pressure-temperature phase diagrams of proteins, in: J. Buchner, T. Kiefhaber (Eds.), Protein Folding Handbook, Vol. 1, Weinheim, Wiley-VCH, 2005, pp. 99–126.

[569] I. Daniel, P. Oger, R. Winter, Origins of life and biochemistry under high-pressure conditions, Chem. Soc. Rev. 35 (2006) 858–875.

[570] R. Winter, D. Lopes, S. Grudzielanek, K. Vogtt, Towards an understanding of the temperature/pressure configurational and free-energy landscape of biomolecules, J. Non-Equilib. Thermodyn. 32 (2007) 41–97.

[571] F. Abe, Piezophysiology of yeast: Occurrence and significance, Cell. Mol. Biol. 50 (2004) 437–445.

[572] A. Krzyzaniak, P. Salanski, J. Jurczak, J. Barciszewski, B–Z DNA reversible conformation changes effected by high pressure, FEBS Lett. 279 (1991) 1–4.

[573] D. Dubins, A. Lee, R. Macgregor, T. Chalikian, On the stability of double stranded nucleic acids, J. Am. Chem. Soc. 123 (2001) 9254–9259.

[574] G. Panick, R. Malessa, R. Winter, G. Rapp, K. J. Frye, C. A. Royer, Structural characterization of the pressure-denatured state and unfolding/refolding kinetics of staphylococcal nuclease by synchrotron small-angle X-ray scattering and Fourier-transform infrared spectroscopy, J. Mol. Biol. 275 (1998) 389–402.

[575] G. Panick, G. Vidugiris, R. Malessa, G. Rapp, R. Winter, C. Royer, Exploring the temperature-pressure phase diagram of staphylococcal nuclease, Biochemistry 38 (1999) 4157–4164.

[576] W. Dzwolak, R. Ravindra, J. Lendermann, R. Winter, Aggregation of bovine insulin probed by DSC/PPC calorimetry and FTIR spectroscopy, Biochemistry 42 (2003) 11347–11355.

[577] G. M. Schneider, Phase behavior and critical phenomena in fluid mixtures under pressure, Ber. Bunsenges. Phys. Chem. 76 (1972) 325–331.

[578] G. M. Schneider, High-pressure investigations on fluid systems. A challenge to experiment, theory, and application, J. Chem. Thermodyn. 23 (1991) 301–326.

[579] G. M. Schneider, Aqueous solutions at pressures up to 2 GPa: Gas–gas equilibria, closed loops, high-pressure immiscibility, salt effects and related phenomena, Phys. Chem. Chem. Phys. 4 (2002) 845–852.

[580] Z. Visak, L. Rebelo, J. Szydlowski, The "hidden" phase diagram of water + 3-methylpyridine at large absolute negative pressures, J. Phys. Chem. B 107 (2003) 9837–9846.

[581] L. Dougan, R. Hargreaves, S. P. Bates, J. L. Finney, V. Reat, A. K. Soper, J. Crain, Segregation in aqueous methanol enhanced by cooling and compression, J. Chem. Phys. 122 (2005) 174514.

[582] Y. Awakuni, J. H. Calderwood, Water vapour adsorption and surface conductivity in solids, J. Phys. D: Appl. Phys. 5 (1972) 1038–1045.

[583] N. Sasaki, Dielectric properties of slightly hydrated collagen: Time-water content superposition analysis, Biopolymers 23 (1984) 1725–1734.

[584] J. Eden, P. R. C. Gascoyne, R. Pethig, Dielectric and electrical properties of hydrated bovine serum albumin, J. Chem. Soc., Faraday Trans. 1 76 (1980) 426–434.

[585] P. R. Gascoyne, R. Pethig, A. Szent-Gyorgyi, Water structure-dependent charge transport in proteins, Proc. Natl. Acad. Sci. U.S.A. 78 (1981) 261–265.

[586] C. Giacomantonio, Charge transport in melanin, a disordered bio-organic conductor, Master's thesis, University of Queensland (2005).

[587] P. Pissis, A. A. Konsta, S. Ratkovic, S. Todorovic, J. Laudat, Temperature- and hydration-dependence of molecular mobility in seeds, J. Therm. Anal. Calorim. 47 (1996) 1463–1483.

[588] A. A. Konsta, J. Laudat, P. Pissis, Dielectric investigation of the protonic conductivity in plant seeds, Solid State Ionics 97 (1997) 97–104.

[589] M. R. Powell, B. Rosenberg, The nature of the charge carriers in solvated biomacromolecules, J. Bioenerg. Biomembr. 1 (1970) 493–509.

[590] G. Careri, M. Geraci, A. Giansanti, J. A. Rupley, Protonic conductivity of hydrated lysozyme powders at megahertz frequencies, Proc. Natl. Acad. Sci. U.S.A. 82 (1985) 5342–5346.

[591] G. Careri, A. Giansanti, J. A. Rupley, Proton percolation on hydrated lysozyme powders, Proc. Natl. Acad. Sci. U.S.A. 83 (1986) 6810–6814.

[592] G. Careri, A. Giansanti, J. A. Rupley, Critical exponents of protonic percolation in hydrated lysozyme powders, Phys. Rev. A 37 (1988) 2703–2705.

[593] J. A. Rupley, L. Siemankowski, G. Careri, F. Bruni, Two-dimensional protonic percolation on lightly hydrated purple membrane, Proc. Natl. Acad. Sci. U.S.A. 85 (1988) 9022–9025.

[594] I. Kovacs, G. Varo, Dielectric dispersion and protonic conduction in hydrated purple membrane, Acta Biochim. Biophys. Hung. 23 (1988) 265–270.

[595] F. Bruni, G. Careri, A. C. Leopold, Critical exponents of protonic percolation in maize seeds, Phys. Rev. A 40 (1989) 2803–2805.

[596] F. Bruni, A. C. Leopold, Hydration, protons and onset of physiological activities in maize seeds, Physiol. Plant. 81 (1991) 359–366.

[597] D. Sokolowska, A. Krol-Otwinowska, J. K. Moscicki, Water-network percolation transitions in hydrated yeast, Phys. Rev. E 70 (2004) 052901–052904.

[598] F. Bruni, G. Careri, J. S. Clegg, Dielectric properties of Artemia cysts at low water contents. Evidence for a percolative transition, Biophys. J. 23 (1989) 932–939.

[599] H. Haranczyk, On water in extremely dry biological systems, Jagiellonian University Press, Krakow, 2003.

[600] P. M. Suherman, P. Taylor, G. Smith, Low frequency dielectric study on hydrated ovalbumin, J. Non-Cryst. Solids 305 (2002) 317–321.

[601] P. M. Suherman, G. Smith, A percolation transition cluster model of the temperature dependent dielectric properties of hydrated proteins, J. Phys. D: Appl. Phys. 36 (2003) 336–342.

[602] D. van Lith, M. P. D. Haas, J. M. Warman, A. Hummel, Highly mobile charge carriers in hydrated DNA and collagen formed by pulsed ionization, Biopolymers 22 (1983) 807–810.

[603] D. van Lith, J. M. Warman, M. P. de Haas, A. Himmel, Electron migration in hydrated DNA and collagen at low temperatures Part 1. Effect of water concentration, J. Chem. Soc. Faraday Trans. 1 82 (1986) 2933–2943.

[604] D. van Lith, J. Eden, J. M. Warman, A. Hummel, Electron migration in hydrated DNA and collagen at low temperatures. Part 2. The effect of additives, J. Chem. Soc., Faraday Trans. 1 82 (1986) 2945–2950.

[605] A. Bonincontro, G. Careri, A. Giansanti, F. Pedone, Water-induced DC conductivity of DNA: A dielectric-gravimetric study, Phys. Rev. A 38 (1988) 6446–6447.

[606] J. M. Warman, M. P. de Haas, A. Rupprecht, DNA: A molecular wire? Chem. Phys. Lett. 249 (1996) 319–322.

[607] A. Szent-Gyorgyi, The study of energy-levels in biochemistry, Nature 3745 (1941) 157–159.

[608] N. Smolin, R. Winter, Molecular dynamics simulations of staphylococcal nuclease: Properties of water at the protein surface, J. Phys. Chem. B 108 (2004) 15928–15937.

[609] F. Merzel, J. C. Smith, Is the first hydration shell of lysozyme of higher density than bulk water? Proc. Natl. Acad. Sci. U.S.A. 99 (2002) 5378–5383.

[610] M. Marchi, F. Sterpone, M. Ceccarelli, Water rotational relaxation and diffusion in hydrated lysozyme, J. Am. Chem. Soc. 124 (2002) 6787–6791.

[611] L. Meinhold, J. C. Smith, Pressure-dependent transition in protein dynamics at about 4 kbar revealed by molecular dynamics simulation, Phys. Rev. E 72 (2005) 061908.

[612] A. Oleinikova, N. Smolin, I. Brovchenko, Origin of the dynamic transition upon pressurization of crystalline proteins, J. Phys. Chem. B 110 (2006) 19619–19624.

[613] V. I. Ivanov, L. E. Minchenkova, The A-form of DNA: In search of biological role (a review), Mol. Biol. 28 (1995) 780–788.

[614] X. J. Lu, Z. Shakked, W. K. Olson, A-form conformational motifs in ligand-bound DNA structures, J. Mol. Biol. 300 (2000) 819–840.

[615] A. Rich, S. Zhang, Timeline: Z-DNA: The long road to biological function, Nat. Rev. Genet. 4 (2003) 566–572.

[616] N. Foloppe, L. Nilsson, J. MacKerell, E. Alexander, Ab initio conformational analysis of nucleic acid components: Intrinsic energetic contributions to nucleic acid structure and dynamics, Biopolymers 61 (1999) 61–76.

[617] H.-L. Ng, M. L. Kopka, R. E. Dickerson, The structure of a stable intermediate in the A-B DNA helix transition, Proc. Natl. Acad. Sci. U.S.A. 97 (2000) 2035–2039.

[618] R. E. Franklin, R. G. Gosling, Molecular configuration in sodium thymonucleate, Nature 171 (1953) 740–741.

[619] M. J. Tunis-Schneider, M. F. Maestre, Circular dichroism spectra of oriented and unoriented deoxyribonucleic acid films: A preliminary study, J. Mol. Biol. 52 (1970) 521–541.

[620] J. Piškur, A. Rupprecht, Aggregated DNA in ethanol solution, FEBS Lett. 375 (1995) 174–178.

[621] I. Brovchenko, A. Krukau, A. Oleinikova, A. K. Mazur, Water clustering and percolation in low hydration DNA shells, J. Phys. Chem. B 111 (2007) 3258–3266.

[622] A. K. Mazur, DNA dynamics in a water drop without counterions, J. Am. Chem. Soc. 124 (2002) 14707–14715.

[623] A. K. Mazur, Electrostatic polymer condensation and the A/B polymorphism in DNA-sequence effects, J. Chem. Theory Comput. 1 (2005) 325–336.

[624] S. Arnott, D. W. L. Hukins, Optimised parameters for A-DNA and B-DNA, Biochem. Biophys. Res. Commun. 47 (1972) 1504–1509.

[625] V. I. Ivanov, L. E. Minchenkova, E. E. Minyat, M. D. Frank-Kametetskii, A. K. Schyolkina, The B to A transition of DNA in solution, J. Mol. Biol. 87 (1974) 817–833.

[626] S. B. Zimmerman, B. H. Pheiffer, Does DNA adopt the C form in concentrated salt solutions or in organic solvent/water mixutres? An X-ray diffraction study of DNA fibers immersed in various media, J. Mol. Biol. 142 (1980) 315–330.

[627] V. B. Zhurkin, Y. P. Lysov, V. I. Ivanov, Anisotropic flexibility of DNA and the nucleosomal structure, Nucleic Acids Res. 6 (1979) 1081–1096.

[628] A. K. Mazur, Titration in silico of reversible B↔A transitions in DNA, J. Am. Chem. Soc. 125 (2003) 7849–7859.

[629] T. E. Cheatham, III, M. F. Crowley, T. Fox, P. A. Kollman, A molecular level picture of the stabilization of A-DNA in mixed

ethanol-water solutions, Proc. Natl. Acad. Sci. U.S.A. 94 (1997) 9626–9630.

[630] A. Oleinikova, N. Smolin, I. Brovchenko, Influence of water clustering on the dynamics of hydration water at the surface of a lysozyme, Biophys. J. 93 (2007) 2986–3000.

[631] A. Oleinikova, I. Brovchenko, N. Smolin, A. Krukau, A. Geiger, R. Winter, Percolation transition of hydration water: From planar hydrophilic surfaces to proteins, Phys. Rev. Lett. 95 (2005) 247802.

[632] D. Rosen, Dielectric properties of protein powders with adsorbed water, Trans. Faraday Soc. 59 (1963) 2178–2191.

[633] S. C. Harvey, P. Hoekstra, Dielectric relaxation spectra of water adsorbed at lysozyme, J. Phys. Chem. 76 (1972) 2987–2994.

[634] G. Careri, A. Giansanti, E. Gratton, Lysozyme film hydration events: An IR and gravimetric study, Biopolymers 18 (1979) 1187–1203.

[635] S. Bone, R. Pethig, Dielectric studies of the binding of water to lysozyme, J. Mol. Biol. 157 (1982) 571–575.

[636] S. Bone, R. Pethig, Dielectric studies of protein hydration and hydration-induced flexibility, J. Mol. Biol. 181 (1985) 323–326.

[637] H. Urabe, Y. Sugawara, M. Ataka, A. Ruppecht, Low-frequency Raman spectra of lysozyme crystals and oriented DNA films: Dynamics of crystal water, Biophys. J. 74 (1998) 1533–1540.

[638] F. Pizzitutti, F. Bruni, Glassy dynamics and enzymatic activity of lysozyme, Phys. Rev. E 64 (2001) 052905.

[639] J. Knab, J.-Y. Chen, A. Markelz, Hydration dependence of conformational dielectric relaxation of lysozyme, Biophys. J. 90 (2006) 2576–2581.

[640] A. Bizzarri, S. Cannistraro, Molecular dynamics simulation evidence of anomalous diffusion of protein hydration water, Phys. Rev. E 53 (1996) R3040–R3043.

[641] A. Bizzarri, C. Rocchi, S. Cannistraro, Origin of the anomalous diffusion observed by MD simulation at the protein-water interface, Chem. Phys. Lett. 263 (1996) 559–566.

[642] A. Oleinikova, I. Brovchenko, A. Krukau, A. Mazur, Anomalous diffusion in the hydration shell of DNA, to be published.

[643] M. Settles, W. Doster, Anomalous diffusion of the adsorbed water: A neutron scattering study of hydrated myoglobin, Faraday Discuss. 103 (1996) 269–279.

[644] C. Rocchi, A. R. Bizzarri, S. Cannistraro, Water dynamical anomalies evidences by molecular dynamics simulations at the solvent-protein interface, Phys. Rev. E 57 (1998) 3315–3325.

[645] M. Schoen, J. H. Cushman, D. J. Diestler, C. L. Rhykerd, Fluids in micropores. II. Self-diffusion in a simple classical fluid in a slit pore, J. Chem. Phys. 88 (1988) 1394–1406.

[646] M. Sega, R. Vallauri, S. Melchionna, Diffusion of water in confined geometry: The case of a multilamellar bilayer, Phys. Rev. E 72 (2005) 041201.

[647] P. Gallo, M. Rovere, Anomalous dynamics of confined water at low hydration, J. Phys.: Condens. Matt. 15 (2003) 7625–7633.

[648] D. van der Spoel, P. J. van Maaren, H. J. C. Berendsen, A systematic study of water models for molecular simulation: Derivation of water models optimized for use with a reaction field, J. Chem. Phys. 108 (1998) 10220–10230.

[649] I. Brovchenko, A. Krukau, A. Oleinikova, A. Mazur, Ion dynamics and water percolation effects in DNA polymorphism, J. Am. Chem. Soc. 130 (2008) 121–131.

[650] R. Kohlrausch, Theorie des Elektrischen Ruckstandes in der Leidener Flasche, Pogg. Ann. Phys. Chem. 91 (1854) 179–214.

[651] G. Williams, D. C. Watts, Non-symmetrical dielectric relaxation behaviour arising from a simple empirical decay function, Trans. Faraday Soc. 66 (1970) 80–85.

[652] Y. Leng, P. T. Cummings, Fluidity of hydration layers nanoconfined between mica surfaces, Phys. Rev. Lett. 94 (2005) 026101.

[653] R. Petig, Protein-water interactions determined by dielectric methods, Annu. Rev. Phys. Chem. 43 (1992) 177–205.

[654] S. Bone, Dielectric and gravimetric studies of water binding to lysozyme, Phys. Med. Biol. 41 (1996) 1265–1275.

[655] P. Tran, B. Alavi, G. Gruner, Charge transport along the λ-DNA double helix, Phys. Rev. Lett. 85 (2000) 1564–1567.

[656] V. P. Tomaselli, M. H. Shamos, Electrical properties of hydrated collagen. I. Dielectric properties, Biopolymers 12 (1973) 353–366.

[657] U. Heugen, G. Schwaab, E. Brundermann, M. Heyden, X. Yu, D. M. Leitner, M. Havenith, Solute-induced retardation of water dynamics probed directly by terahertz spectroscopy, Proc. Natl. Acad. Sci. U.S.A. 103 (2006) 12301–12306.

[658] A. Krukau, I. Brovchenko, A. Geiger, Temperature-induced conformational transition of a model ELP GVG(VPGVG)$_3$ in water, Biomacromolecules 8 (2007) 2196–2202.

[659] H. Reiersen, A. Clarke, A. Rees, Short ELPs exhibit the same temperature-induced structural transitions as elastin polymers: Implications for protein engineering, J. Mol. Biol. 283 (1998) 255–264.

[660] C. Nicolini, R. Ravindra, B. Ludolph, R. Winter, Characterization of the temperature- and pressure-induced inverse and reentrant transition of the minimum elastin-like polypeptide GVG(VPGVG) by DSC, PPC, CD, and FT-IR spectroscopy, Biophys. J. 86 (2004) 1385–1392.

[661] E. Schreiner, C. Nicolini, B. Ludolph, R. Ravindra, N. Otte, A. Kohlmeyer, R. Rousseau, R. Winter, D. Marx, Folding and unfolding of an elastinlike oligopeptide: "Inverse temperature transition," reentrance, and hydrogen-bond dynamics, Phys. Rev. Lett. 92 (2004) 148101.

[662] C. A. J. Hoeve, P. J. Flory, The elastic properties of elastin, J. Am. Chem. Soc. 80 (1958) 6523–6526.

[663] C. A. J. Hoeve, P. J. Flory, Elastic properties of elastin, Biopolymers 13 (1974) 677–686.

[664] R. Ravindra, C. Royer, R. Winter, Pressure perturbation calorimetic studies of the solvation properties and the thermal unfolding of staphylococcal nuclease, Phys. Chem. Chem. Phys. 6 (2004) 1952–1961.

[665] K. Heremans, L. Smeller, Protein structure and dynamics at high pressure, Biochim. Biophys. Acta 1386 (1998) 353–370.

[666] J. Woenckhaus, R. Khling, T. Thiyagarajan, K. C. Littrell, S. Seifert, C. A. Royer, R. Winter, Pressure-jump small-angle X-ray scattering detected kinetics of staphylococcal nuclease folding, Biophys. J. 80 (2001) 1518–1523.

[667] A. Paliwal, D. Asthagiri, D. P. Bossev, E. Paulaitis, Pressure denaturation of staphylococcal nuclease studied by neutron small-angle scattering and molecular simulation, Biophys. J. 87 (2004) 3479–3492.

Index

Adsorption, 17, 21, 29, 51, 54, 62
  isobar, 92
  isochor, 92
  isotherm, 24–32, 92–9, 114–15, 139, 167
  of water, 24–32, 50, 88, 138–9, 152, 189
  preferential, 17, 67, 86, 152
Anomalous diffusion, 197–201, 209–12
Autocorrelation function, 205–8

Bond percolation, 190–1, 230

Capillary:
  condensation, 43–5, 62, 91, 114–18, 171
  evaporation, 60, 63, 91, 114–18
Cluster:
  infinite, 122, 126–9
  physical, 6, 121–2
  spanning, 126–7, 129–30
    at the spherical surface, 132–3, 136
    in lysozyme powder, 174
    near ELP surface, 218, 224
    on DNA surface, 187
    on lysozyme surface, 189
    on Snase surface, 180, 224
Coexistence curve, 1
  of 2D water, 36
  of bulk water, 1–9, 12–14
  of layering transition, 25, 32–41
  of prewetting transition, 21, 47

of water in pores, 44–8, 59, 69, 71, 76–83, 88, 91, 95–108, 118
Compressibility, 5, 7, 10, 29
Conductivity, 138–9, 165–70, 194, 211–14
  protonic, 195
Conformational changes, 159, 224–5
Coordination shell, 8, 142, 230
Correlation length, 4, 20, 22–3, 51, 55, 58, 67, 72–6, 80–3, 86, 93–4, 108–9, 234
Critical exponent:
  of order parameter, 3, 25, 37
    2D, 93, 47
    bulk (3D), 81, 97
    effective, 96
    in percolation problems, 127, 168, 188
    surface, 84, 97
Critical temperature, 2, 128
  hysteresis *see* Hysteresis critical temperature
  of 2D water, 228
  of bulk water, 2–6, 52–3, 77, 83, 225
  of layering transition, 24–5, 28–9, 34–42, 61, 66, 122, 137, 165
  of liquid–liquid transition, 230
  of prewetting transition, 21, 45–6, 61, 64–6
Crossover, 68, 75, 102–3, 169
  3D–2D critical behavior, 93, 109–10
  bulk-surface critical behavior, 72–3, 82–6
  ordinary–normal transition, 69, 85

Dead layers, 23, 24, 45, 50, 105–11, 149
Diameter of the coexistence curve, 4, 83
Dielectric:
  constant, 138–9, 209, 213–14
  properties, 138–9, 208, 214
  spectra, 208
Diffusion coefficient, 14, 149, 201–4, 210, 218
Disorder, 12
  spatial, 196–7, 201, 209–10
  temporal, 196, 199
Disordered pores, 92–4
Disordered water, 227, 229–31
Disordering effect of the surface, 28, 82, 97
DNA, 162, 169, 192, 194
  conformational transition, 158, 162, 170, 181–2
  dynamic transition, 112
  ion mobility, 209–13
  water diffusion, 199–204
  water percolation, 181–9
Domain structure, 12, 93–4, 104–5, 171
Dynamic transition:
  pressure induced, 181, 204–5
  temperature induced, 112, 160–1, 212, 215, 228

Elastin-like peptide (ELP), 162, 216–27

Fluctuations:
  of concentration, 11
  of density, 4, 5, 7, 39
  of interface, 23
Fractal dimension, 128, 137, 186, 189–92
  of largest cluster, 128, 130, 172–5, 191, 224
  of random walk, 169
Fractal structure, 189

Freezing temperature, 1–10, 21–4, 40, 62–6, 111, 234
  of 2D water, 39
  of confined water, 110

Heat capacity, 3–7
  maximum, 6, 92, 235
Hysteresis, 27, 92, 95, 99, 115
  coexistence curve, 92, 96, 99
  critical temperature, 92, 99, 195

Ice:
  2D, 29, 36, 145
  amorphous, 10–11, 148
  cubic, 110
  hexagonal, 1, 10, 110, 148
Interfacial equation, 55
Ising film, 103
Ising model, 10, 12, 18–19, 24, 67–8, 72, 82, 89, 94
  3D, 18, 102
  2D, 24, 37, 46, 93
Ising lattice see Ising model
Ising magnet see Ising model

Layering transition, 24–5, 28–47, 64–6, 111, 121–2, 137, 145, 153, 165, 178, 234
Liquid–liquid:
  coexistence curve, 13, 96
  critical point, 10–14, 69, 112
  phase transition, 10–15, 98, 113, 230
Liquid–vapor:
  coexistence curve, 18, 51, 59
    in pores, 69–70, 76–80, 92, 96, 99–100, 113
    of bulk water, 1–15
    of layering transition, 25
  critical point, 1–9, 14, 46, 93, 97, 99, 117
  critical temperature, 8, 21, 51–3, 233
  see also Critical temperature
Local diameter, 74, 83–5

Index

Local order parameter, 68, 71–7, 80–3
Lysozyme, 112, 156–62, 167–78, 184, 189–99
   powder, 159–60, 167–75, 190, 192, 195
   water dynamics, 202–8

Master curve, 71–4, 80, 83, 133
Mean-square displacement (MSD), 199–204, 209
Metastable states, 7, 27, 32, 92, 118, 151
Missing neighbor, 21, 49–60, 63, 67–76, 83, 86–8, 108, 234
Mixture model, 9, 10

Normal transition, 67–9, 73, 85

Order parameter:
   in bulk fluid, 3–7
   of layering transition, 25, 36–7, 47
   see also Critical exponent of order parameter
Ordinary transition, 67–9, 72, 82, 85
Orientational ordering, 42, 121, 139–42, 148, 230, 234

Phase diagram, 1
   of bulk water, 1–15
Percolation transition of water:
   in bulk, 6, 13–14
   in fully hydrated biosystems, 217–36
   in low-hydrated biosystems, 165–214
   on smooth surface, 28, 39, 121–39
Prewetting, 17–26, 42–8, 61, 64–5, 91
   film, 47
Pseudocritical point, 93

Relaxation time, 206–8

Sequential wetting, 21, 23, 49, 66, 235
Site-bond percolation, 122, 191
Site percolation, 188, 190–1, 230
Strongly bound water, 169, 197–201, 206–8
Supercooled water, 98, 111–3, 230

Temperature of wetting transition, 43, 45, 48–50
   see also Wetting temperature
Tetrahedral ordering, 7–8, 144, 230
Thermal conductivity, 52

Weakly bound water, 138, 206–8
Wetting:
   transition, 18–26, 43–50, 53, 62, 66, 69, 234
   temperature, 18, 20–1, 49, 66
   see also temperature of wetting transition